DIAMOND FILMS AND COATINGS

DIAMOND FILMS AND COATINGS

Development, Properties, and Applications

Edited by

Robert F. Davis

North Carolina State University
Department of Materials Science and Engineering
Raleigh, North Carolina

np

NOYES PUBLICATIONS
Park Ridge, New Jersey, U.S.A.

Library of Congress Catalog Card Number: 92-35011
ISBN: 0-8155-1323-2
Printed in the United States

Published in the United States of America by
Noyes Publications
Mill Road, Park Ridge, New Jersey 07656

10 9 8 7 6 5 4 3 2 1

Library of Congress Cataloging-in-Publication Data

Diamond films and coatings : development, properties, and applications / edited by Robert F. Davis.
p. cm.
Includes bibliographical references and index.
ISBN 0-8155-1323-2
1. Diamonds, Artificial. 2. Diamond thin films. I. Davis, Robert F. (Robert Foster), 1942- .
TP873.5.D5D52 1992
666'.88--dc20 92-35011
CIP

MATERIALS SCIENCE AND PROCESS TECHNOLOGY SERIES

Editors

Electronic Materials and Process Technology

DEPOSITION TECHNOLOGIES FOR FILMS AND COATINGS: by Rointan F. Bunshah et al

CHEMICAL VAPOR DEPOSITION FOR MICROELECTRONICS: by Arthur Sherman

SEMICONDUCTOR MATERIALS AND PROCESS TECHNOLOGY HANDBOOK: edited by Gary E. McGuire

HYBRID MICROCIRCUIT TECHNOLOGY HANDBOOK: by James J. Licari and Leonard R. Enlow

HANDBOOK OF THIN FILM DEPOSITION PROCESSES AND TECHNIQUES: edited by Klaus K. Schuegraf

IONIZED-CLUSTER BEAM DEPOSITION AND EPITAXY: by Toshinori Takagi

DIFFUSION PHENOMENA IN THIN FILMS AND MICROELECTRONIC MATERIALS: edited by Devendra Gupta and Paul S. Ho

HANDBOOK OF CONTAMINATION CONTROL IN MICROELECTRONICS: edited by Donald L. Tolliver

HANDBOOK OF ION BEAM PROCESSING TECHNOLOGY: edited by Jerome J. Cuomo, Stephen M. Rossnagel, and Harold R. Kaufman

CHARACTERIZATION OF SEMICONDUCTOR MATERIALS—Volume 1: edited by Gary E. McGuire

HANDBOOK OF PLASMA PROCESSING TECHNOLOGY: edited by Stephen M. Rossnagel, Jerome J. Cuomo, and William D. Westwood

HANDBOOK OF SEMICONDUCTOR SILICON TECHNOLOGY: edited by William C. O'Mara, Robert B. Herring, and Lee P. Hunt

HANDBOOK OF POLYMER COATINGS FOR ELECTRONICS—Second Edition: by James J. Licari and Laura A. Hughes

HANDBOOK OF SPUTTER DEPOSITION TECHNOLOGY: by Kiyotaka Wasa and Shigeru Hayakawa

HANDBOOK OF VLSI MICROLITHOGRAPHY: edited by William B. Glendinning and John N. Helbert

CHEMISTRY OF SUPERCONDUCTOR MATERIALS: edited by Terrell A. Vanderah

CHEMICAL VAPOR DEPOSITION OF TUNGSTEN AND TUNGSTEN SILICIDES: by John E.J. Schmitz

ELECTROCHEMISTRY OF SEMICONDUCTORS AND ELECTRONICS: edited by John McHardy and Frank Ludwig

(continued)

HANDBOOK OF CHEMICAL VAPOR DEPOSITION: by Hugh O. Pierson

DIAMOND FILMS AND COATINGS: edited by Robert F. Davis

ELECTRODEPOSITION: by Jack W. Dini

Ceramic and Other Materials—Processing and Technology

SOL-GEL TECHNOLOGY FOR THIN FILMS, FIBERS, PREFORMS, ELECTRONICS AND SPECIALTY SHAPES: edited by Lisa C. Klein

FIBER REINFORCED CERAMIC COMPOSITES: by K.S. Mazdiyasni

ADVANCED CERAMIC PROCESSING AND TECHNOLOGY—Volume 1: edited by Jon G.P. Binner

FRICTION AND WEAR TRANSITIONS OF MATERIALS: by Peter J. Blau

SHOCK WAVES FOR INDUSTRIAL APPLICATIONS: edited by Lawrence E. Murr

SPECIAL MELTING AND PROCESSING TECHNOLOGIES: edited by G.K. Bhat

CORROSION OF GLASS, CERAMICS AND CERAMIC SUPERCONDUCTORS: edited by David E. Clark and Bruce K. Zoitos

HANDBOOK OF INDUSTRIAL REFRACTORIES TECHNOLOGY: by Stephen C. Carniglia and Gordon L. Barna

CERAMIC FILMS AND COATINGS: edited by John B. Wachtman and Richard A. Haber

Related Titles

ADHESIVES TECHNOLOGY HANDBOOK: by Arthur H. Landrock

HANDBOOK OF THERMOSET PLASTICS: edited by Sidney H. Goodman

SURFACE PREPARATION TECHNIQUES FOR ADHESIVE BONDING: by Raymond F. Wegman

FORMULATING PLASTICS AND ELASTOMERS BY COMPUTER: by Ralph D. Hermansen

HANDBOOK OF ADHESIVE BONDED STRUCTURAL REPAIR: by Raymond F. Wegman and Thomas R. Tullos

CARBON-CARBON MATERIALS AND COMPOSITES: edited by John D. Buckley and Dan D. Edie

Preface

The synthetic diamonds developed by industrial organizations in the 1950's and 1960's using a combination of high-pressure, high-temperature (HPHT) techniques are used primarily in cutting, grinding and polishing applications. However, a variety of potential applications for this material require thin films and coatings which cannot be produced by these methods. The results of much smaller efforts, simultaneous with the HPHT studies, in the United States and Russia and concerned both with the kinetics of the pyrolysis of various hydrocarbon-hydrogen mixtures and with the mechanisms of nucleation and growth of diamond on diamond seeds at low pressures were considered to be only of "academic" interest. However, the publication of micrographs of highly faceted and continuous diamond films, and confirming analytical data, in the Russian and, subsequently, in the Japanese scientific literature in the late 1970's and early 1980's catalyzed substantial activity in this field. This resulted in the development and use of a variety of deposition and characterization techniques to achieve and study diamond films and coatings.

The purpose of this book is to provide an in-depth examination of selected topics concerned with synthetic diamond. This unique material has been considered from the perspectives of historical development, its properties, the latest in scientific results regarding its growth and characterization and present and future applications.

The eight chapters of this book are authored by experts in their fields. The initial chapter by Yoder provides an applications-related cornerstone for the book. Tables of all the major properties of diamond are given in conjunction with discussions concerned with the importance of these properties to the present and potential employment of this material. Having determined the need for synthetic diamond, the subsequent chapter by Nassau describes the considerable efforts to develop this material with an emphasis on the high-temperature, high-pressure research. The remainder of the book is concerned with diamond materials grown primarily from the vapor phase. Pehrsson, Celii and Butler address the important issue of the chemical mechanisms occurring within the gas phase which ultimately allow the deposition of this material under non-equilibrium conditions on various substrates. The various methods and experimental conditions within which one can obtain the proper gas-phase chemistry and resulting diamond films is described by Koba. Lux and Haubner describe the market potential and current applications of diamond in terms of wear resistant coatings for a variety of products. The chemical, physical and microstructural characterization of the low pressure diamond materials was used initially to convince the skeptics of its existence. Subsequently it was used to determine the nature of the deposited material and to provide information for the optimization of the particular growth method. The chapter by Zhu, Kong and Glass provides an in-depth review of the methods and results of this significant effort. Applications of diamond films and coatings are now being realized. In the final two chapters, Plano, Pinneo and Das have taken the information content in the book full circle with their discussion regarding both the identification of the relevant properties which affect the development of nonelectronic and electronic applications for diamond films and the future research necessary for property enhancement. Extensive lists of reviews, research papers and bibliographies at the end of each chapter also provide resource materials for more detailed study.

Numerous individuals have contributed and assisted in the preparation of this book. I wish to express my appreciation firstly to the authors and their employers who contributed considerable effort, time and facilities for the generation of each chapter. I am also especially indebted to Jeff Glass who provided numerous suggestions, assistance and author contact. I also appreciate the patience and continual encouragement of George Narita, Executive Editor of Noyes Publications, during the completion of all the tasks involved in the editing of this book. Last but not least, I wish to extend my special thanks to my wife, Jan, and daughter, Leona, for their patience and for the time I could not spend with them during this endeavor.

Raleigh, North Carolina
September, 1992

Robert F. Davis

Contributors

James E. Butler
Gas/Surface Dynamics Section
Naval Research Laboratory
Washington, DC

Frances G. Celii
Texas Instruments, Inc.
Dallas, TX

Kalyan Kumar Das
Kobe Steel Research Laboratories
Electronic Materials Center
Research Triangle Park, NC

Robert F. Davis
North Carolina State University
Department of Materials Science and Engineering
Raleigh, NC

Jeffrey T. Glass
North Carolina State University
Department of Materials Science and Engineering
Raleigh, NC

Roland Haubner
Technische Universitat
Institute fur Chemische Technologie anorganischer Stoffe
Vienna, Austria

Richard J. Koba
Plasma Therm, Inc.
St. Petersburg, FL

Hua-Shuang Kong
Cree Research, Inc.
Durham, NC

Benno Lux
Technische Universitat
Institute fur Chemische Technologie anorganischer Stoffe
Vienna, Austria

Kurt Nassau
154A Guinea Hollow Road
Lebanon, NJ

Pehr E. Pehrsson
Gas/Surface Dynamics Section
Naval Research Laboratory
Washington, DC

Michael Pinneo
Kobe Steel Research Laboratories
Electronic Materials Center
Research Triangle Park, NC

Linda S. Plano
Kobe Steel Research Laboratories
Electronic Materials Center
Research Triangle Park, NC

Max N. Yoder
Electronics Program
Office of Naval Research
Arlington, VA

Wei Zhu
North Carolina State University
Department of Materials Science and Engineering
Raleigh, NC

Notice

Contents

1. DIAMOND PROPERTIES AND APPLICATIONS 1
Max N. Yoder
1.0 The Crystalline Structure of Diamond 1
1.1 Crystalline Carbon 2
2.0 Mechanical and Elastic Properties of Diamond 3
2.1 Heat 4
2.2 Elastic Waves 7
2.3 Hardness 7
2.4 Young's Modulus 8
2.5 Fracture Toughness 8
2.6 Coefficient of Friction 9
2.7 Density 10
3.0 Optical Properties of Diamond 11
4.0 Electrical Properties of Diamond 11
5.0 High Temperature and Tribological Applications of Diamond 14
5.1 Graphitization 14
5.2 Corrosion and Oxidation 14
5.3 Heat Exchangers 15
5.4 Tribology and Hardness 16
6.0 Optical Applications of Diamond 19
6.1 Windows 19
6.2 Optical Matching 21
6.3 Fresnel Lens 22
6.4 X-Ray Applications 24
7.0 Electro-Optic and Related Applications 25
7.1 Optically Triggered Switches 25

7.2 Lasers 26
8.0 Biological Applications 26
8.1 Artificial Heart Valves 26
8.2 Scalpels 27
8.3 Marine Fouling 27
References 28

2. SYNTHESIS OF BULK DIAMOND: HISTORY AND PRESENT STATUS 31
Kurt Nassau
1.0 Introduction 31
2.0 The Early History of Diamond Synthesis 32
2.1 Historical Aspects 32
2.2 Early Diamond Claims 33
2.3 The Diamond Problem 34
2.4 P.W. Bridgman and High Pressure Research 36
3.0 The High Pressure Growth of Diamond 40
3.1 The General Electric Diamond Project 40
3.2 H.T. Hall, the First Person to Grow Diamond 46
3.3 The High Pressure Synthesis of Diamond Cut 49
3.4 High Pressure Single Crystal Growth of Diamond 53
3.5 Color in Synthetic Diamond 55
4.0 Other Bulk Diamond Synthesis Processes 57
4.1 High Pressure Carbonado 57
4.2 The Shock Wave Process 58
5.0 The Present Status of High Pressure Diamond Growth 61
References 64

3. CHEMICAL MECHANISMS OF DIAMOND CVD 68
Pehr E. Pehrsson, Frances G. Celii, and James E. Butler
1.0 Introduction 68
2.0 Nucleation 71
2.1 The Two-Step Nucleation Process 76
2.2 Epitaxial Diamond Growth 100
2.3 Nucleation: A Summary 106
3.0 Growth 108
3.1 The Diamond Growth Surface 109
3.2 Characterization of the Gaseous Growth Environment 111
3.3 Models of the Gaseous Growth Environment 119
3.4 Models of Diamond Growth 125
References 133

4. TECHNOLOGY OF VAPOR PHASE GROWTH OF DIAMOND FILMS 147
Richard J. Koba
1.0 Introduction 147
2.0 General Conditions for Diamond Film Deposition 148
3.0 Pyrolysis Methods 150
3.1 Hydrocarbon Pyrolysis 150
3.2 Fluorocarbon Pyrolysis 152
4.0 Heated Filament Method 154
5.0 Plasma Enhanced Chemical Vapor Deposition . . . 158
5.1 Microwave Plasma Enhanced CVD 158
5.2 RF Plasma Enhanced CVD 163
5.3 DC Plasma Enhanced CVD 165
6.0 Other Methods of Diamond Deposition 167
6.1 Physical Vapor Deposition Methods 167
6.2 Combustion 171
7.0 Comparison of Deposition Technologies 173
7.1 Epitaxial Diamond for Semiconductor Device Research 174
7.2 Thick, Free Standing Diamond Structures 174
7.3 Rim Supported Diamond Windows/ Membranes 175
7.4 Cutting Tools 175
8.0 Recommendations and Predictions 176
References 178

5. DIAMOND AS A WEAR-RESISTANT COATING 183
Benno Lux and Roland Haubner
1.0 Introduction 183
2.0 Deposition and Applications of Wear-Resistant Materials 184
2.1 Improvement of Materials by Hard CVD and PVD Coatings 184
2.2 PCD and PCBN, the Superhard High-Pressure Products 189
2.3 Low-Pressure Diamond Layers as a New Challenge 192
3.0 Potential Products and Industrial Applications . . . 197
3.1 In Situ CVD Diamond Deposition (CVD Coating Analogy) 197
3.2 "Bonded" Freestanding Diamond Sheets (PCD Analogy) 205

3.3 Cutting Performance of Low-Pressure Diamond Tools with Al-Si and Other Workpieces . . . 205
3.4 Estimations of Production Costs for Diamond-Coated Cemented Carbide Cutting Tool Inserts . . . 210
4.0 Problems Related to the Fabrication and Adhesion of Low-Pressure Diamond Products . . . 215
4.1 Influence of Substrate Chemistry . . . 215
4.2 Nucleation Enhancement . . . 216
4.3 Adhesion . . . 217
5.0 Technical and Economical Problems Linked with Synthesis Methods, Industrial Production Units and Fabrication . . . 218
5.1 Diamond Synthesis Methods . . . 218
5.2 Industrial Production Units . . . 220
5.3 Comparison of the Two Concepts for Applying Low-Pressure Diamond . . . 222
5.4 Economic Considerations . . . 223
6.0 Market Potentials and Future Prospects . . . 224
6.1 Market Potential . . . 224
6.2 New Superhard Composite Powders and Compacted Products . . . 228
6.3 The Superhard BN Phases . . . 229
7.0 Conclusions and Outlook . . . 230
7.1 The Short-Range Outlook . . . 231
7.2 The Long-Range Outlook . . . 232
References . . . 234

6. CHARACTERIZATION OF DIAMOND FILMS . . . 244
Wei Zhu, Hua-Shuang Kong and Jeffrey T. Glass
1.0 Introduction . . . 244
2.0 Vibrational Spectroscopies . . . 245
2.1 Raman Spectroscopy . . . 248
2.2 Infrared Spectroscopy . . . 252
3.0 Electron Spectroscopies . . . 257
3.1 X-Ray Photoelectron Spectroscopy . . . 258
3.2 Auger Electron Spectroscopy . . . 263
3.3 Electron Energy Loss Spectroscopy . . . 267
4.0 Electron Microscopies . . . 272
4.1 Transmission Electron Microscopy . . . 272
4.2 Scanning Electron Microscopy . . . 287
5.0 Diffraction Techniques . . . 298
5.1 X-Ray Diffraction . . . 298

5.2 Electron Selected Area Diffraction 303
5.3 Low Energy Electron Diffraction 311
5.4 Reflection High Energy Electron Diffraction 315
6.0 Other Analytical Techniques 316
6.1 Secondary Ion Mass Spectrometry 316
6.2 Luminescence Spectroscopy 318
6.3 Rutherford Backscattering Spectrometry 321
6.4 Spectroscopic Ellipsometry 323
7.0 Summary 325
References 327

7. NON-ELECTRONIC APPLICATIONS OF DIAMOND FILMS 339
Linda Plano and Michael Pinneo
1.0 Introduction 339
2.0 Goals of Applied CVD Diamond Research 341
2.1 Identification of Relevant Properties 341
2.2 Applications-Related Characterization of Diamond Films 345
3.0 Research Approaches for Diamond Property Enhancement 351
3.1 Proof of Principle 351
3.2 Commercialization of Prototype Diamond Film Products 358
4.0 Future Non-Electronic Diamond Applications 363
4.1 Mechanical Applications 364
4.2 Passive Thermal Applications 371
4.3 CVD Diamond Optical Applications 374
References 377

8. ELECTRONIC APPLICATIONS OF DIAMOND FILMS AND COATINGS 381
Kalyan Kumar Das
1.0 Introduction 381
2.0 Doping 382
3.0 Contacts 386
3.1 Effect of H Content in CVD Films on Contact Properties 386
3.2 Effect of Surface Preparation on Metal Contacts 387
3.3 Ohmic Contacts 387
3.4 Rectifying Contacts 389
4.0 Etching of Diamond 393
5.0 Diamond SiO_2 Interface 393

6.0 Device Structures 395
7.0 Conclusions 406
References 406

INDEX 411

1

Diamond Properties and Applications

MAX N. YODER

1.0 THE CRYSTALLINE STRUCTURE OF DIAMOND

Sir Isaac Newton was first to propose that diamond was organic. His suppositions were drawn from careful measurements of the indices of refraction. In 1772 the French chemist Antoine L. Lavoisier found that the products of diamond combustion behaved as, and most assuredly were, solely carbon dioxide. The English chemist Smithson Tennant conducted an experiment in which he mixed equal amounts of diamond, graphite, and coal together with equal amounts of saltpeter in separate respective closed chambers. He determined that the ignition and burning of each produced equal amounts of "bound air". Having thus established that all three materials were composed of the same element, their differences remained intriguing until the advent of x-rays and the ingenuity of the famous father and son Bragg duo—Sir William Henry and Sir William Lawrence. They found carbon to embody three separate structures: cubic, hexagonal, and amorphous. Natural diamond actually consists of approximately 99% carbon 12 and 1% carbon 13. As seen later, were it not for this isotropic impurity, the thermal conductivity of diamond would be even higher than its already unmatched value.

From these humble beginnings our knowledge of the properties of diamond have increased markedly. From this knowledge and the recent introduction of artifact diamond films, our ability to efficaciously apply diamond to the technological requirements of the modern age is gaining momentum.

1.1 Crystalline Carbon

Diamond. Gemstones are the most commonly known forms of diamond and are exclusively of the face-centered diamond cubic structure shown in Fig. 1. In addition to the eight corner atoms and six face-center atoms (shown in cross-hatch) of the unit cell, there are four other atoms (shown dotted). These dotted atoms are from adjacent interpenetrating lattices and are spaced one quarter of a cube diagonal from the former. Each carbon atom is tetrahedrally coordinated to four other carbon atoms via sigma bonds emanating from sp^3 hybrid atomic orbitals. The four {111} directions in cubic diamond are in the bond directions. The {111} crystallographic plane consists of six-membered rings of carbon atoms puckered such that adjacent atoms are alternately displaced upward and downward from the plane. The lattice may be envisioned as stacked layers of the six-membered rings with each ring in the "chair" conformation (1). In the cubic form, the layers of chairs are joined by staggered bonds and the six-membered rings between the chairs are also chairs. In any {111} direction the stacking sequence is ABC ABC ABC .... The lattice constant is 3.56 Å and the bond length is 1.54 Å. The Raman spectrographic identifier signal for cubic diamond is at 1332 cm^{-1} for carbon 12 and 1284 cm^{-1} for carbon 13.

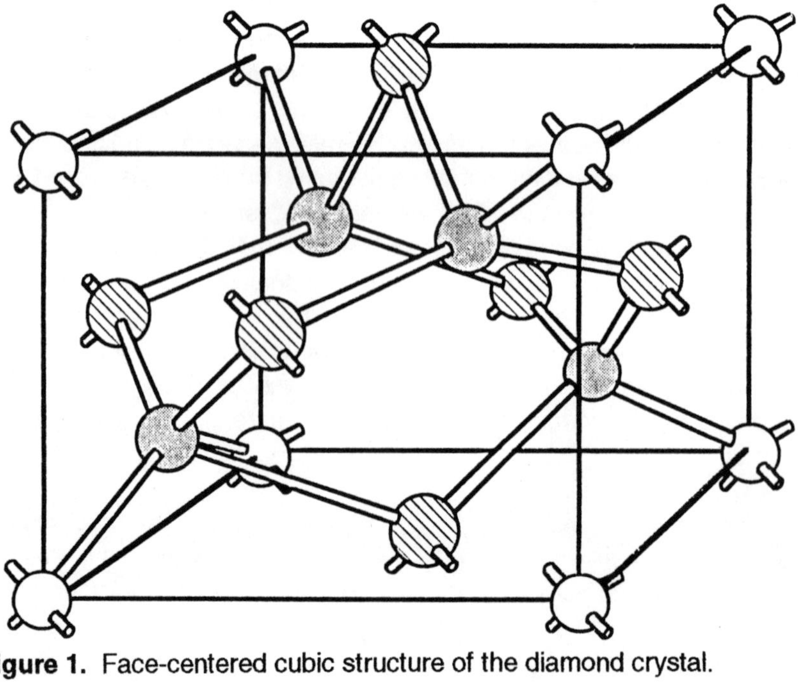

Figure 1. Face-centered cubic structure of the diamond crystal.

Lonsdaleite. If the layers of chairs noted above are joined by "eclipsed" rather than staggered bonds, the six-membered rings joining the layers of chairs conform as "boats" rather than chairs (1). This configuration is known as lonsdaleite and was first found in meteorites. In each plane the arrangement of atoms is the same as in the cubic structure, but the planes are differently joined. This leads to a stacking sequence of AB AB AB .... The eclipsed bonds result in slightly stronger (closer) chemical bonding with lattice constant in the a and c directions of 2.52 Å and 4.12 Å respectively. Nearest neighbor spacing is 1.52 Å, and the material is birefringent. This material is thought to be responsible for a Raman signature of 1315 - 1325 cm^{-1}.

Graphite. The most common form of carbon is graphite. The six-membered, trigonally bonded carbon rings are not puckered as in diamond, but lie in a flat plane. Using sp^2 hybrid atomic orbitals, each carbon atom exhibits three in-plane sigma bonds; the remaining electron forms pi bonds using a p_z atomic orbital. Each plane is staggered from the planes above and below resulting in an AB AB AB ... sequence. In-plane, nearest neighbor spacing is 1.42 Å while the lattice constant in the basal plane (between repeating layers) is 6.707 Å. The in-plane layers are Raman active with a characteristic signature of 1580 cm^{-1}.

Microcrystalline Carbon. Poorly synthesized artifact diamond may contain little long range order resulting primarily in sp^3 bonding, but with sp^2 bonding near the crystallite grain boundaries. Such material is characterized by Raman signatures centered at 1355 cm^{-1}.

Amorphous Carbon. By scaling the Raman signatures of amorphous silicon, Nemanich et al have shown that artifact diamond having a characteristic signature near 1140 cm^{-1} may be sp^3 bonded amorphous carbon and that with a signature near 1500 cm^{-1} may be sp^2 bonded amorphous carbon (2).

2.0 MECHANICAL AND ELASTIC PROPERTIES OF DIAMOND

The structure of diamond—especially its very strong chemical bonding—leads to unique mechanical and elastic properties. Its hardness, molar density, thermal conductivity, and sound velocity are higher than that of any other known material while its compressibility and bulk modulus are lower than that of any other material. It has a very low thermal expansion coefficient 0.8×10^{-6}/°C.

2.1 Heat

Thermal Conductivity. At 20 watts/cm-°C, diamond is unmatched as a thermal conductor. This attribute should not be confused with the specific heat of a material—a property for which diamond is NOT competitive with other materials. It is informative to compare the thermal conductivity of diamond with other materials with which it may be used or with which it may compete. Such a comparison is provided in Table 1. When nitrogen is introduced into diamond in concentrations leading to aggregation such as in type IA diamond, the thermal conductivity is reduced to 30 - 50% of its maximum value. Polycrystalline artifact diamond films of good quality generally exhibit thermal conductivities in the direction of growth of 55% of single crystals. In the plane of growth, 25% of single crystal values is more typical, but these figures are dependant on grain size.

Table 1. Thermal Conductivities Compared

Material	Conductivity (Watts/cm-°C)
Diamond (Type II)	20.00
Boron nitride (theory)	13.00
Beryllium oxide	3.70 - 5.90
Water	5.61
Silicon carbide	4.90
Silver	4.18
Copper	3.80
Aluminum nitride	3.70
Gold	3.11
Aluminum	2.38
Tungsten	1.70
Silicon	1.50
Oil, Transformer	1.36
Iron	0.82
Graphite	0.80 - 2.50
Palladium	0.70
Platinum	0.69
Tin	0.65
Gallium arsenide	0.46
Germanium	0.28
Titanium	0.20
Quartz	0.07 - 0.14
Indium arsenide	0.07

Heat may be transferred in a solid by one of two principal mechanisms. The most common is that of drifting electrons. This mechanism is most prevalent in metals and alloys. Since electron conduction at lower temperatures is critically dependent on scattering mechanisms, the presence of impurities and defects should be avoided if thermal conductivity is to be maximized. At higher temperatures, phonon and lattice scattering predominate as the principal scattering mechanisms in these materials. The second principal mechanism is that of thermal phonons (i.e., the directional wave motion of interacting lattice ions). This mechanism is most prevalent in dielectrics (e.g., diamond) and certain highly disordered alloys. At low temperatures, phonon scattering (i.e., conduction limiting mechanisms) results primarily from impurities and defects in a manner similar to metals. Sample size may also affect thermal conductivity as phonon phase cancellation from sample boundaries may not be insignificant. Boundary scattering certainly affects most of the published data for the thermal conductivity of diamond at low temperatures. Actual values are thought to be considerably higher than those published. At higher temperatures, phonon-phonon scattering predominates, therefore the thermal conductivities of many dielectrics tend to converge at high temperatures. Phonons may also be scattered by isotopic variations. Carbon naturally contains about 1% of the C^{13} isotope. While the mass differential between Si^{28} and Si^{29} is only 3.6%, the mass differential between C^{12} and C^{13} is 8.3%. If diamond were to be synthesized from isotopically pure carbon, its thermal conductivity would probably be increased by a factor of five over that of present material (3).

Molar Heat Capacity. The specific heat capacity of a given mass of any substance is the quantity of heat required to raise its temperature by one degree. The molar heat capacity of any substance is the quantity of heat required to raise the temperature of one molecular weight of that substance by one degree. Being comprised solely of the low atomic mass of carbon, diamond is characterized by a comparatively small molar heat capacity. so it is better employed as a heat distributor than as a heat sink. Comparative values are provided in Table 2.

Thermal Expansion. Expansivity is the ratio of the change in length per degree Celsius to the length at 0°C. It is directly proportional to the molar heat capacity. For solids, the coefficient of volume expansion is approximately three times that of the linear coefficient. Diamond has a value lower than that of Invar outside of the 18 - 100°C range. Table 3 provides comparisons.

Table 2. Molar Heat Capacities at 298 K

Material	Capacity (Joules/moles-°C)
Silver	25.50
Gold	25.41
Copper	24.50
Aluminum	24.37
Silicon	19.85
Graphite	8.54
Diamond	6.19
Water	4.22

BeO 0.74 cal/cc
Sapphire 0.69 cal/cc
Si 0.42 cal/cc

Table 3. Thermal Expansion Coefficients

Material	Coefficient (x 10^{-6}/°C)
Silicon dioxide	0.5
INVAR (18 - 100°C)	0.64
Diamond (300 K)	0.8
Diamond (193 K)	0.4
Diamond (400-1200 K)*	1.5
Silicon	2.6
Aluminum nitride	4.5
Tungsten (polycrystal)	4.5
Sapphire	5.5
Germanium	5.7
Gallium arsenide	5.9
Beryllium oxide	5.9
YAG	7.8
Graphite (in plane)	8.8
Platinum (polycrystal)	8.9
Magnesium oxide	11.0
Gold (polycrystal)	14.1
Graphite (along c axis)	27.3

* Gruneisen's law obeyed in this region

2.2 Elastic Waves

Elastic waves within the volume of solid diamond propagate in one of two modes. The first is a longitudinal wave wherein the direction of displacement at each point is normal to the wavefront. The second is the shear wave wherein the shape of the material may change without a corresponding change of volume. Shear waves exhibit displacement orthogonal to the wavefront and their velocity field has zero divergence. When shear waves combine to form standing waves, linear displacements may result. Table 4 provides comparative data.

Table 4. Comparative Sound Velocities (meters/second)

Material	Longitudinal	Shear
Diamond	18,000	*
Boron	15,000	*
Beryllium	12,890	8,880
Silicon carbide	11,400	*
Silicon	7,500	*
Boron nitride	7,000	*
Tungsten carbide	6,900	*
Aluminum	6,420	3,040
Nickel	6,040	3,000
Aluminum nitride	5,500	*
Tungsten	5,220	2,890
Copper	4,760	2,325
Gold	3,240	1,200
Water	1,498	(n/a)

2.3 Hardness

Diamond is the hardest of all natural materials. As steel is to butter, so is diamond to steel. The hardness of any given material is usually determined by pressing a carefully shaped indenter onto a surface under a load W. A permanent indentation of area A results and the hardness H is described as $H = W/A$ and is a measure of the plasticity of the material. The indenter must be at least as hard as the substance being indented; for measuring diamond it is obvious that only a diamond indenter is useful. Even then, the indenter itself becomes misshapen after several uses and must be replaced. A comparative listing of the hardness of various materials is presented in Table 5.

Table 5. Comparative Hardness

Material	Knoop Hardness (kg/mm^2) (298 K)
Diamond	5,700 - 10,400
Cubic BN	4,500
Boron carbide	2,250
Tungsten carbide	2,190
Titanium carbide	2,190
Aluminum oxide	2,000
Silicon carbide	1,875 - 3,980
Titanium nitride	1,800
Case hardened steel	800
High carbon steel	400
Copper	40

The carbon-nitrogen bond is thought to be stronger than the carbon-carbon bond of diamond. If so, the material that could possibly be harder than diamond (if it could be successfully synthesized) is C_3N_4. The in-plane C=C bonds of graphite are also stronger that the C–C bonds of diamond. There has been speculation that a special bonding of carbon known as C_8 could exploit this bonding configuration for extreme hardness.

2.4 Young's Modulus

The region over which material subjected to a deforming force reverts to its original condition is known as the elastic region. Hooke's law states that the ratio of stress to strain is approximately a constant over a specified region. The ratio of the tensile stress to the linear strain is known as Young's modulus. The value for diamond is unexcelled as shown in Table 6.

2.5 Fracture Toughness

The fracture toughness of any material is calculated as follows (4):

$$K_{IC} = (0.16)EP/Hc^{1.5}$$

where E is Young's Modulus, H is hardness, P is indentation load, and c is the crack length. While this value is necessary to determine the efficacy of diamond in certain protective applications, there is no data currently available for diamond. Should data become available for diamond, the reader wishing a comparative analysis may note that the typical ZnS IR window material will exhibit a crack length of 0.234 mm under a 20 Newton point load.

Table 6. Young's Modulus comparison

Material	Value ($\times 10^{11}$ dynes/cm^2)
Diamond	105.0
Tungsten	34.0
Steel	17.2 - 20.5
Ferrites (most)	15.0
Platinum	14.7
Cast iron	13.8
Copper	12.8
Silicon	11.3
Quartz	7.9 - 10.7
Gold	7.4 - 8.0
Silver	7.1 - 7.8
Aluminum	6.9
Tin	4.1 - 4.5
Lead	1.4

2.6 Coefficient of Friction

The friction between two surfaces is dependant on many variables. Most important among these are the nature of the material itself, the surface finish, the velocity between the moving surfaces, environmental atmosphere, temperature, and degree of cleanliness. The coefficient of friction is

$$f_s = \frac{\text{(magnitude of maximum frictional force)}}{\text{(magnitude of the normal thrust)}}$$

The magnitude of the maximum frictional force required to begin movement between the surfaces is always greater than the magnitude required to ensure continued motion. These coefficients are commonly separated into static and dynamic, respectively. Table 7 provides comparative values of dynamic friction for several materials of interest.

2.7 Density

Of the light elements, diamond exhibits the greatest density; it is, however, not nearly as dense as are materials made of the heavier elements. Table 8 provides comparisons.

Table 7. Friction Comparison

Material	Coefficient
Diamond	0.05
Teflon	0.05
Metals (lubricated)	0.10
Graphite	0.10
Tungsten carbide	0.20
Sapphire	0.20
Nickel (bare)	0.70
Iron (bare)	1.00
Aluminum	1.30
Lead	1.50

Table 8. Density Comparison

Material	Density (gm/cm^3)
Osmium	22.8
Platinum	21.5
Tungsten	19.3
Uranium	18.7
Copper	8.9
Germanium	5.4
Diamond	3.52
Beryllium oxide	2.86
Aluminum	2.7
Quartz	2.65
Boron	2.54
Silicon	2.42
Graphite	2.25
Hydrogen (solid)	0.076

3.0 OPTICAL PROPERTIES OF DIAMOND

Optical properties of greatest interest include index of refraction and optical absorption at a given wavelength and temperature. Table 9 provides relative comparisons of refractive indices at room temperature and in the near IR.

Diamond was originally classified by its optical properties into types I and II. Later the a and b suffix was added. Type I has large amounts of nitrogen included. If the nitrogen is aggregated (as in most natural diamond) then it is type Ia. If it is dispersed (as in most artifact diamond) then it is type Ib. Type II is virtually free of nitrogen. Type IIa is highly insulating and has no detectable boron while the boron in type IIb exceeds the nitrogen. Table 10 provides the data for these and other optically active centers in diamond.

Table 9. Comparative Optical Properties

Material	Refractive Index	
Germanium	4.00 - 4.10	
Silicon	3.42 - 3.50	
GaAs	3.34 - 2.12	
Diamond	2.41 - 2.44	(656-486 nm)
Sapphire	1.58 - 1.83	
Fused silica	1.40 - 1.48	
Crystal quartz	1.52 - 1.69	
Water	1.33 - 1.34	
Calcium fluoride	1.31 - 1.51	
Lithium fluoride	1.11 - 1.45	
Magnesium oxide	1.62 - 1.77	

4.0 ELECTRICAL PROPERTIES OF DIAMOND

As a semiconductor, the properties of diamond (except for electron mobility) are virtually unexcelled. The Hall coefficient of natural diamond extends from 10^{12} at 140 K to 10^5 at room temperature (6). Relaxation time disparities among the three hole bands lead to widely differing values for the density of states depending on the measurement technique used—Hall, Faraday rotation, or cyclotron resonance (7)-(9). The Hall mobility of holes in natural diamond can be as high as 1800 cm^2/volt-sec and that of electrons can reach 2000 cm^2/volt-sec (10). Even higher hole mobilities have been found by magnetoresistance data (11). Hole mobilities in artifact homoepitaxial diamond films grown at 500 °C have been measured as high as 850 cm^2/volt-sec (12). Of semiconductors currently in use, only germanium and indium antimonide exhibit superior hole mobilities; even then the latter materials exhibit poor saturated hole velocities and low dielectric strength. At 10^4 volts/cm electric field strength, the carrier drift velocities of both holes and electrons in natural diamond begins to saturate. For holes it reaches a maximum value of 10^7 cm/sec and for electrons in natural diamond, saturation occurs at 2.0 x 10^7 cm/sec (13)(14). The electrical resistivity of natural diamond is as high as 10^{15} ohm-cm and that of homoepitaxial artifact diamond films grown both by RF immersed plasma methods using carbon monoxide diluted in hydrogen and by thermal CVD using CF_4 in F_2 is equally high (15).

Table 10. Optical Centers in Diamond (5)

Infrared Spectra

Designator	Features cm^{-1}	Features nm	Explanation
Ib	1130	8850	Isolated N atom
B-Aggregate	1185	8400	Even # N atom aggregate
A-Aggregate	1282	7800	Substitutional N pair
Platelet	1360	7300	(unknown)
Boron	900-1330 2808	11000-7500 3562	Substitutional B concentration dependant

Visible and UV Spectra

Designator	Features eV	Features nm	Explanation
ND1	3.150	393	5 electron vacancy (V)
N3	2.985	415	carbon with 3 N neighbors
GR2-8	2.880 3.004	430 413	Neutral V
H4	2.498	496	V + N aggregate
H3	2.463	503	N-V-N
?	1.945	637	N-V
GR1	1.673	741	4 electron V

In natural diamond, the impurity distribution is not homogenous and the measured charge carrier transport values are open to question. Theoretical values are much higher (16).

Substitutional boron is an acceptor in diamond. The activation energy of boron is somewhat dependant upon concentration in that large concentrations may introduce electrically active defects at more shallow levels. Dilute concentrations of boron (e.g., $10^{19}/cm^3$) are generally thought to have an activation energy of 370 meV. Substitutional nitrogen and phosphorous are known to be donors, but the energy level of nitrogen is 1.57 eV below the conduction band minimum rendering it virtually useless for n-type material. Phosphorous (being much larger than carbon) is thought to create electrically active defects. Interstitial lithium and sodium are thought to be relatively shallow donors (17). Experimental evidence indicates that some boron must be present to force lithium into interstitial sites and resultant (but compensated) donor activity (18). Viable n-channel diamond semiconductor devices may require electron injection via a heterojunction from a higher bandgap material or from a degenerately doped (graphitic) ohmic contact.

Referring to Table 11 below, it is readily apparent that the electronic properties of diamond are superior to those of conventional semiconductors. While the electron mobility of diamond is noted to be less than that of GaAs, it: *(i)* is seen to be greater than that of silicon and *(ii)* can be effectively circumvented using clever device design by virtue of its high saturated carrier velocity (19). The bandgap of diamond is noteworthy from a x-ray detector viewpoint in that the background (dark current) from IR and visible light is non-existent.

Table 11. Electrical Comparisons

Property	Diamond	Silicon	GaAs
Bandgap (meV)	5450	1100	1430
Electron mobility#	2000	1500	8500
Hole mobility #	1800	600	400
Breakdown V/cm x 10^6	10	5	6
Resistivity (ohm-cm)	10^{16}	10^3	10^9
High Field electron velocity (cm/sec x 10^7)	2.7	1	1
Dielectric constant	5.7	11	12.5
Work function (eV)	*	4.8	4.7

\# cm^2/volt-sec

* 4.8 eV (p-type). Diamond is thought to exhibit a negative electron affinity on its {111} surface.

The dielectric loss tangent is useful for applications where diamond is used as a substrate for microwave and/or millimeter wave applications. Comparative values for commonly used materials are provided in Table 12.

Table 12. Loss Tangent Comparisons

Material	Loss Tangent x 10^{-4}
Diamond	2
Beryllium oxide	5
Sapphire	10
Gallium arsenide	16
Silicon	100
Epoxy	300

5.0 HIGH TEMPERATURE AND TRIBOLOGICAL APPLICATIONS OF DIAMOND

Friction is usually dissipated as heat and this causes temperature to rise. Diamond has many potential applications at high temperatures that are not associated with friction or tribology, but diamond is certainly compatible with tribologically generated heat as well. The susceptibility of the diamond surface to heat is very much dependant on the environment.

5.1 Graphitization

The onset of graphitization of diamond in an inert atmosphere or in vacuum begins around 1520°C. At 2120°C about 0.3 carats/minute are graphitized. It is not unusual to find the charred graphitic remains of a diamond gemstone set in the still intact platinum setting of a wedding ring subjected to a house fire. The activation energy for this graphitization is orientation dependent. For the {111} surface, three carbon-carbon bonds must be broken and the activation energy is approximately 1060 kJ/mol. On other crystalline faces where only two carbon-carbon bonds are broken, the activation energy is reduced to 730 kJ/mol.

5.2 Corrosion and Oxidation

In the presence of atomic oxygen, a black graphitic-like coating may be detected on the diamond surface at temperatures as low as 625°C. Molten sodium nitrate is known to attack diamond at temperatures as low as 427°C. Oxidizing substances appear to be the only effective way to attack diamond at temperatures below 1000°C. At higher temperatures or under high pressure contact, two other groups of materials will attack diamond. The first of these are the solvents. Molten manganese, iron, cobalt, nickel and the platinum group of metals are true solvents for carbon and diamond. This is the reason that diamond cutting tools are not applicable for cutting ferrous and related materials. The other group of metals attacking diamond are the avaricious carbide formers such as titanium, zirconium, tantalum, and tungsten. Although boiling KOH and other similarly corrosive liquids do not attack artifact diamond films, they can permeate the films and attack the underlying material. To prevent this, the individual diamond grain size should be less than 30 nm. Permeability is also dependant on the method used to grow the diamond films and the quality (i.e., graphitic content) of the films.

5.3 Heat Exchangers

With unmatched thermal conductivity, diamond appears to be the perfect heat exchange material. It may be thought of as a solid state heat pipe, but capable of operating over a much wider range of temperatures than any liquid heat pipe.

Diamond has long been used as an electrically insulating thermal conductor for various electronics applications. Most high power IMPATT diodes today are mounted on diamond (natural or high pressure artifact material). Certain high power traveling wave tubes have their helices mounted in diamond. In these applications, diamond serves the triple roles by offering *(i)* exceptional dielectric strength, *(ii)* low loss tangent, and *(iii)* unmatched thermal conductivity. GaAs microwave and millimeter wave power transistors (both field effect and bipolar) currently must be designed with their active "fingers" spread about five times further apart than would otherwise be necessary if the thermal conductivity of GaAs would be even ten times better. This finger spreading introduces excessive parasitic capacitance and inductance to the devices. Not only is the heat a problem, but the dielectric loss tangent of GaAs in the millimeter wave spectrum is not much better than that of silicon. Recently an epitaxial lift-off technique has been designed for GaAs monolithic microwave integrated circuits (MMIC) wherein the entire epitaxial layer containing the MMIC die is removed from its substrate and attached by Van der Waals forces to any smooth surface (20). When this substrate is diamond, the exceptional thermal conductivity and low loss tangent should render it possible to design and operate the circuit for much increased performance.

Non-electronic heat exchangers can also benefit from diamond coatings. Here the corrosion and erosion resistant properties of diamond are combined with its thermal conductivity. Because of the exceptional corrosion and erosion resistant properties of diamond, the walls of the heat exchanger need not be as thick as would otherwise be necessary as thinning with use is expected to be virtually non existent. The unmatched strength of diamond is another factor conducive to designing with thinner walls. Thinner walls means less weight and this is particularly attractive to space borne applications. Caustic molten salts such as sodium solutions are difficult to contain in conventional heat exchangers. Diamond plated heat exchangers may again render viable liquid sodium based nuclear reactor plants. Such an implementation would not only lead to the improved efficiency of such sodium based plants, but would also address the safety issue that previously "grounded" them.

Fuel cells are a special case of heat exchangers. In addition to the requirement for exceptional thermal conductivity, corrosion, and erosion resistance, fuel cells also demand electrically insulating thermal materials. Again diamond is expected to generate significant performance and efficiency improvements in these devices. Fuel cell reactions employing oxidants at high temperatures are not, however, amenable to diamond (unless overcoated with BN).

5.4 Tribology and Hardness

The tribological properties of the natural diamond surface have been known for some time (21)(22). The coefficient of friction for the natural diamond surface in air at 300 K is approximately 0.1—equivalent to that of teflon. It is, however, somewhat dependant on load, orientation, and geometry. In all directions along the {111} surface, it can be as low as 0.05. On other faces, it is somewhat more direction dependant, but does not generally exceed 0.15 in air. The low coefficient of friction is generally attributed to the termination of the diamond surface in hydrogen (there is some indication that diamond getters hydrogen at temperatures below 600°C). In the atmosphere, there appears to be ample hydrogen present to provide for renewed hydrogen termination of the diamond surface as the rubbing removes it. In a vacuum (e.g., outer space), however, this is not the case. Under such conditions, the coefficient of friction may approach unity. Termination in fluorine offers some advantages. Should fluorine atoms be removed from the diamond surface, it is energetically more favorable for them to return to the diamond surface than to combine to form F_2. Such an energetically favorable situation is NOT the case for hydrogen atoms leaving the diamond surface.

Magnetic Disks. Head crashes on magnetic disks can prove to be disastrous. Normally a read/write head "flys" above a hard platter magnetic storage disk and is kept in this flying position by fluid dynamic principles dependant upon the rotation speed of the platter. It is removed to a position not used for data storage prior to powering down the system. In situations of unexpected power outage, the platter speed can be slowed below that necessary to sustain flying; upon contact with the rotating platter, the read/write head can physically damage the rotating media and destroy information. To be effective in protecting such a surface, the media must be *(i)* very hard, *(ii)* conformally applied, *(iii)* possess a low coefficient of friction, and *(iv)* not exceed 25 nm thickness. While diamond easily meets the first three of these conditions, the latter can be troublesome. The condition itself derives from the magnetic field strength and resulting signal-to-noise ratio. While the flying height may be reduced to compensate for a thicker protective film, tolerances must be improved. For the films to be conformal (no greater roughness than the underlying media and of uniform thickness), the diamond grain size must be reduced to approximately 5 nm or less.

Wire Dies and Water Jet Nozzles. Natural diamonds are currently used as wire dies and water jet nozzles. Hole diameters from 0.01 to 0.5 mm are typical. Variation of roundness of 0.005 mm is typically specified. Natural diamonds not only contain heterogeneous impurities and microcracks that lead to anisotropic wearout, but the single crystal itself is anisotropic in its hardness (wear resistance). Since fine grain polycrystalline diamond films are much less anisotropic, they will exhibit better wear properties than do the single crystal dies now in use. Hole cutting for such dies typically uses a yttrium aluminum garnet (YAG) laser although the etching process described in Sec. 6.2 (Moth's Eye "Coating") may also be used. Nozzles for high speed water jets are required to withstand pressures of up to 4000 kg/cm^2. Polycrystalline artifact diamond nozzles not only provide isotropic wear, but the inherent strength of diamond permits a less massive (lighter weight) nozzle—important in many automatic control water cutting operations. The annual world wide market for diamond dies exceeds a quarter million.

Guides for high speed wire cutting electrical discharge machines will also benefit from artificial diamond. Not only is friction reduced, but the thermal conductivity of the diamond guide distributes the generated heat to an outer (larger) surface where it can more efficiently dissipated.

Cutting and Grinding Tools. Over 10 million diamond grinding wheels, grinding stone dressers, and cutting-saw segment elements are produced each year. Presently these are mostly sintered from diamond powders and chips.

Sintered diamond powders are also used for cutting tool bits to provide isotropic wear properties. Their density is, however, considerably less than that afforded by polycrystalline diamond films. The annual market for diamond cutting tools is currently 200,000 and growing at a rate of 7%. Adhesion of the sintered diamond to the tool base can be a problem—especially with diamond cutters set into high speed circular saw blades. A costly electroplated bonding is usually required to insure the safety of the blade. Diamond tool bits are preferentially considered for their hardness (long life) and the ability to be honed to a very fine edge. The latter attribute is particularly attractive in micromachining operations where the profile of the tool edge is transferred to the finished product. Inasmuch as a diamond edge can be fabricated to a minimum radius of 2 nm, its maximum roughness can be less than 1 nm. The roughness of the tip edge is calculated as follows:

$$R_k = [(R_s)^2 + 2R_sR_f \cos(w) + (R_f)^2]^{0.5} / \sin(w)$$

where:
R_k = tip edge roughness,
R_s = cutting surface roughness,
R_f = flank roughness, and
w = wedge angle.

Conformally coated diamond films offer production cost benefits. Materials most amenable to machining by diamond cutting tools include aluminum and its alloys, copper and its alloys, various fluorides and chlorides (glasses), NaCl, ZnS, ZnSe, quartz, sapphire, polycarbonates, plastics, Si_3N_4, SiC, and even Ti and WC.

Needle-like Applications. An early (ca 1947) widespread use for diamond was that of the phonograph needle. These needles are classified by their conical, ellipsoidal, and solid type tips. Each requires precision brazing, cutting, and polishing. Conformally deposited diamond films offer potential means of decreasing production cost.

Various surface profilometers use diamond tips as do tools to scribe and separate the die of integrated circuits and to serve as indenters for hardness testers. The use of conformal diamond coatings would largely eliminate the cutting, lapping, and polishing required by present manufacturing techniques.

Anvils. In modern high pressure research the diamond anvil is an essential tool. While pressures of up to 4 megabar are known, 0.5 megabar pressures are more typical. Both synthetic (high pressure produced) and natural diamond is currently in use. Artifact films have not yet been produced to thicknesses useful for this application.

6.0 OPTICAL APPLICATIONS OF DIAMOND

Optical applications of diamond are both passive and active. The earliest of these was, of course, passive. The passive applications generally exploit the low absorption coefficient, the high thermal conductivity, the corrosion resistance, the extreme hardness and even the extremely low coefficient of friction.

6.1 Windows

A celebrated early use of a diamond window was for the IR emission sensor of the Venus explorer. Another proposed optical application of diamond exploits not only its very low absorption coefficient, but also its very high thermal conductivity (23). This particular combination of properties permits diamond to be used as a window for high power lasers such as the free electron laser (FEL). Conventional materials (a) absorb too much of the laser energy, (b) may become nonlinear under the intense energy levels impinging, and (c) simply can not dissipate the heat from the center of the window or lens to the outer periphery where a cooling heat sink may be attached. The use of diamond not only reduces absorption by several orders of magnitude, but the absorbed energy is quickly conducted to the window periphery where it may be dissipated by various heat sink and water cooling techniques. Optical mirrors are also similarly vulnerable to high energy laser beams. Diamond mirrors or diffraction gratings have been calculated to reduce the length of a 2500 meter free-electron laser to but 15 meters and that of a 60 meter FEL to less than 2 meters by reducing the need to diverge the beam to lower power density. Mirror diameter is proportionately reduced.

The combination of its low absorption coefficient and its extreme hardness suggests the use of diamond coatings or films for the protection of other materials that are not so endowed. One such example is that of infrared (IR) windows. Germanium windows are widely used today as they are isotropic and exhibit very little distortion. They are, however, comparatively soft and subject to microparticle abrasion and rain droplet erosion. This vulnerability is greatly exacerbated when such windows are moving at high velocities with respect to their environment—such as on an aircraft or missile. Diamond may be used to advantage in this application either as a coating for conventional windows or as stand alone windows. In the latter application, the extreme strength of diamond may be used to create larger apertures without requiring excessive thickness of the window material.

The marine environment presents rather demanding requirements for optical windows. Periscope windows must not only resist corrosion from the salt water, but also require strength to encounter extremely high pressure differentials and must resist the attachment of barnacles and other marine life. Even more, when a periscope window emerges from its underwater environment, the tribological properties of diamond ensure that water droplets do not adhere to its surface and obscure vision. While it is not known why the smooth diamond surface resists the attachment of marine life, it is thought that its teflon-like coefficient of friction combined with its extremely short lattice constant prevent tentacles (however small) from obtaining a "toe hold". Unlike germanium or silicon, the unterminated diamond surface has a propensity to form pi bonds among its adjacent carbon atoms. Once formed, it is energetically difficult to break these bonds. With the diamond surface reconstructed solely with these pi bonds, other prospective chemical bonds (almost always of lower energy than the pi bonds on the diamond surface) can not chemically bond to its surface.

Stripper Foils. To obtain positively charged ions for use in high energy accelerators, electrically conductive stripper foils are placed in the path of a collimated beam of negative ions. In passing through the stripper foil, the ions are stripped of their electrons thereby creating positive ions. By virtue of the positive potential applied to the stripper foil, the nascent positive ions are repulsed—thereby further accelerating them. Conventional stripper foils are made of hard carbon films and their thickness per unit available area is typically between 2 and 10 micrometers/cm^2. Desired properties for these films are electrical conductivity (to act as an accelerating anode), low atomic number (to provide low cross sectional absorption to the beam), and strength (to enable the thinnest possible films). They must also provide useful operating lifetime. P-type semiconducting diamond would appear to be ideally suited for such an application. The first known attempt at such utilization provided mixed results (24). Subsequent to that work diamond films exhibiting thickness to area factors of 0.3 micrometers/cm^2 have been fabricated by several different facilities and would undoubtedly exhibit better performance than did those originally tested.

6.2 Optical Matching

Diamond has a refractive index of 2.4. This is lower than that of most semiconductors, but higher than that of typical dielectrics. A interface coating whose optical thickness is a quarter wavelength of the incident wave and whose refractive index is suitably low may be used to reduce any reflection caused by the diamond index of refraction. If the index of refraction of diamond is n_3 and that of the media to which diamond is to be matched (e.g., air or vacuum) is n_1 then the refractive index n_2 of the coating material must be the geometric mean of the two:

$$n_2 = (n_1 n_3)^{-1/2}$$

To match the diamond to air interface where n_1 is 1.00 and n_2 is 2.4, the coating index must be 1.55. If so, the reflectivity of a normal incident wave would be zero. From Table 9, it is seen that crystalline quartz would be ideally suited for this purpose. Unfortunately, coating diamond with quartz or virtually *any* other material would severely reduce resistance to microparticle abrasion, erosion, corrosion, and other superlatives for which diamond was chosen in the first place. Fortunately, there is a solution shown in the section below (Moth's Eye "Coatings").

The magnitude of the reflection of an optical coating of an underlying material having a higher index of refraction is given by (25):

$$R_1 = [(n_1^2 - n_0 n_2)/(n_1^2 + n_0 n_2)]^2$$

where n_0, n_1, and n_2 are the respective refractive indices of air (vacuum), the coating film, and the underlying substrate. IR detectors such as silicon, germanium, the II-VI materials and the lead salts typically have refraction indices much greater than that of diamond, therefore diamond becomes a coating material of choice not only because of its mechanical and chemical durability, but also because of its refractive index. As an example, the surface reflection of a silicon detector at 0.6 nm is approximately 30%, but this can be reduced to about 6.6% by a single quarter wave coating of diamond film. Greater than 90% transmittance on silicon is typically achieved by 0.25 nm diamond films over a spectral range of 0.1 to 2.5 nm. There is an imperceptible difference in transmittance over a temperature range from 15 to 300 K. An even greater improvement is engendered for longer wavelength detectors having higher refractive indices. Silicon solar cell efficiency has been increased by as much as 40% while that of germanium cells can be improved by up to 88% (26). While diamond-like carbon (amorphous diamond with predetermined hydrogen concentration) coatings may be used in certain applications, absorption is generally greater and many of the superlative mechanical properties of diamond are lost.

Moth's Eye "Coating". The effective refractive index of any material may be reduced by rendering the material somewhat porous providing that the porous voids are a very small fraction of a wavelength of the incident light beam. This procedure was taken further by Allen Cox for use in silicon (27) and adapted for diamond by this author as shown in Fig. 2. Small rectangularly-shaped depressions are etched (by a process to be described below) into the diamond surface engendering a surface structure akin to that of a moth's eye. Design dimensions for such a structure are:

$$l < \text{wavelength}/n_2$$
$$d = 0.25 * \text{wavelength}/(n_1 n_2)^{1/2}$$
$$w = (0.5 * l)^{1/2}$$

To increase the effective bandwidth over which the procedure is effective, the walls of the depression may be made curved rather than straight. This same technique may be readily adapted to diamond coatings for IR detectors. The approach may be used as a single layer coating or as a double layer coating wherein the diamond next to the detector material is a continuous film and superimposed upon it is the moth's eye diamond superstructure. Using these techniques, surface reflection may be minimized.

Although diamond is not etched by boiling acids or other common etchants, it is perhaps the easiest of all materials to etch. Using nitrogen dioxide in the presence of a 50 eV noble gas ion beam and with the diamond held at near room temperature, etching rates of 200 nm/min are possible (28). Very high aspect etch ratios (e.g., deep, narrow troughs) are possible as the etchant residue is gaseous carbon suboxides. For requirements wherein etch depth must be controlled to within one atomic layer, a variation of the process known as atomic layer etching is possible (29).

6.3 Fresnel Lens

Although an expendable mandrel may be used with low temperature diamond deposition processes to construct a concave or a convex diamond surface, a simpler approach is warranted in certain lens applications. A comparatively crude Fresnel lens may be made of the diamond surface by depositing concentric circular rings of a opaque metal as shown in Fig. 3.

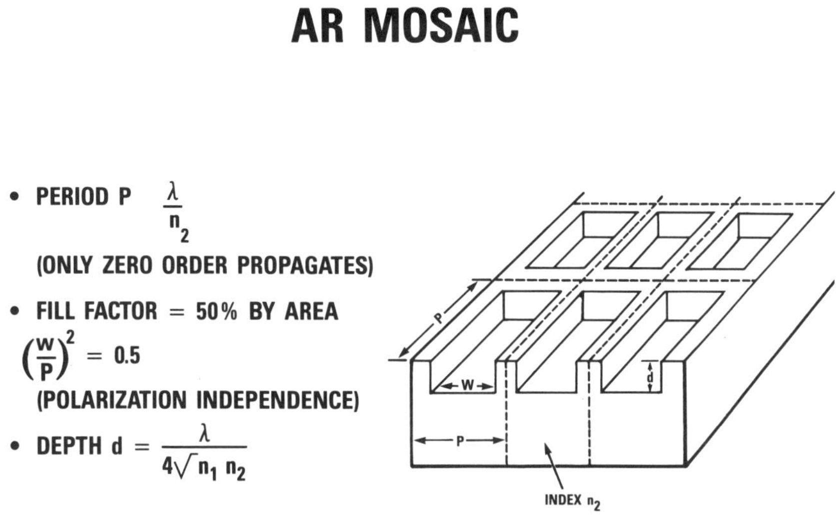

Figure 2. Moth's eye anti-reflection coating of a diamond window.

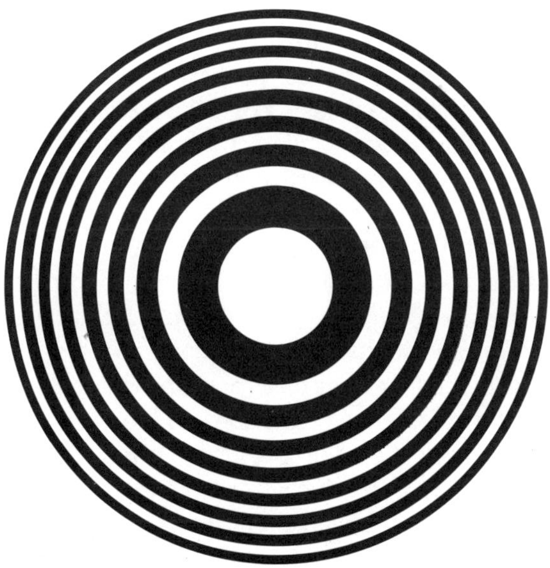

Figure 3. Pattern for diamond fresnel lens.

The positioning of the opaque rings is such that from a point source of light located on axis and a focal distance away from the surface, alternate half-wave zones are in phase with each other. To accomplish this, the radii of these zones are made proportional to the square root of their integer number. A lens made in this manner will, however, pass only one half of the incident light at most. If the opaque regions were replaced by etched depressions in the diamond, then the entire incident beam (less material absorption and reflection losses) could be transmitted if the depressions were of such a depth so as to introduce a half-wave wavefront advance. Superposition of the moth's eye anti-reflection "coating" on the Fresnel etching will further improve the transmissivity of the lens.

6.4 X-ray Applications

X-ray applications of diamond are currently limited to windows and masks. An x-ray window must withstand the atmospheric pressure to vacuum differential; it must be strong and impermeable. It must also be characterized by a low absorption coefficient at short wavelengths. Since diamond is the strongest of all contending materials, it can be made very thin such that absorption is decreased. Being thin, however, increases the possibility of permeability. To maintain the pressure differential and impermeability, the crystallite size in these diamond films must generally be kept smaller than 30 nm. The permeability is typically ascertained and characterized by a helium leak test. While the x-ray absorption of diamond is not as low as that of boron, diamond is much stronger; the window thickness can therefore be made much thinner and this more than compensates for the absorptivity. Figure 4 illustrates such a window manufactured by Crystallume.

X-ray Targets. Semiconducting diamond may also be considered as a target for the generation of x- rays. K shell electrons can be ejected from the carbon atoms by bombarding the diamond crystal with energetic electrons. In the process, K shell electrons are ejected as Auger electrons. Electrons from the outer L shell transition into the inner K shell and, in the process, emit photons in the x-ray spectrum. Diamond is useful as an x-ray target because its thermal conductivity is so very large and because of its shorter wavelength. While conventional targets require rotation to uniformly distribute the heat induced by the energetic electron bombardment, the thermal conductivity of diamond accomplishes this heat distribution process without mechanical rotation.

Figure 4. Diamond x-ray window. *(Courtesy of Crystallume, Inc.)*

7.0 ELECTRO-OPTIC AND RELATED APPLICATIONS

7.1 Optically Triggered Switches

The ability to generate extremely short pulses of extremely high currents while withstanding extremely high electric potentials and fields is required of certain impulse radar systems. Since intense optical pulses of tens of femtoseconds are comparatively readily available, optical triggering of these high power switches is the method of choice. Other factors to consider are the dielectric constant and breakdown strength of the semiconductor being triggered, the mobility and high field velocity of the charge carriers, the thermal conductivity, and its refractive index (to most efficiently absorb the optical trigger). Tables 1, 9, and 11 show that diamond exceeds all other semiconductors in *each* of these factors. A final consideration is the ability of a diamond semiconductor to absorb the optical trigger. With a bandgap of 5.5 eV, it is obvious that ultrapure diamond would be transparent to the wavelength of ultra short pulse lasers. Fortunately optical absorption centers do exist in diamond as shown in Table 10.

7.2 Lasers

Diamond is not a direct bandgap material, therefore lasing action via the traditional semiconductor band-to-band recombination mode is not possible in the bulk material. Diamond lasers have, however, been demonstrated (30). These have been electron beam pumped with optical generation from impurity centers. Properties of interest when considering the suitability of a laser host material are many. Thermal conductivity is important to insure that all parts of the laser are at equal temperature so as not to distort the wavefront. Here diamond is unexcelled. Thermal expansion coefficient is also important to insure that mirror-to-mirror spacing remains stable to within a fraction of a wavelength. Here again diamond is unexcelled. When pumped with extremely large amounts of power, stress is built up within the crystal and resultant strain can adversely affect mirror-to-mirror spacing. In this respect, diamond has an unexcelled Young's modulus. For high power lasers, the strength of the material (to resist stress) is also an important parameter. At 272 kg/mm^2, diamond is the strongest of all optical materials. Another parameter of interest in selecting a laser host is the reciprocal of the ratio of linear extensional strain to the concomitant lateral contractional strain—otherwise known as Poisson's ratio. For diamond, this value is 0.2 whereas for other candidate optical materials it is generally in the range of 0.24 to 0.35. A critical factor for a laser host material is the manner with which the host crystal interacts with the dopant impurity ions within the crystal. More specifically, it is the manner in which the host crystal fields lift the degeneracy of the electrons in the impurity. This is known as crystal field splitting and values for impurities in diamond are generally unknown. Even more questionable is the size of the impurity ions. The rare earth ions traditionally used in pumped lasers are too large to be inserted into the diamond lattice without defect-producing stress. While smaller impurities (e.g., the transition metals) may be considered, their use is largely unproven.

8.0 BIOLOGICAL APPLICATIONS

8.1 Artificial Heart Valves

Diamond is extremely hydrophobic. Because of this, biological tissue (nominally 98% water) does not readily adhere to diamond, so bacteria and virus are thought not to readily cling to the diamond surface. The combination of the hydrophobic nature of diamond, its non-toxicity, its corrosion-free properties, and its hardness combine to render it an ideal candidate for the coating and/or fabrication of artificial heart valves. Not only should valves so fabricated not corrode, erode, or otherwise wear out, but the hydrophobic nature of the surface suggests that the major problem of conventional artificial heart valves—providing nucleation sites for blood clotting—may be overcome. Current recipients of conventional artificial heart valves must receive blood thinning medication for the duration of their lives. This not only requires careful periodic monitoring, but also restricts activity such as pregnancy.

8.2 Scalpels

Carefully cut and polished natural diamond inserts are currently in widespread use as surgical scalpels. These scalpels are not only efficacious by virtue of their hydrophobic properties noted above, but also because of their sharpness in comparison to surgical steel scalpels. The sharpness to which an edge may be polished is fundamentally dependant upon the hardness of the material. With a radius of curvature some 30 times smaller than the steel equivalents, the diamond scalpels introduce much less tearing of the tissue. As a result, healing is much faster and scar tissue formation is much less. Although cutting can be much faster when using a diamond scalpel, the much smaller radius of curvature on the diamond cutting edge provides much, much less "spring back" force to the surgeon's hand. The "touch factor" requires significant readjustment for the surgeon. With the advent of artifact diamond coatings, the cost and availability of diamond scalpels should be favorably affected. Current applications for diamond surgical instruments include those for such microsurgical procedures as radial keratotomy, cataract extraction, and neurosurgery. Special diamond knives are used in coronary bypass surgery (anastomosis of the coronary artery). Such a knife eliminates the need for changing different knives and scissors during coronary surgery. Larger diamond knives are in use for extracting excess "gristle" in the hand in patients with Dupuytrens syndrome (failure of the hand to fully close).

8.3 Marine Fouling

Hundreds of millions of dollars are spent worldwide each year overcoming the costs of barnacles on ships. These salt water creatures attach themselves to the hulls of ships with their glues of repetitive cross-linking protein strands. Their attachment significantly increases ship drag resulting in much increased fuel costs in addition to the costs of placing the ships into drydock to scrape and refinish the hulls. Although arsenic based paints have proven partially effective in discouraging barnacle attachment, they are not entirely effective and the yearly costs stated above are considerable. While significant technical and economic problems remain to be solved, it is not inconceivable that diamond plating of ships and other underwater objects can be made to be cost effective.

ACKNOWLEDGEMENT

This chapter was prepared by a U.S. Government employee. Pursuant to Title 17, U.S. Code, Section 105, it is not subject to copyright.

REFERENCES

1. Angus, J. C., Buck, F. A., Sunkara, M., Groth, T. F., Hayman, C. C. and Gat, R., *MRS Bull.* XIV:10 (1989)
2. Nemanich, R. J., Shroder, R. E., Glass, J. T. and Lucovsky, G., *Proc. of the 19th Int. Conf. on the Physics of Semiconductors,* (W. Zawadzki, ed.) p 515, Inst. of Physics, Polish Acad. of Science, Warsaw (1988)
3. Seitz, R., *The End of the Age of Sand,* Presented at the Gorham Advanced Materials Institute, Marco Island, FL 15-17 October (1989)
4. Anstis, G. R., Chantikul, P., Lawn, B. R. and Marshall, D. B., *A Critical Evaluation of Indentation Techniques for Measuring Fracture Toughness: I, Direct Crack Measurements,* Vol 64 No. 9 pp 533-538 (1981)
5. See G. Davies, *Chem. Phys. Carbon* 13:1-143 (1977) for comprehensive listings
6. Williams, A. W. S., Lightowlers, E. C. and Collins, A. T., *J. Phys. C* 3:1727 (1970)
7. Dean, P. J., Lightowlers, E. C. and Wight, D. R., *Phys. Rev.* 140:A352 (1965)
8. Prosser, V., *Czech. J. Phys. B* 15:128 (1965)

9. Rauch, C. J., *Phys. Rev. Lett.* 7:83 (1961)

10. Koronova, E. A. and Shevchenko, S. A., *Sov. Phys. Semicond.* 1:299 (1967)

11. Russell, K. J. and Leivo, W. J., *Phys. Rev. B:* 6:4588 (1972)

12. Markunas, R. J., private communications

13. Canali, C., Nava, F., Gatti, E., Kozlov, S. F., Manfredi, P. F., Manfredotti, C. and Quirini, A., *Nucl. Instrum. Methods* 160:73 (1979)

14. Reggiani, L., Bosi, S., Canali, C., Nava, F. and Kozlov, S. F., *Solid State Commun.* 30:333 (1979)

15. Rudder, R. A., Posthill, J. B. and Markunas, R. J., *Elect. Lett.* 25(18):1220 (1989) and private communications relating thereto

16. Ferry, D. K., *Phys. Rev. B:* 12:2361 (1975)

17. Bernholc, J., Antonelli, A. and Kajihara, S., presented at the *Fourth SDIO/IST-ONR Diamond Technology Initiative Symposium,* Crystal City, VA (12 Jul 1989)

18. Geis, M. (private communications)

19. Yoder, M., "High Power Diamond Traveling Wave Amplifier," U.S. Patent 4,929,986

20. Yablonovitch, E., Kapon, E. Gmitter, T. J. Yun, C. P. and Bhat, R. *IEEE Photonics Tech. Lett* 1:41 (1989)

21. Enomoto, Y. and Tabor, D., *Proc. R. Soc. Lond.* A373:405-417 (1981)

22. Tabor, D., *The Properties of Diamond,* Chapter 10, (J. E. Field, ed.) Academic Press, London (1979)

23. Seitz, J. R., "Laser Systems With Diamond Optical Elements," U.S. Patent 3,895,313

24. Phelps, A. W. and Koba, R., *Proc. First Int'l. Symp. on Diamond and Diamond-Like Films,* The Electrochemical Society , Vol 89-2 (1989)

25. Hovel, H. J., *Semiconductors and Semimetals* (Willardson and Beer, eds.), Vol. II, p. 204, Academic Press, New York (1975)

26. Altshuler, A., *Diamond Technology in the USSR,* Delphic Associates Press (1989)

27. Cox, A., Honeywell, Inc. (used by permission)

28. Geis, M. W., Rathman, D. D., Ehrlich, D. J., Murphy, R. A. and Lindley, W. T., presented at the SDIO/ONR Diamond Technology Initiative Workshop, Lexington, MA (1987)

29. Yoder, M. N., "Atomic Layer Etching," U. S. Patent # 4,756,794

30. Rand, S. C. and DeShazer, L. G., *Optics Ltrs.* 10(10):481-3 (1985)

2

Synthesis of Bulk Diamond: History and Present Status

Kurt Nassau

1.0 INTRODUCTION

Before the advent of modern science, alchemists sought the philosopher's stone which would turn lead into gold and provide eternal youth. Had the alchemists known that diamond was a form of carbon, there is little doubt that the philosopher's stone would have acquired the additional task of converting charcoal into diamond. Interestingly enough, two of these three tasks, the transmutation of lead into gold and the conversion of charcoal into diamond can now be achieved, albeit not on an economical basis; only the secret of eternal youth continues to remain beyond our grasp.

Attempts at diamond synthesis began almost as soon as it was recognized that diamond is the high pressure form of carbon. Unfortunately, this phase of the story, described briefly in Sec. 2.1, is filled with claims both sincere as well as fraudulent which could not be verified by other investigators. It was not until the 1955 announcement from the General Electric Research Laboratories (1) that a process was described in detail which could be duplicated by other scientists and thus met the criteria for acceptance as a valid achievement.

Until quite recently, work on the synthesis of bulk diamond has been inspired by the aura attached to its gemstone use. This same driving force led to the synthesis of ruby and sapphire by Verneuil at the end of the 19th century (2), the laboratory duplication of emerald by Chatham in the 1930's (3) followed by the 1955 diamond announcement from GE (1). With the successful laboratory duplication of opal by P. Gilson in 1972 (4), the five major gemstones have now yielded up their secrets. These and other aspects of man-made gemstones have been summarized in Ref. 5.

Before diamond could be duplicated, high pressure techniques had to be perfected. The two outstanding innovators in this field are P. W. Bridgman and H. T. Hall, and brief biographical sketches and outlines of the various high pressure techniques are given. Details of the high pressure GE process are reviewed together with a discussion of the shock wave and sintering techniques (excepting thin film growth, however), as well as the current status of availability and prospects for the future.

2.0 THE EARLY HISTORY OF DIAMOND SYNTHESIS

2.1 Historical Aspects

It is now 35 years since the first successful synthesis of diamond, and 20 years since the duplication of gem size diamonds. This is a sufficiently long period so that, without ignoring the tremendous excitement engendered at the time, one can nevertheless now look back at these achievements with some of the balance of the historian and try to obtain a better perspective.

Certain principles have been developed over the years for evaluating scientific accomplishments and attributing priorities. For example, the claim of a successful experiment cannot be fully accepted until it has been duplicated by other experimenters, following the published descriptions of the technique. It is obviously the claimant's fault if inadequate descriptions do not permit a successful duplication, and the claims then have to be discounted.

If more than one experimenter claims priority in an achievement, then independent proof of achievement at the time claimed is essential

Patent applications and/or publication of reproducible experimental details, examination of the product by independent experimenters, deposition of the "secrets"' in a sealed document (as was done by Verneuil) (2), all have been used to establish priority of inventions. The patent system is a protection for the inventor and his firm, so that it is not necessary to resort to keeping one's secret to protect an invention. In fact, in today's sophisticated technology, there are essentially no discoveries which one could hope to keep as a trade secret; merely the knowledge that something can be done is sufficient encouragement for others to follow the hidden footsteps and duplicate the secret process, usually in a surprisingly short time. There are several points in the history of diamond synthesis where these principles of establishing achievement and/or priority are relevant.

2.2 Early Diamond Claims

In 1797 Smithson Tennant demonstrated that diamond was a form of carbon; this is readily shown by burning a diamond in oxygen, resulting in carbon dioxide gas with just a minute trace of ash remaining. Since diamond has a higher density, 3.52 gm/cc, than other forms of carbon, e.g., the 2.25 gm/cc of graphite, it became immediately plausible that pressure (which produces a smaller volume and therefore a higher density) might convert other forms of carbon into diamond.

Over the course of the following one and one-half centuries there were many experimenters who claimed to have succeeded in this quest. Perhaps the best-known of these were J. B. Hannay and Henri Moissan.

J. B. Hannay reported success in 1880 at making diamonds (6) by sealing organic material with lithium into iron tubes and heating them to red-heat. The majority of his tubes exploded, but some that did not were claimed to contain diamonds.

H. Moissan was a contemporary of Verneuil in Paris. Just before the turn of the century he used the known solubility of carbon in solid iron to attempt diamond synthesis by quenching such a solution from white heat into water. The pressure generated by the contraction of the outer parts of the iron was claimed (7) to produce diamonds.

Over a thirty year period C. Parsons attempted to duplicate these and other diamond growth experiments. He was at first misled by insufficiently detailed study of the product, but finally came to conclude (8) that neither he nor anyone else had succeeded.

It would be tedious to go through the list of other claimants, and quite pointless, since we can discount all their claims on three points: *(i)* none of their processes has proved to be reproducible, which alone is conclusive; *(ii)* from our present knowledge of the conditions necessary for diamond formation we can deduce that none of these early experimenters reached such conditions; and *(iii)* detailed examination techniques have shown that the specimens claimed to be synthetic diamonds were either other hard substances, such as spinel $MgAl_2O_4$ or moissanite SiC, many of which crystallize in the cubic system, or were indeed diamonds, but of the natural kind. Indeed Moissan's widow said to Parsons who reported (9) this to Bridgman that one of Moissan's assistants had introduced natural diamond chips "to please the old man" and cut short the tediously long experiments! Hannay's diamonds have recently been examined by cathodoluminescence by Collins (10) who concluded that these were natural and not synthetic diamonds.

The only other claim that requires comment is that of H. Liander and E. Lundblad at Allemanna Swenska Elektriska Aktiebolaget at Västerås, Sweden. This work was begun in 1942 and ultimately led to some very limited success. It was not described publicly until years after the GE work had been announced and, accordingly, priority is not usually attributed to this work. Details have been given elsewhere (11).

2.3 The Diamond Problem

What then is the difficulty in making diamond in the laboratory? We can isolate three major problems. First, there is difficulty in achieving the compact, strongly-bonded structure of diamond, unique in being the hardest material known to man; to do this requires extreme high pressure. Second, even when such a high pressure has been achieved, a very high temperature is needed to make the conversion from other forms of carbon to diamond proceed at a useful rapid rate. Third, when diamond is thus obtained, it is in the form of very small grains; to achieve large single crystal diamond involves yet another set of constraints.

The difference in structure between graphite and diamond is shown in Fig. 1. Diamond has tetrahedral bonding in which all bonds are essentially purely covalent sp^3 and equivalent, all 1.54 Å in length. The only weakness in this structure lies in the octahedral planes, where diamond crystals can be cleaved with relative ease. Graphite, however, has hexagonal sp^2 sheets as shown in Fig. 1, with strong, short carbon-carbon distances of 1.42 Å in length in the sheet layers, but the bonds between the sheets are 3.35 Å long. These long bonds are so weak that even a small force can break them, resulting in the lubricating properties of graphite. Most other forms of carbon, such as charcoal, consist of disordered fragments of the graphite structure.

Figure 2 shows how the state of carbon depends on the temperature, plotted horizontally, and the pressure, plotted vertically. Pressures are given on the left in bars and atmospheres (these are almost but not quite the same) and on the right in pounds force per square inch (psi). Much study by many investigators was required to establish the details of this figure, and a description has been given by Bundy (12).

In the bottom region of Fig. 2, at ordinary pressures all the way up to the melting curve near 4000°C, graphite is the stable form of carbon. The only reason diamond is able to exist on the surface of the earth as a metastable form is because the temperature is too low for its bonds to break and re-form as graphite. If, however, a diamond is heated to over 1500°C

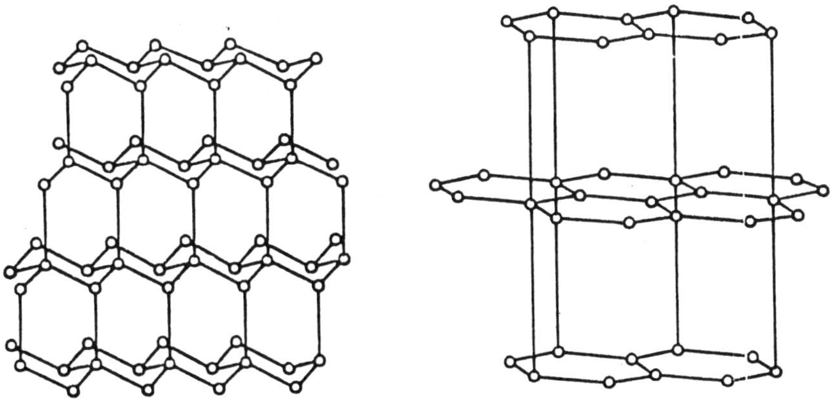

Figure 1. The structures of diamond (left) and graphite (right).

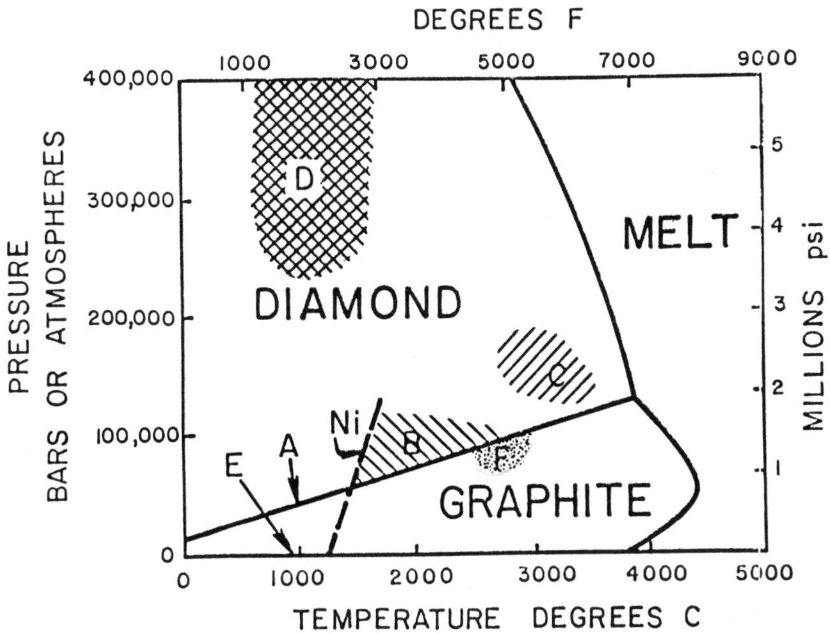

Figure 2. Phase diagram of carbon. Diamond can also exist metastably in the graphite region and graphite in the diamond region. After Bundy (12).

at ambient pressure in the absence of air, it will convert to graphite and remain as graphite.

Only above the line marked A in Fig. 2 is diamond the stable form of carbon. Again, it is possible for graphite to exist metastably above line A if conditions are not appropriate for its conversion to the diamond which is stable there. In Fig. 2, there are four regions of particular interest where diamond has been made from other forms of carbon, marked B, C, D, and E.

The region marked B represents conditions under which most synthetic diamond has been prepared by the General Electric process as first announced in 1955 (1) with details in 1959 (13). This can be done at pressures as low as about one million psi and temperature below 2000°C if a metallic solvent is used. The line marked Ni in Fig. 2 shows the melting curve of nickel in the presence of carbon, and to the right of this line the molten nickel can dissolve graphite, for which it has a high solubility, while at the same time crystallizing out diamond, for which its solubility is much less. At one time it was thought that the metals used, such as Ni, acted as a catalyst, but it is now clear that the metals serve purely as solvents (14) and that the growth process is analogous to the travelling zone solvent technique.

At higher temperature and pressures, such as in region C, diamond has been crystallized without any metal solvent as described in 1962 by Bundy (12)(15). This is, however, a more difficult region to work in than region B. In region D, using much higher pressures but much lower temperatures, diamond is also produced by the shock wave technique as described in Sec. 4.2. Thin film growth at low to ambient pressures, as described elsewhere in this volume, occurs in region E of Fig. 2, and the sintering technique for carbonado in region F is described in Sec. 4.1.

2.4 P. W. Bridgman and High Pressure Research

The difficulty in diamond research has always been the simultaneous achievement of sufficient high pressures and temperatures, and the work of many experimenters over the years contributed toward the solution of the problems involved.

Percy Williams Bridgman (16) (Fig. 3) was, without doubt, the most prolific and well-known worker in the field of high pressure. In addition to many awards and prizes, he received the Nobel Prize which was awarded to him in 1946 "... for the invention of an apparatus to produce extremely high pressures, and for the discoveries he made thereby in the field of high pressure physics."

Figure 3. Dr. P. W. Bridgman. *(Photograph courtesy of Harvard University News Office.)*

He was born in Cambridge, Massachusetts, on April 21, 1882. His father, Raymond L. Bridgman, was a newspaper correspondent and the author of books on public affairs. Percy attended public schools in Newton, Massachusetts, but all of his subsequent education and career took place in the town of his birth, for he obtained his B.A. in 1904 (summa cum laude), his M.A. in 1905, and his Ph.D. in 1908, all from Harvard, and stayed on there as research fellow, instructor, and up the professional ladder, ending with professor emeritus which he attained in 1954. The day after his death on August 20, 1961 in his eightieth year, Harvard received the index he had compiled for the collection of his complete papers in seven volumes, published in 1964 (17).

The thesis he proposed for his graduate work was on the effect of pressure on the index of refraction, and he remained with various aspects of high pressure for the balance of his long scientific career. Even in his thesis work he immediately encountered the restricting limitation of experimental technique, and began a series of improvements which revolutionized the high pressure field. When he began his work, he had to use a screw compressor turned by a six foot wrench to achieve less than 100,000 psi. Yet, by the end of his career, he was able to achieve over six million psi. One of his first inventions was a self-sealing high-pressure packing which functioned without leaking and permitted hydraulic operation, the so-called "Bridgman unsupported area seal" still in full use today.

The limit of his technique, the "Bridgman Anvil" is illustrated in Fig. 4. The tapered anvils A and B are made of cobalt-bonded tungsten carbide, one of the strongest materials known. This material has a stress limit of less than one million psi, but this can be safely exceeded by the technique of "massive support," namely by using a small taper on the anvil surfaces and supporting the outside of the anvils against rupture by shrunk-on steel rings C and D. The sample E (admittedly of only small size) is surrounded by a ring F of pipestone (catlinite); this has the interesting property that under high pressure it squeezes out into a thin gasket which adheres well to the anvil and does not extrude further.

Bridgman's hydraulic seal, massive support, and non-extruding gasket are essential parts of all modern high pressure apparatus. Although he could reach over six million psi, a pressure more than adequate for diamond synthesis, this eluded his grasp because of temperature limitations; in the 1000 to 2000°C region he could still only reach one half million psi at most.

As a scientist Bridgman tended to work alone, with rarely more than two graduate students. He designed and built almost all his own equipment, and left the task of collecting data mostly to assistants. As described by J. C.

Slater (18), one of his students, even in a time of great cooperative research projects he remained an old-fashioned physicist, with traits of honesty and straightforwardness combined with good judgment and deep theoretical insight. The simplicity of his apparatus and the mastery of his manipulative techniques were reflections of his mechanical genius (16). Some of his equipment is now at the Smithsonian Institution in Washington, D.C.

As a result of his teaching assignments he became interested in the logic and philosophy of physics, and a number of his later articles (of a total of over 260), and most of his thirteen books deal with this subject.

He married Olive Ware in 1912 and had two children. His hobbies included music, chess, handball, gardening, mountain climbing, and photography which he pursued with the same concentration and perfectionism as his technical work (16). He was president of the American Physical Society in 1942 and a member of the National Academy of Sciences and of the American Academy of Arts and Sciences.

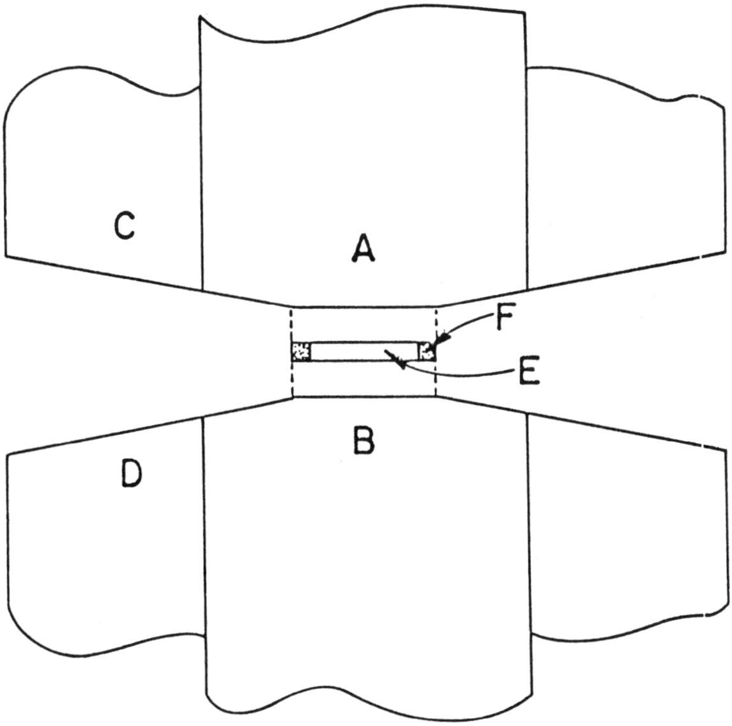

Figure 4. The Bridgman anvil.

3.0 THE HIGH PRESSURE GROWTH OF DIAMOND

3.1 The General Electric Diamond Project

Already in 1941, there had been agreement between the General Electric, Carborundum, and Norton Companies and P. W. Bridgman to investigate diamond synthesis, but this effort ended prematurely because of the war. Nevertheless, graphite at almost half a million psi had been heated by a thermite reaction to 3000°C for a few seconds (19)-(20). Some work continued at the Norton Co., but was apparently unsuccessful and never published.

In 1951 a new high pressure diamond group was formed at General Electric under H. A. Nerad. At first there were F. P. Bundy and H. M. Strong, both physicists, who were shortly joined by H. T. Hall and R. H. Wentorf, Jr., both physical chemists, and later by two engineers, J. E. Cheney and H. P. Bovenkerk. Modifications of the Bridgman anvils first by Bundy, and then by Strong raised the pressure capability to almost one million psi with high temperatures, but it was a series of novel designs by Hall, culminating in his "belt" apparatus (21) of 1953 which permitted success. This apparatus was later shown to be capable of reaching three million psi and 5000°C simultaneously (19), permitting stable conditions to be maintained for an hour or more.

The belt utilized the Bridgman concepts of massive support and nonextruding gaskets (now made of pyrophyllite, a hydrous aluminum silicate, coated with a thin dusting of hematite (Fe_2O_3) to further increase the friction), but added two new concepts. By using strongly tapered pistons, A and B in Fig. 5, a large piston motion became possible without squeezing the pyrophyllite gasket too thin; the compressed gasket in turn provided additional strength by giving support to the pistons. This arrangement permitted a very large sample volume S. The separate "belt" C, also supported the outside of the sample volume and interacted with the two tapered pistons to complete the arrangement. Both the pistons A and B and the belt C were made of cobalt-bonded tungsten carbide and were supported by several rings of special steel such as D, E, and F shrunk on to place the tungsten carbide into compression.

The sample S was contained in a small graphite tube which bridged the gap between the two anvils A and B. Since the pyrophyllite is an insulator, an electric current passed from one anvil to the other through the graphite tube and heated the sample inside it. Figure 6 shows a belt being loaded into a large press by Strong (below) and Wentorf.

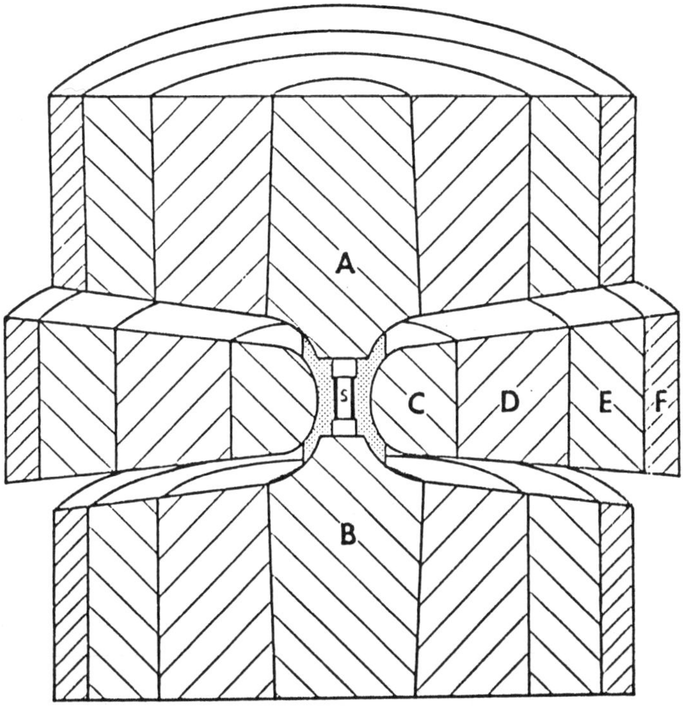

Figure 5. The Hall Belt; the dotted region surrounding the sample S is the pyrophyllite gasket (simplified).

Not only the apparatus design, but also the chemistry and physics were being investigated, and during this period there were several occasions when diamond synthesis was believed to have been achieved. However, none of these promising experiments could be repeated, and thus did not meet the essential test of reproducibility discussed above in Sec. 2.1.

It was finally on December 16, 1954 that Hall obtained diamonds, and during a series of twenty-seven runs over the next fifteen days, eleven more successfully produced diamonds. On December 31, an independent experiment by H. Woodbury and two weeks later three more runs performed in Hall's absence all successfully produced diamond (19)(20).

These initial experiments were performed in a leaky old 400 ton Watson Stillman press and needed only a few minutes at the growth conditions. They have since been widely reproduced elsewhere. In the first run, the iron sulfide FeS (troilite) had been added to the graphite, based on the diamond-troilite association in the Canon Diablo meteorite (Arizona Meteor Crater). During the series, Hall found that iron was necessary, but sulfur was not. An interesting personal account has been given by Hall (22).

Figure 6. A belt being loaded into the largest hydraulic press of General Electric by Strong (below) and Wentorf (above). *(Photograph courtesy of the General Electric Co.)*

Over the next few months the details of the process were worked out by the group and the first announcements appeared under the names of Bundy, Hall, Strong and Wentorf in 1955 (1). A historic group portrait is that of Fig. 7, which shows the original GE diamond team of Bundy, Strong, Hall, Wentorf, Nerad, and Cheney around a model of an ultra-high-pressure hydraulic press.

Figure 7. The original GE diamond team. Left to right: F. P. Bundy, H. M. Strong, H. T. Hall, R. H. Wentorf, Jr., A. J. Nerad and J. E. Cheney. *(Photograph courtesy of the General Electric Co.)*

Francis P. Bundy received his Doctorate degree from Ohio State University in 1937, and stayed on as a faculty member until 1942. After four years at Harvard he joined General Electric in 1946, where he worked on the physics of high pressure and various other physics problems until his retirement. Herbert M. Strong, who obtained his Ph.D. from Ohio State University, worked for 10 years with the Kendall Co. and also joined General Electric in 1946. His main areas of interest have been crystal growth and the properties of diamond. Since his retirement in 1973 he has been working as a research associate at Union College in Schenectady, New York.

Robert H. Wentorf, Jr. obtained his Ph.D. in Chemistry at the University of Wisconsin in 1952; at that time he joined General Electric and has since worked on a variety of chemical and physical problems. In 1957 he achieved the high pressure growth of cubic boron nitride crystals, a chemical analog of diamond. H. Tracy Hall's biography is given in the next section.

Patent applications were begun immediately; based on the filing data given within these patents, the scheme of Table 1 can be constructed. As new information arose, there were several abandonments and re-filings, and the iron, nickel, and tantalum of the first claims (stated to be catalysts but now believed to act as solvents) (14)—were quickly joined by other elements and alloys. Much later Bundy (12)(15) showed that a solvent or catalyst was not necessary, and also (23) that even a simple Bridgman anvil could be used to grow tiny diamonds, once the required conditions were known.

Table 1. The Five Original GE Diamond Patents

	I.	II.	III.	IV.	V.
Patent Applicant	Hall	Hall, Strong, Wentorf	Hall	Bundy	Strong
2/14/55	file 488050	file 488116			
8/29/55			file 530935		
1/10/57		refile 635505			
4/29/57				file 655884	file 655885
1/6/58	refile 707432	refile 707435		refile 707436	refile 707433
6/21/60	pat. granted				
8/20/60		pat. granted	pat. granted	pat. granted	pat. granted
U.S. Patent No.	2,941,248	2,947,610	2,947,608	2,947,611	2,947,609
Subject of Patent	Belt Apparatus	Diamond growth using one of 11 metals	Diamond growth with 3 metals using resistance inflection	Diamond growth using platinum at higher temperatures	Diamond growth using using some alloys

Full details of the processes and patents were held up because of a government secrecy order (22) and details of the process and the belt were first published by Bovenkerk, Bundy, Hall, Strong, and Wentorf in 1959 (13) and Hall in 1960 (21), respectively. Meanwhile commercial production had started at GE in 1957. A description of the full process appears in Sec. 3.3.

Another historic photograph is that of Fig. 8. The occasion here was a special lecture given at the Royal Institution in London, England. Ever since the famous Michael Faraday instituted his Evening Discourse series over a century ago, there has been a continuous series of experimental demonstrations for the public. For the May 2, 1974 occasion, the hydraulic press of Fig. 8, weighing some two tons was shipped specially from Schenectady to London together with all accessories. In front of a large public audience, the first demonstration was the burning of a diamond (from a ring donated by Lady Parker, wife of the director of the Royal Institution) in a flask containing liquid oxygen to show that diamond was indeed composed of carbon.

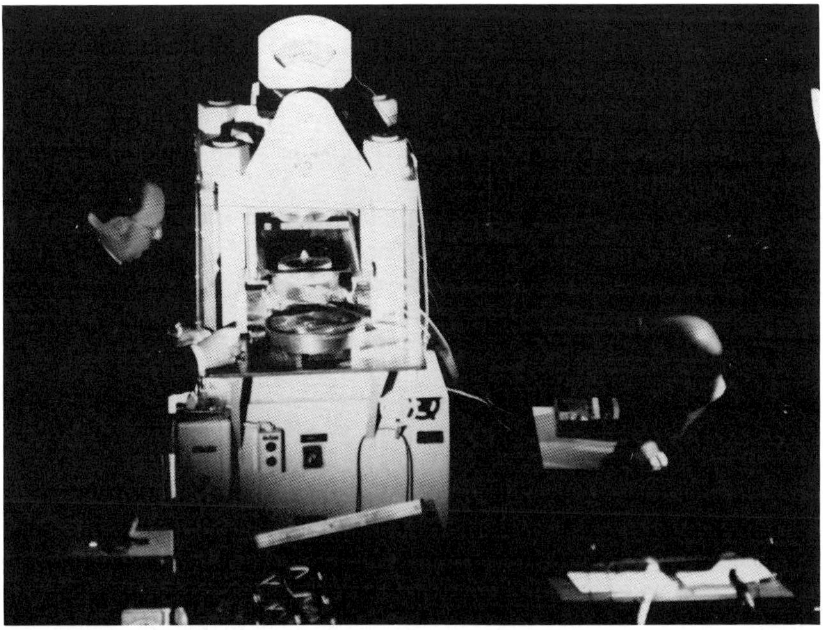

Figure 8. Diamond synthesis demonstration, Royal Institution, London, May 2, 1974. *(Photograph courtesy of the General Electric Co.)*

The highlight of the demonstration was the actual performance of diamond synthesis as in Fig. 8 by Roy Tuft (left), who has performed many of the high-pressure experiments at GE over the years. The audience could view the details of the operations by projected television, while A. M. Bueche, GE's vice president for research and engineering (right), described the processes occurring. At the end of the demonstration Lady Parker was presented a ring containing a faceted man-made diamond to hold as curator for the Royal Institution (and to compensate her for the diamond she had donated).

Faceted synthetic diamonds made by a modification of this technique, described in Sec. 3.4, were donated to the Royal Institution of London and the Smithsonian Institution in Washington, D.C; the latter received a colorless synthetic diamond, the first to have been faceted, weighing about one third of a carat, together with a one carat uncut crystal and a yellow and blue faceted stone on June 17, 1971 (24).

3.2 H. T. Hall, the First Person to Grow Diamond

Based on accounts by Hall himself (19)(21) as well as on a consistent account by G. G. Suits, then Director of Research of the General Electric Research Laboratory (20), we can conclude that Hall was indeed the first person to grow diamond by a revealed, reproducible process. The American Chemical Society awarded Hall its 1972 gold medal for creative invention (25) with an award citation stating "... for being the first to discover a reproducible reaction system for making synthetic diamonds from graphite, and for the concept and design of a super high pressure apparatus which not only made the synthesis possible, but brought about a whole new era of high pressure research ..."

Howard Tracy Hall (Fig. 9) was born on October 20, 1919 in Ogden, Utah, where he grew up on a farm with his parents Howard and Florence Tracy Hall. His scientific training was at the University of Utah where he received his B.S. in 1942, M.S in 1943, and Ph.D. in 1948 under Professor H. Eyring. In 1944 he enlisted and spent two years in the Navy as an ensign, studying electronics at Bowdoin College, MIT, Harvard, and at the Honolulu Naval Base. During his studies he also worked as a chemist at the U.S. Bureau of Mines in Salt Lake City and also as a photographer. It was during his Master's thesis work under Professor V. J. Beard that he first made contact with the problems of diamond synthesis.

Immediately after completing his graduate work, Hall joined the General Electric Research Laboratory. His work in their diamond group was described in the previous section. In 1955, a few months after his success, he left GE and became Professor of Chemistry at Brigham Young University in Provo, Utah, as well as Director of Research for the university until 1967, when he became Distinguished Professor of Chemistry.

Figure 9. H. T. Hall in front of his tetrahedral press. *(Photograph courtesy of H. T. Hall)*

Immediately on leaving GE he faced the problem of not being able to use in his work the high pressure belt apparatus he had just invented for GE (22). However, such is his inventive ability that within two years he had perfected a new device, the tetrahedral press (26), which did not conflict with the belt patent and with which he could attain temperatures and pressures in the same range as those produced by the belt.

In the place of two anvils and a belt, the tetrahedral press uses four anvils which form the walls of a tetrahedral space when they meet. Such a press can be seen in Fig. 9, and details are shown in Fig. 10. The anvils are again made of cobalt-bonded tungsten carbide and supported by massive shrunk-on steel rings. The anvils are shaped so that when they push onto a pyrophyllite tetrahedron that is 25% larger than the final space, some of the pyrophyllite is initially squeezed out to form a gasket, which then permits the pressure to rise as the four pistons push further inward.

For heating purposes a graphite tube A is positioned across the pyrophyllite tetrahedron as shown in Fig. 10; this makes contact with two metal tabs BC and DE, which in turn each contact two of the four anvils. In this way an electric current is passed through the graphite tube to heat a sample inside it.

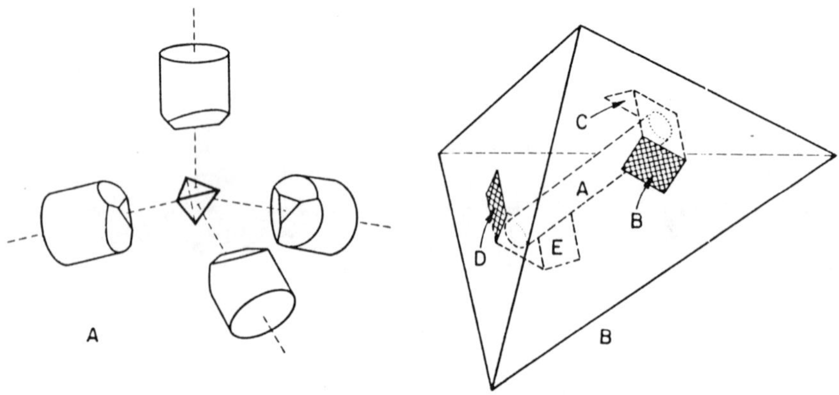

Figure 10. The Hall tetrahedral press: (a) detail; (b) the pyrophillite tetrahedron with graphite heating tube A, and metal contact tabs B - E.

There have since been modifications of the tetrahedral arrangement, such as the use of this geometry in a single-acting press and the addition of extra anvils to form a six anvil "cubic" press; however, the extra complexity of the additional anvils brings some disadvantages. There have also been modifications of the belt apparatus. At present, some twenty years later, Hall's belt and tetrahedral press, or modifications thereof, are nevertheless still the primary tools for ultra high pressure research and manufacturing.

As early as 1958 (27), Hall had proposed the potential usefulness of a sintered diamond compact, equivalent to the naturally-occurring carbonado (see Sec. 4.1 below). In 1966 he founded a company, Megadiamonds, with B. J. Pope and M. D. Horton to manufacture such a synthetic carbonado for commercial tool use (28). He was president and held other offices until 1975.

Among Hall's many prizes and awards is the 1970 Chemical Pioneer Award from the American Institute of Chemists. In the acceptance speech for this award (22), Hall gave a personal account of his view of the circumstances surrounding the first diamond growth at GE.

Hall married Ida Rose Langford in 1941 and has seven children. The fact that six of these are, or are married to, scientists or engineers does seem to tell its own story.

For many years Hall has had his own private machine shop, where he has built and operated his own high pressure equipment and performed research in energy conservation. He is very active in the Mormon Church.

Hall has published a book and close to one hundred research articles; there are also some twenty patents. He retired from teaching in 1978 to be able to devote more of his time to research.

3.3 The High Pressure Synthesis of Diamond Grit

Most manufacture of bulk diamond is designed to produce diamond grit for cutting, grinding, and polishing operations. Single crystal growth will be described in the next section. Wentorf has given a useful review of the high pressure process (29).

Curiously enough, almost any form of carbon can be used as the starting material. One report (30) lists over twenty substances used with success. Some, such as naphthalene (moth flakes), first form graphite, which then converts to diamond. Others, such as paraffin wax or sugar, convert directly to diamond. In five minutes at 2000°C and two million pounds per square inch even peanuts yielded 60% diamond!

The carbon, usually in the form of graphite, is mixed with a metal which serves as a solvent (because of an earlier interpretation, the term catalyst is still frequently used). The metals most often mentioned are manganese, iron, cobalt, nickel, palladium, platinum, and alloys of these metals. Nickel, for example, at 57,000 bars and 1500°C can dissolve about four weight percent of carbon from graphite, which is metastable under these conditions (see Fig. 2). However, since diamond is the stable form, the limiting solubility is that of carbon from diamond, which is only about 3.6 weight percent. Thus the difference of 0.4 weight percent cannot remain in solution and must crystallize out as the form of carbon stable under these conditions, namely diamond (29).

The graphite is packed into a pyrophylite tube with nickel end-plates, as shown in Fig. 11A, and placed between metal and disks into the belt apparatus of Fig. 5 and into a hydraulic press such as that of Figs. 6 or 8.

After raising the pressure so that the sample is at, say, one million pounds per square inch, an electric current is passed from one piston of the press through the metal ends and the graphite to the other piston. Since the graphite has the highest electrical resistance in the current path, it will heat up; the current is adjusted to reach the desired temperature, say 1800°C.

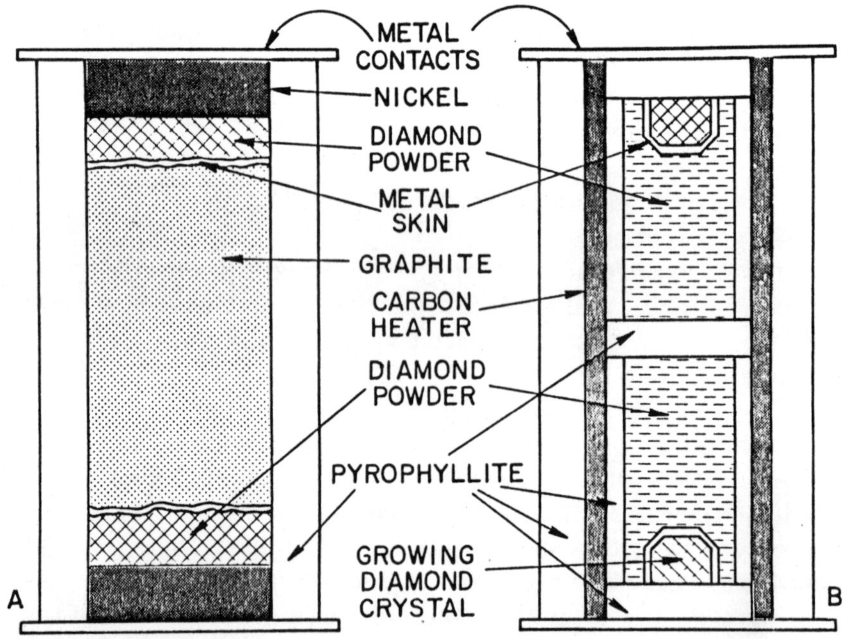

Figure 11. Sample configuration for the growth of diamond grit (A) and single crystal diamond (B).

Under these conditions the nickel melts, dissolves the graphite and begins to crystallize out diamond. The film of metal separating the graphite from the diamond may be less than 0.1 mm thick and movement of the carbon through the metal is, accordingly, quite rapid. In only a few minutes all the graphite will convert to diamond as the nickel moves through it from both ends toward the hotter center.

If the pressure were to be released at this point, the diamond would rapidly convert back to graphite; therefore the pressure is not released until after the electric current has been turned off and the sample has cooled somewhat. All that is then required is to break open the tube and dissolve away the metal in acid, thus releasing the diamond crystals. These are separated by size, usually by selective settling in a heavy liquid where larger particles settle more rapidly.

Depending on the exact conditions, namely, metal solvent used, form and nature of the graphite, temperature, pressure, and time at temperature, the result will be diamond crystals of various sizes and differing mechanical properties, such as friability. A group of crystals, each about ¼ mm across, is shown in Fig. 12. Grit up to ½ mm in size can be made without the necessity of crushing; this gives synthetic diamond its large advantage over crushed natural diamond grit.

Figure 12. Type MBS-70 diamond grit grown by GE, crystals are about ¼ mm across.

The shape of the crystals depends on the temperature, and one study (31) found that there is a change from cubic (100) at 1300°C to mixed cubic and octahedral at 1400 - 1550°C to octahedral (111) at 1600°C, as illustrated in Fig. 13, as the temperature is raised at a fixed pressure of 900,000 psi. The rhombohedral (110) habit was also observed, but only if conditions were in a narrow region close to the diamond-graphite boundary line (Fig. 2).

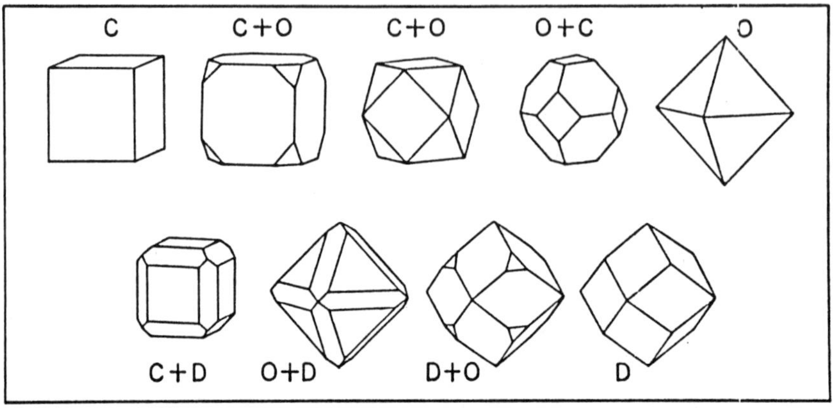

Figure 13. Common habits of high pressure grown synthetic diamond; C = cube, D = dodecahedron, O = octahedron.

The manufacture of cutting blades with GE diamonds has been described (32)(33). For the cutting of rocks, glass, etc. blocky grit sintered in a metal matrix is best. For the shaping of tungsten carbide and hard metals, friable grit in a resinoid matrix is preferred. For compounds (pastes) used for polishing, the grit is typically formulated with olive oil, lithium stearate (a soap), and some amorphous silica (a thickener) for an oil based diamond compound; polyethylene glycol, potassium hydroxide in methanol, amorphous silica, and a little water are used for a "water soluble" diamond compound.

The reason why synthetic diamond grit outlasts natural diamond grit in most uses lies in the ability to control the size and shape of the man-made crystals. By contrast, natural diamonds must be powdered in a mortar to yield grit, a process which involves severe shock and attendant cracking. Many of the resulting natural grit particles will thus contain cracks, which represent points of weakness and can lead to failure of the grains and their loss from the cutting blade. There is no difference in hardness between natural and synthetic diamond.

3.4 High Pressure Single Crystal Growth of Diamond

Whenever one can grow small crystals, it would seem only logical that by growing a little longer one could obtain larger crystals. It is well known in the field of crystal growth, however, that if large and high quality crystals are desired, they must grow very slowly and under much more carefully controlled and very stable conditions. The same is true for the growth of large, industrial or gem grade synthetic diamond crystals. And when the growth of even small crystals is as complicated and difficult as it is for diamond, then rapid progress can not be expected.

It was not until 15 years after the initial announcement from GE, that the achievement of large, gem quality crystals was announced in 1970, with details published by Wentorf, Strong, and Chrenko in 1971 (34)-(36). Details are also given in the excellent review by Wentorf (29).

The general arrangements and procedures are similar to those described in the previous section, except that now seed-crystals are used. These may be thin slices of natural diamond or of previously-grown synthetic diamond. The feed material can be either diamond grit (synthetic or natural) or graphite; the latter rapidly converts to diamond in the first few minutes. The use of grit makes it easier to control the process since there is no shrinkage from the initial graphite-to-diamond conversion.

The arrangement used is shown in Fig. 11B. Because more heat is lost at the end discs, the center of the carbon tube is hotter than the ends where the seed crystals are located. There is thus a temperature gradient, resulting in a higher solubility of diamond in the molten metal solvent in the hotter center of the tube than at the slightly cooler seed crystals. This produces a steady transport of carbon from the grit in the center to the seeds at the ends of the tube. Since the feed diamonds float in the molten metal, they tend to settle on the growing upper seed.

In this way it takes about one week to grow a 5 mm, 1 carat, high quality diamond crystal such as that of Fig. 14. If an attempt is made to grow more rapidly, flaws appear in the crystal, and spontaneous nucleation causes other crystals to grow at the expense of the seed. It is this necessity for slow, controlled growth which makes gem diamond growth so difficult.

The morphology of these large crystals is dominated by the usual cube (100) and octahedral (111) faces. However, dodecahedral (110) and the more complex (113) and (117) faces also occur, particularly if the crystals are pure and therefore colorless. As many as forty faces have been observed on a single crystal (36).

Twins and other unusual features have been observed, and a rare "fiveling" twin is shown in Fig. 15. This has (111) twin planes at each of the

Figure 14. Large synthetic diamond crystal (front) grown from synthetic diamond grit (back). *(Photograph courtesy of the General Electric Co.)*

Figure 15. "Fiveling" twinned synthetic diamond crystal. *(Photograph courtesy of General Electric Co.)*

four re-entrant angles; the fifth re-entrant is only a mismatched meeting.

A group of six crystals weighing about one carat each were faceted in 1970-71 by Lazar Kaplan & Sons of New York City. After the usual cutting losses, the faceted gems weighed from 0.26 to 0.46 carats. Detailed gemological testing was performed on some of these by Crowningshield (37). The conclusion was that these diamonds closely resembled natural ones, and at least one of them would have been accepted as natural at the time. There were, however, sufficient differences so that with experience one could expect to distinguish them from natural diamonds, particularly by using ultraviolet fluorescence and absorption spectroscopy. Colors ranged from a remarkable "F" on the GIA scale (where "D" is the best, colorless, rating) up to "J"; the best clarity grade was a quite respectable "VS_1." The current status of single crystal diamond growth is discussed in Sec. 5.

3.5 Color in Synthetic Diamond

Natural diamonds are classified into four groupings: type Ia diamonds contain platelets, usually attributed to nitrogen conglomerations; type Ib are yellow and contain dispersed nitrogen atoms; type IIa are colorless and essentially free of impurities; and type IIb are blue and semiconducting.

All but the first of these four types have been duplicated in the synthesis process. Essentially colorless type IIa crystals result from the absence of impurities, but special conditions must be used to ensure that all nitrogen is eliminated (29), since as little as a few nitrogen atoms per hundred thousand carbon atoms will result in a strong yellow type Ib crystal. At much over one nitrogen atom per thousand carbon atoms the color deepens to a green, but growth becomes much more difficult, and a significant amount of nitrogen weakens the crystal (29)(36).

In band terminology (38) nitrogen forms a deep donor (n-type, about 4 eV) and Fig. 16 shows (39) how the yellow color results from light absorption at the blue end of the spectrum unlike pure diamond, which has essentially no absorption in the visible region of the spectrum.

The only other elements which have been found to enter the diamond structure readily (29) are aluminum, beryllium, boron, and lithium, all of which produce semiconductivity, yielding shallow levels (38). Of these, only boron was found to produce a blue color, needing only a few boron atoms per million carbon atoms; at one boron atom per ten thousand carbon atoms an opaque blue-black results. Only in this way was it possible to prove (40) that natural blue semiconducting diamonds such as the "Hope" owe their color to a few parts per million boron and not the much higher level of aluminum which is also present and to which the color had previously been attributed. Boron is a shallow acceptor in diamond, (p-type, about 0.4 eV). The spectrum in Fig. 16 shows that the origin of the blue color lies in the absorption of light at the red end of the spectrum.

These absorptions are all very intense, since the transitions causing them are fully allowed. The impurities do not always enter uniformly, depending on growth conditions, and Fig. 17 shows color zoning in a slice from a boron-doped diamond; similar effects have been observed in some yellow crystals.

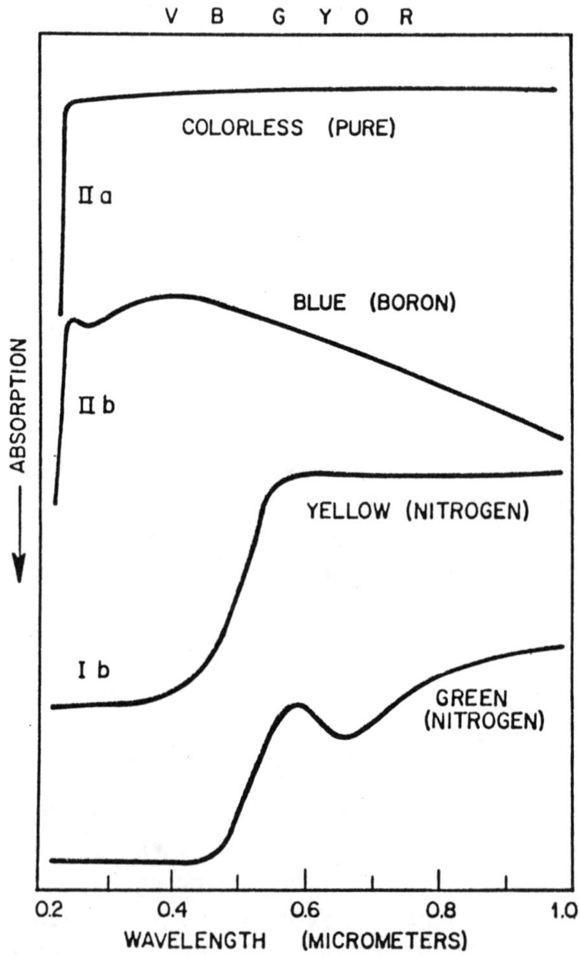

Figure 16. Absorption spectra of synthetic diamond.

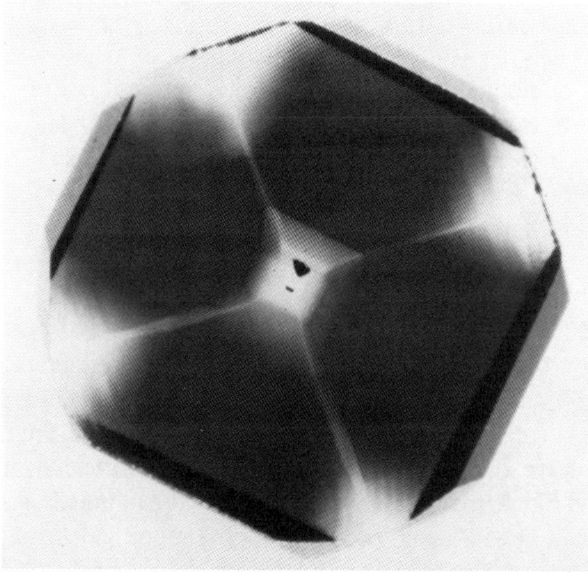

Figure 17. Zoned blue synthetic GE diamond crystal; slice is 5 mm across. Diagonal (100) directions are very low in boron and almost colorless. *(Photograph courtesy of the General Electric Co.)*

4.0 OTHER BULK DIAMOND SYNTHESIS PROCESSES

4.1 High Pressure Carbonado

There is a natural polycrystallite form of diamond which is generally called carbonado. This is usually black and is much tougher than single crystal diamond. Because it consists of many small diamond crystals tightly bonded together, the diamond [111] cleavage which can so easily sever a single crystal does not weaken carbonado; if a cleavage crack starts in one of the crystallites, it cannot propagate through other crystallites, since the orientations are random. Although carbonado may contain significant impurities, there is strong direct carbon-to-carbon bonding between the individual diamond grains.

As early as 1958 Hall and Kistler (41) discussed the possibility of a synthetic carbonado for use as a construction material for high pressure equipment. Today, polycrystalline cemented or sintered diamond is made in several forms by different processes in the USA, in the USSR and elsewhere. A brief discussion is given in the Wentorf diamond review (29).

A close approach to carbonado can actually be an as-grown compact obtained directly from the high pressure apparatus. This consists of grains of diamond cemented together by the solvent metal, with perhaps a little grain-to-grain diamond bonding. The grain-to-grain bonding can dominate under special conditions. It appears that this is achieved at GE by using fairly high pressures and temperatures with boron carbide added to act as a binder and promoter of carbon migration. This is marketed by GE under the name "Compax" described as a "polycrystalline aggregate of Man-Made diamond on carbide substrate."

In 1966 Hall, together with M. D. Horton and B. J. Pope, formed the Megadiamond Corp. to manufacture a synthetic carbonado (42)(43). Their technique appears to use diamond powder under conditions close to or in the graphite stability region; with a very short heating time, little or no graphite is formed. With a relatively low overall pressure in the region marked F on Fig. 2, the concentration of pressure at the diamond-diamond contact points nevertheless produces high enough localized pressure for contact growth. A filler may be used to fill voids, but probably does not add to the strength. Fine powder of either natural or synthetic diamond can be used in the manufacture of carbonado. Diamond powder recovered from cutting and grinding wastes can also be employed.

Synthetic carbonado is used for wire-drawing dies (where its polycrystalline nature produces a more uniform wear pattern than does the usually-used single crystal diamond which tends to wear into an oval). The major application is for metal-machining cutting tools and related uses such as grinding wheel dressing tools. Where employed in this way, carbonado is usually more durable than any other material, even than single crystal diamond, particularly under high stress conditions.

Since it is almost impossible to machine carbonado to shape, it is manufactured in individual pieces close to the desired dimensions. The General Electric "Compax" product is made in thin pieces which are bonded to tungsten carbide tools, while the "Mega-diamond" tools are made as solid pieces to be inserted into tool holders; single pieces over twenty carats have been produced. Other carbonado products include the "Syndite" of DeBeers as well as a similar product made in the USSR.

4.2 The Shock Wave Process

In 1961 DeCarli and Jamieson (44) used an explosive charge to generate an estimated 300,000 atmosphere (4½ million psi) pressure for a millionth of a second by a shock wave in graphite. The result was a small amount of tiny particles of diamond. There has been much subsequent work, with patents by Allied Chemical and DuPont. Workers at DuPont have included G. R. Cowan, B. W. Dunnington, A. H. Holtzman (45) and, more recently, L. C. Trueb (46).

If the instantaneous pressure during such a shock wave exceeds about five million psi at, say, 1000°C, then a curious thing happens. Any metastable graphite still present will melt immediately but this molten carbon will immediately form solid diamond; this is a necessary result from phase diagram considerations not shown in Fig. 2. The only drawback is that, since the pressure drops back down to normal much more rapidly than does the temperature, most of the diamond formed will convert back to graphite.

To avoid this, a mixture of graphite and a metal such as copper is pressed into a steel tube which is surrounded by a hollow tube of explosive with a gap between as shown in Fig. 18. The force of the explosion shock wave is intensified in this "shaped charge" configuration. The copper rapidly absorbs the heat and cools off the formed diamond so that it does not revert to graphite. Pressures used are in the four to seven million psi region with a temperature near 1000°C (see D in Fig. 2). The graphite is at these conditions only for about one fiftieth of a second.

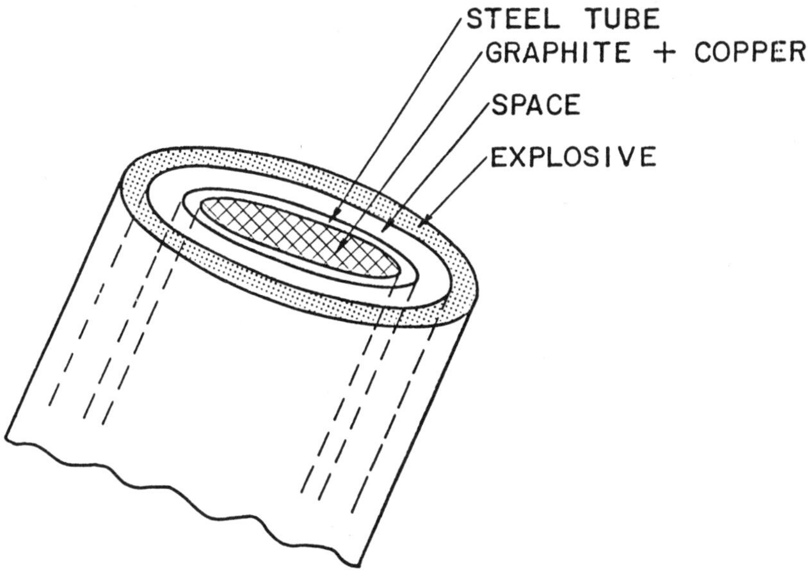

Figure 18. Arrangement for the explosion process for making synthetic diamond.

This process is used on a commercial scale by DuPont at its Gibbstown, NJ facility. As in the GE process, the size of the particles and their structure (friability, etc.) can be controlled by varying the conditions used (nature of metal and explosive, relative amounts, geometry, etc.). After dissolving away the copper with acid, particles from about 50 microns down to less than one micron in size can be obtained and separated, usually by heavy liquid settling.

The structure and cutting characteristics of the shock wave diamond (Fig. 19) have been described (47)(48) and appear to be somewhat different from those of natural or GE process diamonds. The term "diamond abrasive machining" has been used (49). Apparently there is present a mixture of the diamond (cubic) and lonsdaleite (hexagonal) forms of carbon. The structural relationship between these is similar to that of the zinc sulfides: sphalerite (cubic) and wurtzite (hexagonal). Unfortunately, lonsdaleite has been, and often still is, referred to as "hexagonal diamond;" since diamond means cubic carbon, this would then read "hexagonal cubic carbon!" Lonsdaleite was first noted in man-made diamond, and has subsequently been identified in natural meteoritic carbonados (50)-(52).

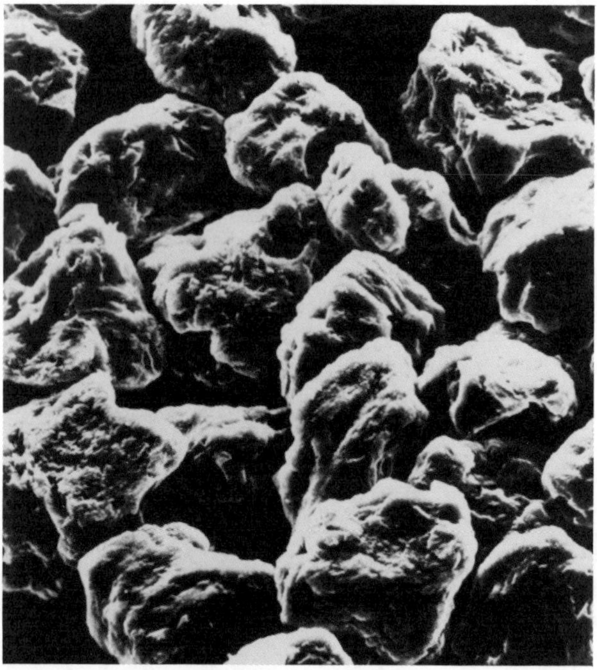

Figure 19. Shock-wave produced synthetic diamond made by DuPont. *(Photograph courtesy of DuPont)*

5.0 THE PRESENT STATUS OF HIGH PRESSURE DIAMOND GROWTH

High pressure growth of grit diamond by the General Electric process, or minor modifications of it, is being conducted on a world wide basis. In fact it is estimated that about ninety percent of all abrasive diamond grit used is now man-made. In the USA, there is GE itself which manufactures at its Worthington, Ohio facility probably between one third and one half of the world's production, an estimated three hundred million carats (sixty-six tons) of grit per year. Diamond grit costs from one to several dollars per carat, depending on type.

The second largest producer is probably the DeBeers organization who purchased from GE the rights to manufacture diamonds in South Africa and Europe with major production facility on the East Rand in South Africa as in Fig. 20, and additional production in Shannon, Ireland, and at the ASEA facility in Robersfjord, Sweden. Smaller production occurs in Japan and West Germany and probably elsewhere. In the USSR diamond research is conducted in Moscow with large production facilities in Poltowa, Erevan, and Kiev.

Estimates for the much smaller production of shockwave-produced grit and synthetic carbonado do not seem to be available.

The question of the commercial possibility for gem grade synthetic diamond is an interesting one. In the past, synthetics have become commercial as lower cost substitutes for natural gemstones when their production costs fell to less than perhaps ten percent of that of the equivalent natural material. This stage has not yet been reached for synthetic diamond. It may well be that in a few more years the synthetic products could come into use if natural diamonds continue to rise in price. The ability to control color, particularly blue, would be an important factor, although irradiated natural diamond might then be a competitor.

In the absence of such gem production, large single crystal diamonds are now manufactured, particularly by Sumitomo in Japan, some for experimental laser windows and some for high power semiconductor heat sinks, for both in up to 5 mm x 5 mm slices. The power handling capacity of a semiconductor device depends on the ability to dissipate its heat, and for this the best heat conducting material known is type IIa diamond. At room temperature this conducts heat more than four times as efficiently as copper. Synthetic diamonds can be grown consistently to be as good as the best of the relatively rare type IIa natural diamonds in this respect.

Figure 20. A production line in DeBeers synthetic diamond facility in the Rand, South Africa. *(Photograph courtesy of DeBeers)*

Equipment capabilities have increased greatly over the years. Belts now routinely have working volumes estimated at over three inches long and well over one inch in diameter. Most diamond grit production is carried out in such equipment in presses of up to 10,000 ton capacity, although some diamonds are also believed to be produced in tetrahedral presses. Individual diamond single crystals up to 17 mm across and weighing over 11 carats have been reported; smaller single crystals are grown in layers, with several crystals per layer, c.f. Fig. 11B, by Sumitomo in Japan. (Note that a Russian claim of a 2 kg crystal turned out to be not synthetic diamond but cubic zirconia, a diamond imitation) (53). Essentially all such material is yellow from the nitrogen content, and may also be magnetic from included Ni (54). Gemological studies have been reported for the Sumitomo commercial production (55) and for the GE experimental product (56).

The 1966 review by Klement and Jayaraman (57) gives a good overview and many references on high pressure research techniques and phase relations and structures of solids at high pressures. This has been updated in the 1976 review by Pistorius (58).

Looking ahead, the present ninety percent of the world's industrial diamond consumption which is now satisfied by synthetic diamond can certainly be expected to increase even further. If the price of natural gem diamond were to increase continuously, as has happened historically, then the time must come when synthetic gem diamond will become competitive. In view of the strong general preference for the natural over the man-made product, the first break-through may come in the use of the blue synthetic diamond; this color natural gem is extremely rare and expensive. Irradiation of certain diamonds can also produce such a color (59), although not of quite the same hue and intensity as the man-made and the best natural blue.

Success in the ability to significantly increase the size of gem grade diamonds by epitaxial single crystal overgrowth by the low pressure technique would certainly have a profound effect. Since there have been no real breakthroughs in this direction despite over thirty years of intensive effort, the prospects are perhaps not too encouraging.

Finally, there is the question: Would the commercial availability of synthetic diamond affect the value of the natural gemstone? We can only attempt to answer this question by examining the past experience with ruby, emerald, sapphire, opal, etc.; in all previous instances, the value of the gems which have been duplicated by man were not affected in the slightest by the availability of the synthetic counterpart.

ACKNOWLEDGMENTS

The authors are grateful for assistance from, or interviews granted by, many of the investigators involved, including H. Bovenkerk, F. P. Bundy, R. C. DeVries, H. M. Strong, R. Tuft, A. W. Urquart, and R. H. Wentorf, Jr., all of General Electric Co., Schenectady, New York and Worthington, Ohio; G. R. Cowan and H. Straw of DuPont Co., Gibbstown, New Jersey; H. T. Hall of Brigham Young University, Provo, Utah; and B. J. Pope of Megadiamond Corp., Provo, Utah. We are grateful to Harvard University for permission to reproduce Fig. 3; to GE for Figs. 6, 7, 14, and 15; to H. T. Hall for Figs. 8 and 9; to DuPont for Fig. 19, and DeBeers for Fig. 20.

REFERENCES

1. Bundy, F. P., Hall, H. T., Strong, H. M. and Wentorf, R. H. Jr., *Nature* 176:51(1955)
2. Nassau, K. and Nassau, J., *Lapidary Journal* 24:1284,1442,1524 (1971); Nassau, K., *J. Crystal Growth* 13:12 (1972)
3. Nassau, K., *Lapidary Journal*, 30:196,468 (1976); *J. Crystal Growth* 35:211 (1976)
4. Darragh, P. J. and Peridrix, J. L., *Gemmology* (London) 15:215 (1975)
5. Nassau, K., *Gems Made by Man*, Gemological Institute of America, Santa Monica, CA (1987)
6. Hannay, J. B., *Proc. Roy. Soc.* (London) 30:188,450 (1880); *Nature* 22: 255 (1880)
7. Moissan, H., *C. R. Acad. Sci.*. Paris 118:320 (1894); 123:206,210 (1896)
8. Parsons, C. A., *Phil. Trans.* A220:671 (1919)
9. Bridgman, P. W., *Scientific America*, 193:42 (November 1955)
10. Collins, A. T., *Industrial Diamond Review* 1975:434 (1975)
11. Nassau, K., *J. Gemmology* 19:660 (1985)
12. Bundy, F. P., *Science*, 137:1057 (1962)

13. Bovenkerk, H. P., Bundy, F. P., Hall, H. T., Strong, H. M., and Wentorf, R. H. Jr., *Nature* 184:1094 (1959)

14. Litvin, Y. A., *Inorganic Materials* 4:144 (1968)

15. Bundy, F. P., *J. Chem. Phys.* 38:631 (1963)

16. Kemble, E. C., Birch, F. and Holton, G., in *Dictionary of Scientific Biography*, (C. C. Gillispie, Ed.,) Scribner, New York, 2:457-461 (1970)

17. *Collected Experimental Papers of P. W. Bridgman* (H. Brooks, F. Birch, G. Holton, and W. Paul, eds.) Harvard University Press, Cambridge, 7 volumes (1964)

18. Slater, J. C., *Colloid Science*, 7:199 (1952)

19. Hall, H. T., *Proceeding of the 1957 Conference on Carbon*, Pergamon Press, New York, pp. 75-84 (1957)

20. Suits, C. G., *The Synthesis of Diamond*, presented to American Chemical Society, Rochester on 11/3/60; also 24 page publication, General Electric Co., Schenectady, New York (1960)

21. Hall, H. T., *Rev. Sci. Instr.* 31:125 (1960)

22. Hall, H. T., *The Chemist*, 47:276 (1970)

23. Bundy, F. P., *Science* 146:1673 (1964)

24. Anon, *Lapidary Journal*, 25:754 (1971)

25. Anon., *Chem. and Eng. News* 50(12):45 (1972)

26. Hall, H. T., *Rev. Sci. Inst.* 29:267 (1968)

27. Hall, H. T., *Ann. Rev. Phys. Chem.* 9:395 (1958)

28. Anon, *Gems and Gemology* 13:292 (1971)

29. Wentorf., R. H. Jr., *High Pressure Research* 4:247 (1974)

30. Wentorf., R. H. Jr., *Phys. Chem.* 69:3063 (1965)

31. Yamaoka, S., Komatsu, H., Kanda, H. and Setaka, N., *J. Cryst. Growth* 37:349 (1977)

32. Weaver, H. B., *Lapidary Journal* 17:783 (Nov. 1963)

33. Morrow, W., *Lapidary Journal* 17:922 (Dec. 1963)

34. Wentorf., R. H. Jr., *J. Phys. Chem.* 75:1833 (1971)

35. Strong, H. M. and Chrenko, R. M., *J. Phys. Chem.* 75:1838 (1971)

36. Strong, H. M. and Wentorf., R. H. Jr., *Naturwiss* 59:1 (1972)

37. Crowningshield, R., *Gems and Gemology* 13:302 (1971)

38. Nassau, K., *The Physics and Chemistry of Color*, Wiley, New York, pp. 155, 169 ff (1983)

39. Chrenko, R. M., Strong, H. M. and Tufts, R. E., *Phil. Mag.* 23:313 (1971)

40. Chrenko, R. M., *Phys. Rev.* B7:4560 (1973)

41. Hall, H. T. and Kistler, S. S., *Annual Reviews of Phys. Chem.* 9:395 (1958)

42. Hall, H. T., *Science* 169:868 (1970)

43. Horton, M. D., Pope, B. J. and Hall, H. T., *Diamond - Partner in Productivity*, 3rd Internatnational Diamond Technical Symposium, Industrial Diamond Assoc. of America, Inc., Moorestown, NJ, p. 20 (1974)

44. DeCarli, P. S. and Jamieson, J. C., *Science* 133:182 (1961)

45. Cowan, G. R., Dunnington, B. W. and Holtzman, A. H., U.S. Patent 3,401,019 (9/10/68)

46. Trueb, L. C., *J. Appl. Phys.* 42:503 (1971)

47. Sinkankas, J., *Lapidary Journal* 21:1338,1450; 22:346 (1968)

48. Trueb, L. F., *Lapidary Journal* 22:664 (1968)

49. West, W. W., *Diamond - Partner in Productivity*, 3rd International Diamond Technical Symposium, Industrial Diamond Assoc. of America, Inc., Moorestown, NJ, p. 197 (1974)

50. Bundy, F. P. and Kasper, J. S., *J. Chem. Phys.* 46:437 (1967)

51. Hanneman, R. E., Strong, H. M. and Bundy, F. P. *Science* 155:995 (1967)

52. Frondel, C. and Marvin, U. B., *Nature* 214:587 (1967)

53. Koivula, J. I. and Kamerling, R. C., *Gems and Gemology* 25:45 (1989)

54. Rossman, G. and Kirschvink, J. L., *Gems and Gemology* 20:163 (1984)

55. Shigley, J. E., Fritsch, E. et al., *Gems and Gemology* 22:192 (1986)

56. Shigley, J. E., Fritsch, E. et al., *Gems and Gemology* 23:187 (1987)

57. Klement, W. Jr., and Jayaraman, A., *Progress in Solid State Chemistry*, 3:289-376, Pergamon Press, New York (1966)

58. Pistorius, C. W. F. T., *Progress in Solid State Chemistry*, Vol. II, pp. 1-120, Pergamon Press, New York (1976)

59. Nassau, K., *Gemstone Enhancement*, Butterworths, Boston, p. 124 ff (1984)

3

Chemical Mechanisms of Diamond CVD

Pehr E. Pehrsson, Frances G. Celii, and James E. Butler

1.0 INTRODUCTION

The growth of diamond from the vapor phase begins with nucleation, the initial formation of a crystal, followed by homoepitaxial growth, the deposition of additional carbon on the surface to increase the bulk dimensions. In this chapter we review the current understanding of diamond chemical vapor deposition (CVD) growth environments and the mechanisms of diamond deposition from the vapor phase. While there are numerous chemical vapor deposition techniques which successfully produce diamond, as described in Ch. 4, we discuss the underlying principles which are operative in all these methods. We limit the discussion to the chemical vapor deposition processes that produce predominantly the diamond phase of carbon.

The identification of the critical species and mechanisms of the diamond CVD process will assist in the understanding and technological improvement of the process and the material. The CVD growth environment can be quite complex and varied, as presented by the generalized view in Fig. 1. The reactants, methane and hydrogen in this example, enter a high temperature or energetic region in which the gas is activated by a plasma, a hot filament, or a combustion flame front. A myriad of chemical reactions are initiated which proceed on timescales varying from 10^{-9} sec, i.e. on every gas kinetic collision, to the gas residence time in the reactor (10^{-4} to 10 sec). The species participating in these reactions are transported by forced flow, diffusion, and convection throughout the reactor, eventually reaching the substrate or being exhausted from the reactor. Near the substrate, species may diffuse through a stagnant flow region, called the boundary layer. On the substrate surface, various processes may occur, such as adsorption, desorption, surface and bulk diffusion of species, as well as chemical reactions which may be advantageous or deleterious to diamond growth. The relative importance of the various surface processes is a complex function of the surface temperature, the structure of the exposed surface, and the surface composition.

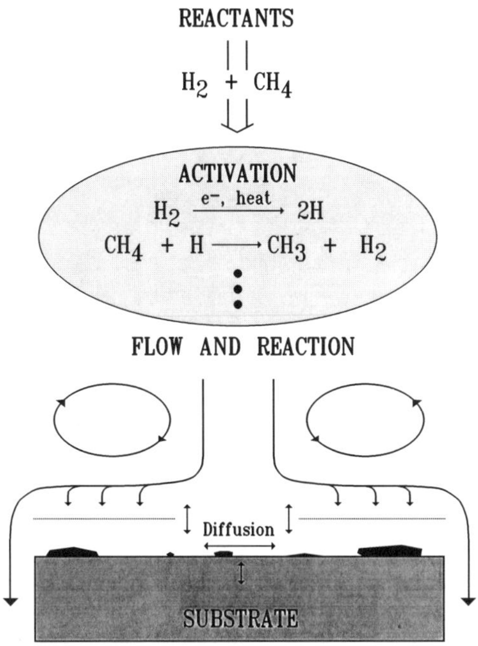

Figure 1. Generalized schematic of the chemical processes occurring in a CVD diamond reactor.

The analysis of diamond CVD can be assisted by an appreciation of the unique aspects of carbon chemistry and the deposition of a *metastable* material, while remembering the lessons of "traditional" crystal growth, in which the chemical potential of the condensing species is the principal driving force for nucleation, growth, and achievement of crystal perfection. This driving force concept also applies to diamond CVD, but requires enhancement and modification to account for the selective deposition of the thermodynamically metastable diamond phase as opposed to the stable graphite phase.

Diamond CVD differs from many other crystal growth processes in several important ways. First, carbon atoms can form different types of chemical bonds using sp, sp^2, and sp^3 hybridization of its valence electrons. This characteristic results in the diverse nature of carbon-based materials and chemistries. Diamond is a crystalline form of carbon comprised solely of sp^3 bonds and is thermodynamically metastable relative to the sp^2 bonded graphite crystalline phase at the temperatures and pressures used in the CVD process. Second, it is probable that the gaseous and surface species in the growth environment will be molecular, not atomic, due to the affinity of carbon to form strong bonds and the prevalence of molecular and atomic hydrogen. Third, diamond CVD occurs at substrate temperatures which are low compared with the CVD of other high quality crystalline materials, when each is scaled to other characteristic temperatures or material properties (e.g., Debye temperature, melting point, bulk diffusion). The diamond CVD growth temperature, 600 to 1100°C, is only a fraction of the characteristic Debye temperature, Θ_D (= 1590°C for diamond) (1), the temperature above which the heat capacity of the solid becomes independent of the material, and a measure of the softening of the lattice (2). By comparison, CVD of other high quality crystalline materials occurs at temperatures several times their Θ_D. For example, the substrate temperature during CVD growth of GaAs is 750°C (3) while Θ_D is 75°C (4). Hence, defect annealing and bulk diffusion processes are expected to be significantly slower for the case of diamond CVD than for other crystalline electronic materials at their optimal growth temperature.

Our review of the growth process is divided into two areas: *(a) nucleation,* including mechanisms on diamond and non-diamond substrates, and *(b) growth,* including the gaseous environment and the surface mechanisms which add material to an existing diamond lattice. This division is somewhat arbitrary; growth of a new layer on a diamond surface requires the nucleation of the next layer, while nucleation on a non-diamond surface necessitates the growth of the nuclei beyond a critical size. We find this organization useful because it reflects the natural sequence of events, and because the optimal environments for nucleation and growth appear to be different. For this discussion, other chapters of this book are relevant background, such as knowledge of the different techniques and growth environments used to grow diamond (Ch. 4), and the resultant quality and nature of the deposited material (Ch. 6). Since the vast majority of diamond deposition techniques involve a dilute mixture of a hydrocarbon (typically methane) in hydrogen as reactants, an understanding of the deposition process in these environments will be the main focus of this review. Other diamond growth chemistries, such as the addition of oxygen or halogen containing species, will also be discussed at the appropriate places.

2.0 NUCLEATION

The control of nucleation is critical to obtaining the grain size (5), orientation (6), transparency (7)(8), adhesion (9) and other properties required for optimal use of CVD diamond in most applications. For example, smooth pinhole-free polycrystalline films for optics (8) and x-ray lithographic masks (10) require high nucleation density and uniform crystal size on appropriate substrates. CVD diamond semiconductor devices require low defect density, single crystal heteroepitaxial diamond (11) or diamond boules (12), and spatially-selective nucleation (13). Wear and cutting applications require uniform nucleation over large and/or irregularly shaped substrates such as tool bits (14).

In this section we review the current understanding of CVD diamond nucleation, and how it can be controlled. In examining CVD diamond nucleation we consider; *(a)* general kinetic and thermodynamic aspects of nucleation and growth, and relevant experimental parameters; *(b)* the substrate surface on which nucleation occurs, and how it is modified by the CVD environment; *(c)* the microscopic nucleation and initial growth processes for both untreated and treated substrates; and *(d)* the effect of nucleation on the nature of the diamond deposit, including hetero- and homoepitaxy.

In order for nucleation to proceed, an incipient nucleus must form and be stable long enough for lattice growth to occur. The development of a critical nucleus, i.e., that size above which it is more probable that a nucleus will survive and grow than that it will decay, depends on the system free energy changes accompanying particle formation under growth conditions. The change in the system free energy must be negative for the process to occur (15)(16), where the system includes the substrate, growing particles, and the gas-phase reservoir of species contributing to lattice growth. The bulk contribution to the free energy change of a material upon incorporation of additional atoms is negative because of the formation of bonds to other atoms in the lattice, whereas the corresponding surface free energy change is positive due to bond strains and unsatisfied valencies at the lattice surface. Nuclei are therefore unstable (the net system free energy is positive) prior to development of the critical dimension because of the large surface free energy contribution resulting from the high surface to volume ratio, as shown in Fig. 2. The nucleus free energy increases with the addition of atoms until the critical radius is attained, at which point the net free energy reaches a maximum and then declines. Formation of critical size nuclei is therefore thermodynamically unfavorable, and occurs because of statistical fluctuations in the numbers of atoms forming temporary aggregations on the substrate surface (17). These random fluctuations in the nucleus size are affected by the coverage and mobility of growth species on the substrate surface, as in a Langmuir-Hinshelwood growth mechanism (18)(19), and on the direct flux of species from the gas phase to reactive surface sites, as in an Eley-Rideal mechanism (20). The relative contributions of these mechanisms depend in turn on growth species supersaturation and transport across the boundary layer, desorption, substrate surface structure, and diffusion of growth species (e.g., carbon) into the substrate bulk (16). Nuclei which do not attain the critical size dissociate or are absorbed into the substrate or other nuclei.

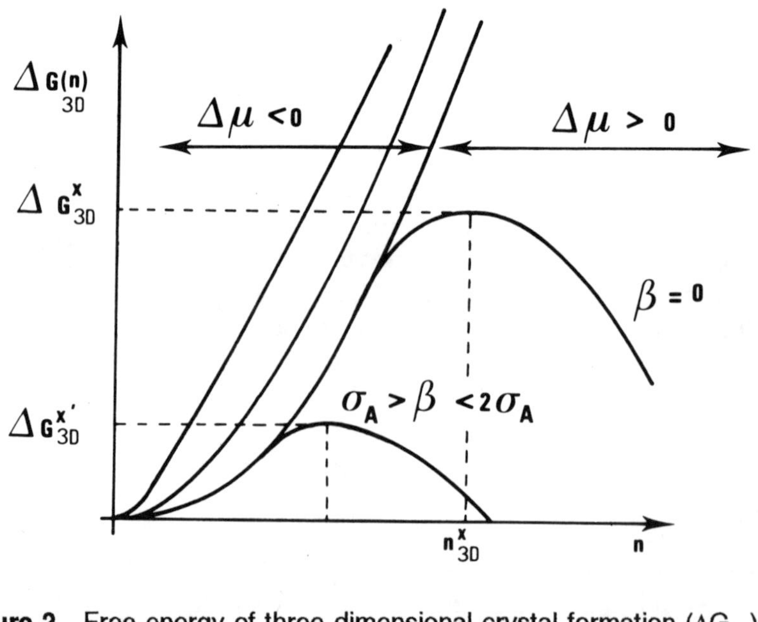

Figure 2. Free energy of three dimensional crystal formation (ΔG_{3D}) at constant supersaturation ($\Delta\mu$) as a function of the number of its particles *n*, and the adhesion energy (β) between the substrate and the deposit. σ_a is the work per unit surface area of the deposit. If $\Delta\mu > 0$, the curve shows a maximum. The barrier is lowered when the adhesion energy, β, is positive (18). *(Reproduced with permission.)*

Diamond nucleation during CVD occurs as a result of some combination of suppression of graphite nucleus formation, stabilization of diamond nuclei and surfaces with respect to graphite nuclei and surfaces, and preferential etching of sp^2 carbon (19)(21)-(26). In order to obtain these conditions, most current diamond CVD growth environments contain large amounts of molecular hydrogen and a supersaturation of atomic hydrogen resulting from some gas-phase activation technique such as exposure to a hot filament. Molecular hydrogen may reduce the partial pressure of gas phase species favorable to graphite nucleation, such as polycyclic aromatic hydrocarbons (27)(28). In addition, the supersaturation of atomic hydrogen probably causes hydrogen-termination of the substrate, diamond and diamond nucleus surfaces, thereby reducing the number of high-energy unsatisfied surface valencies. As a result, the diamond surfaces or small diamond clusters may be stabilized under low-pressure CVD conditions, where bulk diamond is thermodynamically metastable with respect to bulk graphite (although the difference in the heats of formation of bulk graphite and diamond are small) (22). Machlin (23) proposed that the strong pseudomorphic effects of the substrate on the stability of succeeding layers might encourage diamond formation, and even cause layers of graphitic carbon to transform to diamond. It has also been suggested that surface vacancies lower the surface energy of diamond relative to graphite (29). Non-diamond substrates could also influence the stability of various carbon forms via their bonding at the interface. Other proposed effects of atomic hydrogen supersaturation include graphite nucleation inhibition, and selective etching of graphitic nuclei (19)(21)(30)(31). Graphite nucleation can lead to planar defect formation in diamond, reduction in the diamond growth rate (25) and degradation in the quality of the deposited material.

Extensive theoretical and parametric analysis of the interplay between the carbon and hydrogen supersaturation, temperature, and other kinetic and thermodynamic factors influencing graphite and diamond critical nucleus formation was presented by Fedoseev and Deryagin (19)(21) and subsequently by Badzian (25). Gas mixtures with relatively low C/H ratios (32)-(35) and high substrate temperatures (8)(36)(37) (high end of the diamond growth temperature range) generally favor growth of high quality, low defect diamond (25), at the expense of lower nucleation densities and larger grain sizes (probably tied to the lower nucleation density) (37). Oxygen added to the feed gas reduces the C/H ratio and enhances sp^2 carbon etching, and therefore yields crystals with lower defect density.

Other studies have shown that the reactor parameters influencing diamond nucleation, growth rate and morphology include the substrate material and temperature, filament temperature or plasma power, the feedgas composition, the carbon and atomic hydrogen supersaturations, and the flow rate (21)(25)(38). Less work is available on optimization of those parameters specifically to control nucleation. One report found the optimal nucleation temperature was 925°C, as compared to 1000°C for the best crystal quality growth (5). The effects of the substrate type, substrate and filament temperatures, filament to substrate distance, C/H ratio, and the gas flow rate on growth rate and nucleation density were also reported (39). The parameters can be optimized to shift the dynamic balance between nucleation, growth and etching in the complex diamond CVD environment to yield desired properties at different stages of a growth. For example nucleation can be encouraged early in the growth, followed by emphasis on the growth rate or crystal quality (40).

The nature and timing of the nucleation process strongly affects the ultimate diamond deposit. Although the nucleation and growth rates of isolated crystals are initially substrate dependent, the growth rate of continuous polycrystalline films is not. This suggests that the formation and growth of isolated crystals early in the growth is affected by the evolution of the substrate in the CVD diamond growth environment. In particular, during the earliest stages of growth, the diamond nucleation density is influenced by the balance between the carbon flux to the substrate surface, etching of deposited carbon, carbon diffusion into the substrate bulk, and carbon migration on the substrate surface (34)(41)(42). These factors control the concentration of surface carbon species and their arrival rate at existing diamond nuclei. Upon completion of the induction period, i.e., the period between the first exposure of the substrate to the growth environment and the beginning of rapid diamond crystal growth, the diamond deposition rate becomes similar for most pretreatments resulting in heavily nucleated substrates (43). For example, growth rates are the same for diamond grown from diamond seeds and from diamond crystals spontaneously nucleated on a non-diamond surface (44). Growth then appears to be controlled by the arrival rate at, and mobility of, carbon-containing species on the diamond surface (43). Hydrocarbon species (41) and even possibly carbon atoms (45) and diamond clusters (46) appear to be mobile on the diamond surface at CVD temperatures. Such behavior is consistent with growth of the diamond lattice by species diffusion to a growing ledge, which is supported by observations of layer growth on the (100) diamond face (34).

If, after some growth, the entire substrate surface becomes covered with crystals and so forms a polycrystalline film, the crystal size and morphology are limited by impingement on each other, and thus correlate with the nucleation density (5)(8). Such continuous polycrystalline films often contain high densities of small, uniform crystals, suggesting that nucleation occurred early during the growth process. Nucleation then ceased, while crystal growth proceeded by carbon diffusion on the substrate or crystal surfaces and addition to active sites on the existing crystals. Crystals on substrates exhibiting low nucleation densities are isolated and tend to be larger, often with a wider size distribution (16)(47). Such behavior is consistent with continued nucleation later in the growth, and growth limited by the maximum carbon diffusion distance on the substrate. Although the crystal size is usually inversely related to the nucleation density, it may also be affected by other factors such as the initial defect density or buildup of strain in an adjacent diamond crystal (48). For example, growth of linear chains of diamond crystals unrestrained by growth of adjoining crystals has been observed (37). This was interpreted as evidence of a self-propagating diamond lattice defect site for renucleation. Since all of the crystals in the chains were similar in size regardless of the sequence in which they formed, it was suggested that the crystal growth process was self-limiting. Self-propagating diamond nucleation was also observed with clustered diamond crystals on sharp surface features of unpolished Si substrates (49).

As discussed above, the substrate and its pretreatment prior to growth have a large influence on nucleation and the initial stages of growth. Both natural diamond and many non-diamond materials have been used as CVD diamond substrates. Nucleation on most non-diamond substrates is enhanced by scratching with diamond powder (or less effectively, with SiC or cubic boron nitride [c-BN]) (34). Carbide-forming non-diamond substrates on which diamond has been deposited include both single and polycrystalline material: Si(100) (50), α- and β-SiC (on-axis and off-axis) (51), carbide-forming transition metals and carbides (36)(44)(52)(53), Si-coated Fe (54) and WC/Co composites (35). The carbide formers exhibited the highest nucleation density and growth rate (44)(52)(55)(56), and the shortest incubation period, while nucleation rates on single crystal substrates were lower than on the corresponding polycrystalline ones (44). Nucleation on non-carbide forming substrates is comparatively poor or nonexistent unless carbide-forming contaminants are present (44)(55)(56). Diamond growth has been reported on non-carbide forming materials, such as single and polycrystalline Cu (51), Ag, Au (44)(55)(56), Cu/Ni alloys (51), as well as the oxides SiO_2 (52)(57), and Al_2O_3 (52). In general, scratched carbide-formers nucleate better than scratched non-carbide formers (44)(55)(56), although there are some apparent exceptions, such as Cu (52), which, because of its high malleability, may more efficiently trap diamond/carbon particles during scratching. Compared to diamond-scratched substrates, unscratched SiC or other metal carbides do not nucleate diamond significantly better than unscratched Si (58). Combustion flame growths on unscratched Mo and Mo_2C yielded equivalent, low-level nucleation densities (40). Davis et al. (51), reported high nucleation on polished on-axis β-SiC and scratched Si(100), and low nucleation on off-axis β-SiC, on-axis α-SiC and polished Si(100), but provided no quantification of the nucleation densities. Clearly, the nucleation process is complex, and any proposed mechanism(s) must be able to account for the above observations. Despite the utility of general thermodynamic and kinetic principles or experimental parameter regimes for understanding nucleation, elucidation of the microscopic mechanisms is needed. These proposed mechanisms should be valid for the variety of growth techniques, and provide an overall framework in which to understand the large body of experimental observations now in the literature.

2.1 The Two-Step Nucleation Process

A single nucleation mechanism appears to be inadequate to describe all of the current data on nucleation. Rather, the observations are consistent with variations of a two-step nucleation process. In all known CVD diamond growth processes, the first step occurs when carbon species impinge on the substrate from the gas phase and diffuse into the bulk. The rate of carbon diffusion into the substrate bulk decreases as the carbon-rich surface zone of the substrate forms a barrier, e.g., by formation of a carbide layer with a low carbon diffusion coefficient, as shown in Fig. 3 (59). When the substrate surface becomes relatively impervious to further carbon diffusion, the surface carbon concentration increases and eventually exceeds the level necessary for the second step of diamond nucleus initiation/stabilization. As discussed later in Step 2: Nucleation, the second step seems to occur in a variety of ways, depending on the presence or absence of surface features and diamond seeds. On unscratched samples, nucleation appears to occur when carbon moieties on the carbon-saturated substrate surface form nucleation sites at surface flaws. On diamond-scratched samples, the carbon species diffuse to, and are incorporated into, the existing "nucleation" sites, which are probably diamond fragments.

Step 1: Substrates and Interfacial Evolution. A carbon diffusion barrier appears to be a necessary, but not sufficient, precondition for rapid diamond nucleation on unscratched substrates. For example, nucleation leading to continuous film formation does not immediately follow initial carbide formation on unscratched substrates (60), but rather occurs only after a prolonged induction period and at a comparatively low density compared to scratched substrates (43). By contrast, the development of a carbide diffusion barrier may be adequate to allow growth to proceed on diamond-scratched (and presumably already nucleated) substrates. For example, rapid diamond growth immediately followed the achievement of the maximum carbide layer thickness on diamond scratched Mo (5).

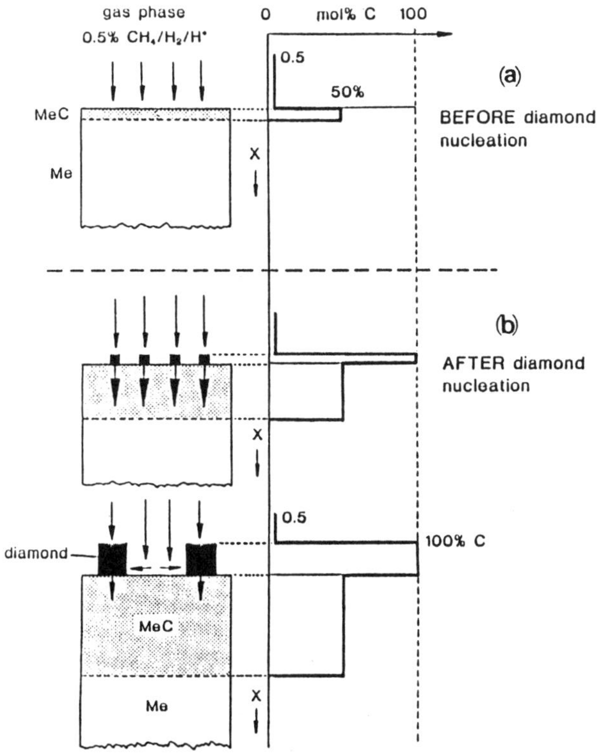

Figure 3. Carburization of refractory metals: showing carbon concentration gradients within a substrate during carburization (a) before and (b) after diamond nucleation (59). *(Reproduced with permission.)*

A correlation between the onset of nucleation/growth and the bulk carbon diffusion coefficients of unscratched carbide-forming substrates (59) was reported for filament-assisted chemical vapor deposition (FACVD) growths. This correlation, illustrated in Fig. 4, suggests that the induction period is related to carbon surface saturation and the attainment of a critical surface carbon concentration necessary for the comparatively low-probability

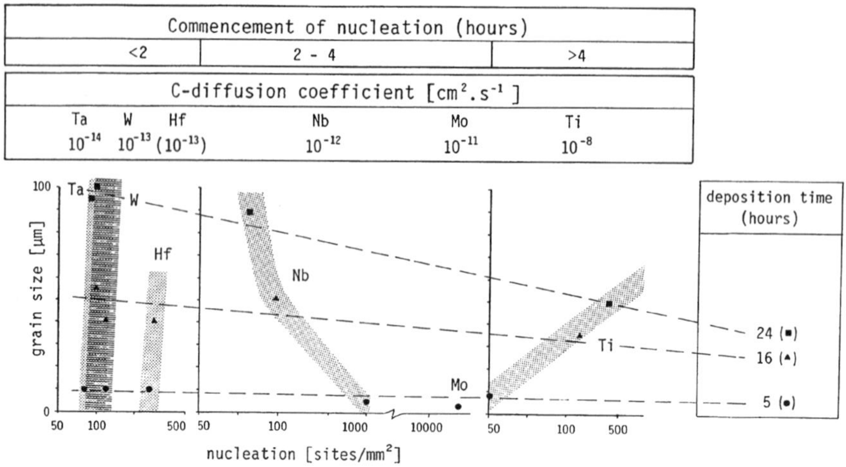

Figure 4. Changes in diamond particle size and nucleation density with respect to deposition time for various refractory metal substrates. The correlation between the onset of nucleation and the carbon diffusion coefficients is also indicated (59). *(Reproduced with permission.)*

of formation and survival of stable nuclei on unscratched substrates, or rapid growth of existing nuclei on scratched substrates. Carbide-forming substrates with low carbon diffusion rates (Hf, Ta, W) exhibited comparatively rapid nucleation. The crystals observed on these substrates later in the growth process were uniform in size, suggesting that nucleation ceased after growth begins, presumably because any additional surface carbon species diffused on the surface until they were incorporated into existing diamond crystals. Growth on Nb, which has a moderate carbon diffusion rate, exhibits a longer induction time due to the longer time needed to achieve a carbide layer thick enough to impede diffusion of surface carbon into the bulk. Denser and more rapid nucleation occurred on unscratched Mo than on W, Nb, Mo, Ta or Hf, possibly because Mo formed a subcarbide during the early part of the growth, and therefore more rapidly formed the diffusion barrier necessary to generate a high surface carbon concentration. Growth on titanium, which has the highest carbon bulk diffusion rate, exhibited a long induction period. A broader size distribution was also observed on Ti, which results from a low initial nucleation density. The crystals which first nucleated were probably too low in number to absorb all the carbon flux to the surface via diffusion, thus additional nucleation continued. However, another report claimed initial nucleation on TiC of one particle micron^{-2} (10^8 particles cm^{-2}) with no subsequent nucleation, only an increase in the size of the existing crystals (61).

The effect of the substrate in promoting nucleation and subsequent growth is altered if an interlayer develops between the diamond crystal and the original substrate surface (18). Non-carbide formers such as Ni or Pt (62) developed hydrocarbon or graphitic carbon overlayers or a carbon-rich surface zone due to carbon diffusion into the outer layers. X-ray photoelectron spectroscopy (XPS), electron loss spectroscopy (ELS) and low energy electron diffraction (LEED) (62)(63) analysis of short growths on clean Ni(100) substrates revealed the formation of a 0.5 monolayer coverage of carbon with a c(2x2) pattern characteristic of the saturated Ni surface. During longer growths, islands of azimuthally oriented graphitic carbon later developed on top of the initial carbon layer. Additionally-deposited carbon lost orientation with respect to the surface, as revealed by the loss of the LEED pattern. Finally, diamond grew on the disordered layer, also without any ordering influence from the substrate. Another growth attempt (64) on single crystal Ni yielded oriented graphite domains, although verification of diamond growth conditions was not provided.

Diamond deposition was also attempted on both scratched and unscratched polycrystalline Pt foils (65). Following the disappearance of surface oxygen and copper contaminants, a 6 Å thick layer of carbon (graphitic and some other form) was deposited, as seen in Fig. 5, and was subsequently replaced by a uniform, 1 - 3 monolayer thick hydrocarbon film. Diamond eventually grew on the scratched sample, although it was unclear if it nucleated on the hydrocarbon or on residual islands of graphitic carbon. Titration experiments with CO revealed that virtually none of the original Pt remained exposed prior to diamond growth, so the diamond probably had little if any contact with the Pt substrate. No diamond grew on the unscratched substrate.

The influence of the substrate evolution and the species covering it prior to growth was demonstrated by the effects of the O/C feedgas ratio on diamond growth in a FACVD reactor (66). The feedgas was 0.5% CH_4 in H_2, diluted with of a 5% O_2 in He gas mixture to attain O/C ratios of 0, 0.5, 0.75, and 1.0. Diamond-scratched Pt substrates exposed to O/C ratios of 0.5 or less displayed the sequence of surface cleaning, development of a graphitic carbon layer, a hydrocarbon layer and finally diamond. Samples exposed to O/C ratios of 0.75 developed the graphitic and then hydrocarbon layers, but no diamond. At O/C = 1.0, the stabilized surface was cleaned to very low levels of carbon, (perhaps graphite islands or diamond grit scratching residue). Samples briefly exposed to O/C = 1 and thereafter to O/C = 0.75 developed a thicker layer of graphitic carbon, probably localized at stable sites generated by the scratching. The fact that no diamond was deposited for O/C > 0.75 suggests that the graphitization step was necessary but not sufficient for nucleation. Exposure of samples with existing diamond particles to O/C = 1 resulted in continued growth, which suggested that high oxygen levels affected the nucleation step more than the growth step, possibly by etching the graphitic carbon and/or hydrocarbon deposits necessary for nucleation. The exact nature of the hydrocarbon layer and its role in the nucleation process is of great potential importance.

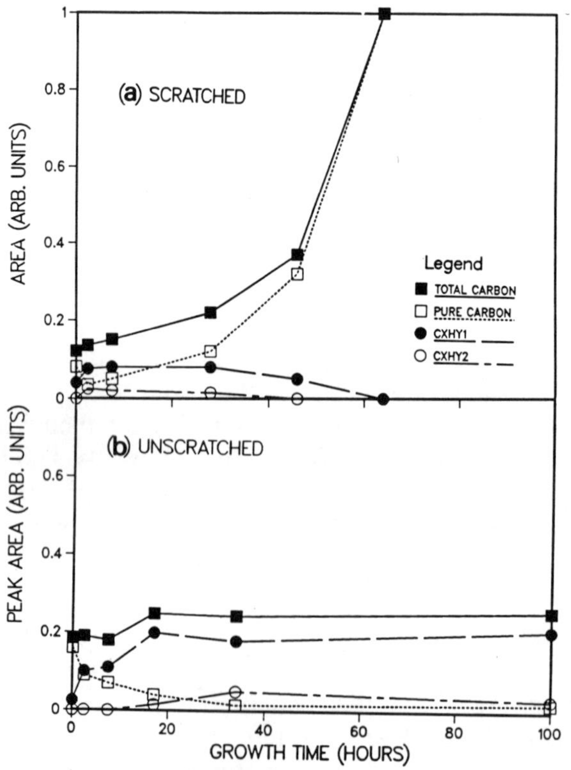

Figure 5. Integrated C(1s) XPS peak areas versus time for (a) scratched and (b) unscratched Pt foils following diamond CVD growths of varying duration. CXHY1 and CXHY2 are peaks attributed to unidentified hydrocarbon species (65). *(Reproduced with permission.)*

Carbide-forming substrates also evolve and develop surface layers between the original substrate material and diamond deposits in the CVD diamond growth environment. A variety of analytical techniques have been used to identify carbide layers on, for example, Si (60)(67), and other metals (5)(32)(53)(59). A Mo carbide layer was detected using x-ray diffraction (XRD) within a minute of growth initiation on Mo. Diamond was first detected with XRD after twenty minutes of growth, although diamond nanocrystals were observed within one minute with scanning electron microscopy (SEM) (5). XRD (32) on W substrates revealed the formation of W_2C and then WC at the interface prior to diamond growth. XRD was used to identify the development of TiC, HfC, NbC, TaC, Ta_2C, Mo_2C, and WC layers on the corresponding metal substrates (53)(59).

XPS and electron loss analysis of diamond grown on Si for varying periods of time revealed that the evolution of the Si surface (68) in the FACVD environment proceeded through three stages, shown in Fig. 6. Within the first fifteen minutes the native oxide and adventitious carbon contaminants were removed, presumably etched by the atomic hydrogen; subsequently a layer of SiC formed on the Si surface; finally the carbide layer was covered with diamond. Williams (67) detected a 50 nm thick SiC layer using high-resolution transmission electron microscopy (HRTEM) and selective area diffraction, although the film was observed only when the CH_4 concentration was low (0.3%). At higher (2%) concentration, no SiC layer was detected. Transmission infrared spectroscopy was used to detect 10 and 20 nm thick SiC layers (5)(39) growing on Si substrates prior to complete diamond film coverage. The SiC layer thickness varied with the C/H feed-gas ratio, with a maximum at 0.3% CH_4 in H_2. Spectroscopic ellipsometry combined with selective chemical etching identified the formation of a damaged Si layer topped by a 40 - 70 Å SiC layer on an unscratched Si substrate after five minutes in a microwave environment (60) using a 0.5% CH_4 in H_2 mixture. The equilibrium thickness was attained in less than five minutes; much less than the induction period for growth on an unscratched sample. The formation of a carbide layer was therefore not the rate-determining step for nucleation. It was suggested (60) that densification and crystallization of the carbide led to development of a microstructure conducive to nucleation. Alteration of the carbide layer might have occurred as the temperature of the substrate achieved equilibrium with the plasma and annealed out damage introduced in the first minutes of carbon exposure. Another report noted that a SiC layer continued to grow after the onset of diamond nucleation on a scratched Si substrate, and that carbide layer growth ceased once the diamond film was complete.

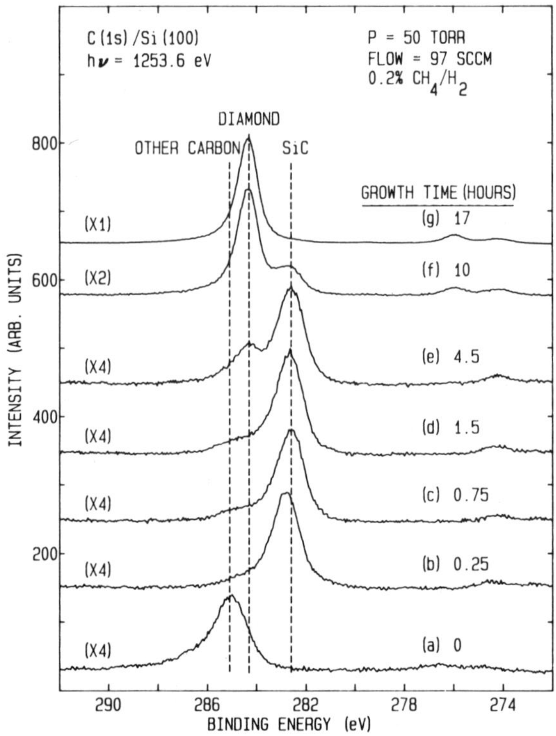

Figure 6. XPS spectra of the C(1s) region for different growth times on a Si substrate. Spectra b-e show the formation of a carbide layer (282.7 eV) followed at later times by deposition of a diamond layer (284.3 eV), shown in spectra f-g (68). *(Reproduced with permission.)*

Using surface-enhanced Raman scattering (SERS) (69), a very thin non-diamond carbon layer was detected on Si substrates in situ in a microwave plasma system prior to development of a diamond film. The extreme surface sensitivity of SERS permitted detection of very small amounts of diamond and non-diamond carbon. The non-diamond carbon was present on the original substrate surface which had been masked during growth, and was also present on the diamond coated region (although it was unclear whether the diamond constituted a continuous film). Graphite pedestals were detected under diamond crystals grown on scratched Si (70), although the point at which they formed during the growth process is unknown.

Hot-filament electron-assisted CVD (EACVD) growths yielded higher nucleation density and lower crystal quality on scratched Mo and Ta (>$1x10^8$ cm^{-2} vs. <$2x10^7$ cm^{-2}) than on scratched Ni, W, or Si(100) (53). The high diamond crystal quality on Ni, Si and W was attributed to, respectively, a good lattice match, structural parameter match, and the high surface energy of the tungsten carbide interfacial layer. However, the crystal quality was judged by the Raman single phonon peak width and may be due instead to the large, isolated crystals present on those substrate surfaces. The higher nucleation density and high sp^2 carbon component on the Mo and Ta substrates was attributed to the larger lattice and surface energy mismatches with diamond.

Kamo (36) reported differences in the nucleation sequences observed on diamond-scratched single crystal WC and Si substrates. During the first stage of growth on Si, the number of nuclei increased but the nucleus size was fairly constant; during the second stage the nucleation rate decreased and the crystal size increased. On WC the number of nuclei and the crystal size both increased during the first stage; thereafter the nucleation rate declined and the crystal size increased. The difference in nucleation behavior can be explained by the longer time needed for development of a SiC layer. The carbon mobility would be reduced on an unsaturated Si surface because of competition between SiC formation and carbon diffusion to existing diamond crystals. Since the WC already is carbon-saturated, no such delay would occur. An inverse correlation between the nucleation and growth rates was also observed, i.e., a higher nucleation rate resulted in a lower growth rate. The formation of diamond nuclei and the growth of existing crystals clearly depend on the evolution of the substrate. In many (if not all) of the substrates examined to date, the original substrate surfaces and the species on them are altered by the CVD environment prior to heavy nucleation and growth of the diamond film. When the substrate is scratched or otherwise pre-seeded with diamond nuclei, the carbon-saturated substrate apparently often functions as little more than a chemically inert platform on which carbon species physically concentrate and contribute to diamond growth by addition to the existing nuclei. When there are no pre-existing diamond nuclei, surface features may function as high energy sites for spontaneous formation of nuclei. These different pathways are discussed below.

Step 2: Nucleation. Diamond nucleation apparently occurs in various ways: *(a)* nucleus formation on appropriate surface features on carbon-saturated surfaces; *(b)* gas phase nucleation; *(c)* growth on diamond fragments or other sp^3 carbon residue deposited by scratching, or *(d)* growth on diamond or other "seeds" deposited by alternative methods, e.g., evaporation, sputtering or deposition in a reactor using a carbon-rich feed-gas mixture. Although the exact microscopic nature of the specific nucleation events described above is experimentally undetermined at this time, we summarize below some of the evidence for each. The relative importance of these alternatives is unclear because of uncertainties in the reported nucleation densities, which can vary strongly with experimental conditions, and are often described only qualitatively in the literature. Comparisons of nucleation densities are misleading if long induction periods are confused with low ultimate nucleation densities. Claims of low nucleation densities should be based on long growths, preferably until a sample has developed a continuous film.

Nucleation on Unscratched Substrates. Nucleation in many non-diamond deposition systems occurs at steps, ledges, screw dislocations and other defects due to enhanced bonding at high energy intersecting surfaces with high densities of unsaturated bonds and low coordination numbers (16)(52)(71). For example, diamond nucleation occurred on features lithographically etched on unscratched Si (72), as shown in Fig. 7. Singh (52) reported enhanced nucleation on Si etched with HF/HNO_3, although other attempts to enhance nucleation by creating etch pits on unscratched Si with acids (43), H^+ (43) or other reactive gases (73) were unsuccessful. Likewise, attempts to generate large numbers of nucleating defect sites via implantation with surface saturation levels of C^+ (74) or Ar^+ (43) resulted in no nucleation enhancement on unscratched substrates, and even suppression on diamond-scratched substrates. Furthermore, arrays of 0.1 - 0.3 micron diameter holes created with a focused Ga^+ ion beam in a Si substrate resulted only in the deposition of non-diamond carbon in the depressions, although diamond growth conditions were not verified (75). These various results suggest that diamond nucleation on unscratched substrates requires the existence of certain types of substrate features or defect sites, and that the topography of the features is important.

One group recently claimed to have obtained nucleation of diamond films by implantation of single crystal and polycrystalline copper substrates with carbon, followed by exposure to nanosecond pulses of a 308 nm laser (76). The laser was assumed to rapidly melt the outer zone of the substrate, resulting in epitaxial regrowth of the copper and then diamond. The presence of diamond on the copper substrates was confirmed with Raman scattering, Rutherford backscattering, Auger spectroscopy, and x-ray diffraction. Such nucleation on unscratched substrates, if confirmed, could be technologically important.

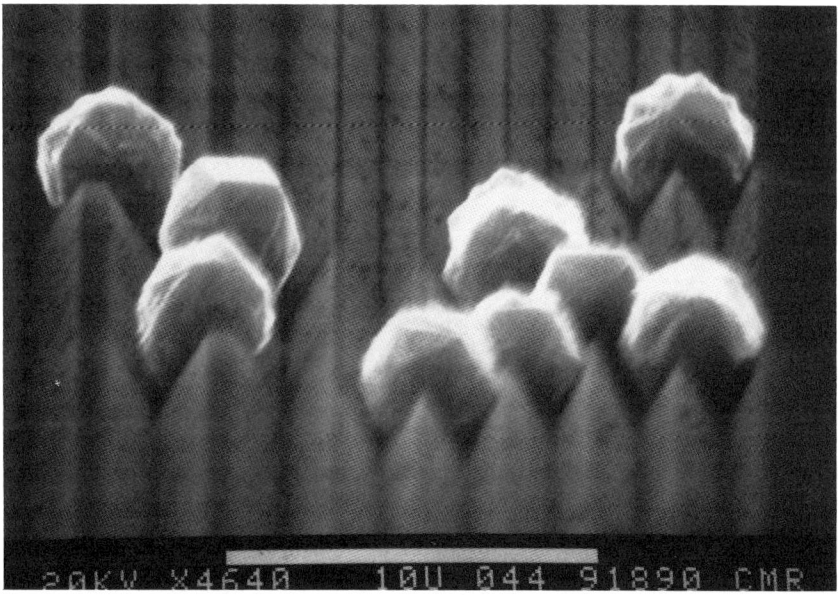

Figure 7. Nucleation of diamond particles on ridges chemically etched in silicon. The centers of the three crystals to the left lie on either side of the top of the ridge. The particles are seen to grow outward from the ridge-edge, indicating these are the nucleation sites (72). *(Reproduced with permission.)*

Wear tracks produced by dragging a sapphire stylus across Si substrates under controlled loads exhibited increased (1×10^7 cm^{-2}) nucleation densities leading to continuous diamond films, while the surrounding substrate was devoid of crystals (77). The nucleation enhancement was more efficient and occurred under a broader range of conditions than for comparable tracks made with a diamond stylus. Scanning Auger microprobe analysis of the wear tracks produced by both diamond and sapphire styli rubbed at 50 - 500 gm loads on Si substrates contained elevated C/Si ratios (77). Since none of the wear tracks made with single passes of the diamond stylus produced diamond crystals during CVD growth, the buildup of carbon in the wear tracks was not sufficient to induce nucleation. However, the sapphire tracks contained a broad size distribution of dense Hertzian stress cracks as well as both sharp-edged and more rounded, microscopic Si debris concentrated at the ends of the tracks and in some of the cracks. The sharp-edged debris and some of the smaller cracks were associated with nucleation enhancement, while the rounded debris and the largest cracks usually had little effect. The high surface to volume ratio of the small particles probably leads to rapid carburization and formation of a surface carbon diffusion barrier, hence a higher surface carbon concentration. The sharp edges function as nucleating sites by stabilizing potentially nucleating carbon species on the substrate surface at multiple high-energy unsatisfied valencies (dangling bonds). The largest nucleation enhancement was caused by pressing the stylus vertically into the substrate, possibly creating a strained region or very small micro-cracks.

Alteration of a Mo foil by bending or damage with a steel tool prior to growth was reported to increase nucleation compared to the undamaged foil (72). The growth temperature of CVD diamond is much lower than the Mo melting point, so the defects induced by the bending presumably were not annealed during growth and functioned as high surface energy sites for nucleation. The Si melting point is much closer to CVD growth temperatures, hence some potentially nucleating defects may be annealed or etched during growth.

Varying degrees of diamond nucleation enhancement have been achieved by depositing interfacial layers of BN, SiC, a-C:H, diamond-like carbon (DLC), or amorphic sp^3 bonded carbon onto substrates prior to diamond growth (40)(58)(70)(78). The enhancement may reflect stress minimization by improved lattice or thermal expansion coefficient matches, which may in turn be reflected in improved adhesion. Substantial nucleation density enhancement and induction period reduction was achieved for an oxyacetylene torch growth by first depositing a layer of DLC on the substrate prior to diamond growth (40). The DLC layer was created by briefly lowering the O/C ratio to 0.75 from the usual growth condition at 0.9. One possible explanation for the enhancement by DLC was thought to be its large numbers of nucleating defect sites, and a high H content which provides atomic hydrogen to the growth surface (40). A separate attempt (58) at nucleation enhancement with DLC layers in a microwave plasma reactor was unsuccessful.

Meilunas (5) compared the nucleation rates and crystal morphology for diamond films grown on Si in a microwave plasma when CH_4 was admitted either during or after the sample achieved growth temperature. Growth with CH_4 present during warmup produced a continuous diamond film with 0.5 - 2 micron grain size. Exposure of the substrates to CH_4 only after growth temperature was attained resulted in a much lower density and larger, more well-faceted crystals. The low surface temperature during the initial nucleation stage may have caused deposition of some carbon species which increased the density of nuclei or reduced the etching of substrate sites.

The use of X-ray Absorption Near Edge Spectroscopy (XANES) to probe local atomic structures revealed that material grown using filament-assisted deposition at a 7% CH_4/H_2 feedgas ratio showed a local diamond environment (79) even while exhibiting little or no diamond Raman signature. The discrepancy between XANES and Raman results may be explained by assuming that the material consists of nanocrystalline diamond embedded in an amorphous matrix, the latter of which caused no sharp features in the XANES spectrum but which can be identified in the Raman spectrum. Nanocrystalline diamond domains could function as nucleation sites for continued growth. Indeed, films grown in a microwave plasma with 5% CH_4/H_2 ratios were etched in an oxidative plasma to reveal a spidery labyrinth of diamond filaments (80). Another microwave study of growth with a 5% CH_4/H_2 mixture yielded diamond nanocrystallites smaller than 10 nm (81). The deposition of high densities of nanocrystalline diamond seeds on the substrate is consistent with the previously discussed oxyacetylene torch results (40), in which a nucleating layer was first deposited by initially using a fuel-rich (carbon-rich) mixture.

Other layers of material can be deposited inadvertently on the substrates and affect the nucleation. For example, some deposition chambers are subject to wall etching and subsequent deposition of wall material onto the substrate. Badzian found that Si particles etched from the reactor walls and carburized to β-SiC catalyzed diamond nucleation on (0001) graphite, which normally exhibited no nucleation (82). He has also described B, Si and Ti as catalysts for nucleation and growth of diamond films (56). However, Singh (39) examined the effects of the tube diameter and wall material on nucleation and growth, and observed no difference between walls of Si, SiO_2, Al_2O_3, pyrex, stainless steel, and Al.

The filament stabilization time during filament-assisted CVD must also be considered in analyzing the nucleation rate and incubation period. During the early stages of growth using a virgin W or Ta filament, most of the carbon results in filament carburization rather than deposition at the substrate (83)(84). The carbon flux to the surface is therefore reduced, possibly slowing down both nucleation and growth. The filament surface composition and reactivity also vary with the temperature and gas phase carbon concentration (83). The choice of filament materials (W, Ta, Re, Mo, or Pt) also affects nucleation (52). A thin film of W_2C, deposited on the substrate (85) during the thirty minute filament carburization period, suppressed diamond nucleation, possibly by covering surface nucleation sites. The use of Mo filaments resulted in the deposition of Mo_2C in the chamber because of the high vapor pressure of Mo at the filament temperatures used (52). Because of their low melting point, Pt filaments were heated only to 1350°C, (below the temperature necessary for significant conversion of H_2 to H) resulting only in graphite deposition on the sample (52).

Biased hot-filament and microwave plasma electron-assisted CVD (in which the substrate is biased with respect to the filament or plasma respectively) have been shown to yield denser nucleation and more rapid growth than conventional hot-filament reactors, depending on the size and magnitude of the bias (86)-(88)(90). Moustakas noted that the carburized tungsten filament is a much better electron emitter than tungsten, and that the use of a bias in conjunction with a carburized filament eliminated the influence of the substrate on nucleation and growth (84). Electron bombardment may enhance the rate of hydrogen desorption from the bonds at the diamond or interfacial surface and thereby accelerate carbon addition to the diamond face. This may result in increased numbers of reactive hydrocarbon fragments near the surface, or may enhance sp^2 carbon etching at the interface (89)(90). Another study demonstrated extremely high (10^{10} cm^{-2}) diamond nucleation density on unscratched Si substrates (90). The substrates were negatively biased and exposed to a brief period of deposition with a microwave plasma operated in a 10% CH_4 / H_2 mixture. The authors stressed the importance of enhanced formation of nucleus-generating species on and above the substrate by the accelerated ions, and improved nucleus/substrate bonding caused by the increased carbon supersaturation. However, excessive ion acceleration limited nucleation by sputtering both the substrate and incipient nuclei. Alternatively, Banholzer has asserted that much of the effect of biasing is a result of enhanced electron-beam heating of the substrate (91).

Speculation on the nature of the specific species (excluding diamond fragments) responsible for nucleating diamond on unscratched substrates has focused on the molecular analogs of diamond crystals. Matsumoto (92) suggested that sp^3 bonded hydrocarbon cage compounds, such as those shown in Fig. 8, could function as nuclei, with adamantane leading to single crystal diamond, and tetracyclododecane and hexacyclopentadecane giving twinned crystals. These molecules possess 6-membered carbon rings in either the chair form, the boat form or a combination of the two. It was noted (92) that the multiply-twinned particles commonly observed with CVD diamond are also found in other materials systems when the deposit/substrate interaction is weak or there is gas-phase nucleation. The 5-fold twins also occur for other systems, such as Au on rocksalt, for which there is no corresponding molecular nucleator (93). Boat-boat bicyclodecane and related multiply-twinned compounds (89)(94)(95) were also suggested as potential diamond and Lonsdaleite nucleators based on the presence of both boat and chair hexagonal carbon rings, Fig. 9, and the overall molecular structure. On the other hand, 3,4,9,10-perylene tetracarboxylic acid dianhydride (PCDTA), which has neither chair nor boat conformation, was reported (96) to strongly enhance nucleation. Perhaps PCDTA merely functioned as an additional source of surface carbon because of its low vapor pressure. Although the stability of the complex hydrocarbon cage compounds is not high at normal diamond growth temperatures, the effects of carbon and atomic hydrogen supersaturation may stabilize them long enough to form the "critical nucleus" necessary for continued diamond growth (15). The frequent occurrence of regularly-shaped, twinned diamond crystals is cited as evidence that the twin nucleus forms from a high concentration of carbon-containing compounds adsorbed on the sample surface (44).

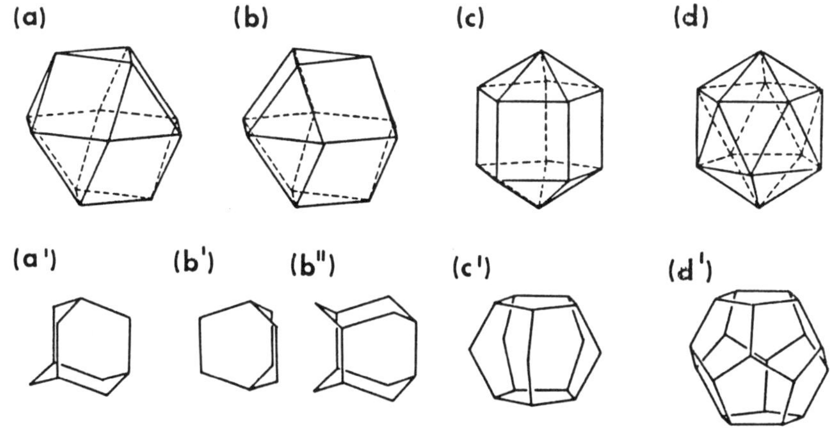

Figure 8. Examples of twinned diamond particles (a - d) and corresponding hydrocarbon cage compound molecular analogs and proposed nuclei for single and twinned diamond crystals, a' - d' (92). (a = cubo-octahedron, b = twinned cubo-octahedron, c = decahedral-Wulff-polyhedron, d =i cosahedron, a' = adamantane, b' = bicyclo [2.2.2] octane, b" = tetracyclo [4.4.0.1.1]dodecane, c' = hexacyclo [5.5.1.1.1.0.0] pentadecane, d' = dodecahedrane). *(Reproduced with permission.)*

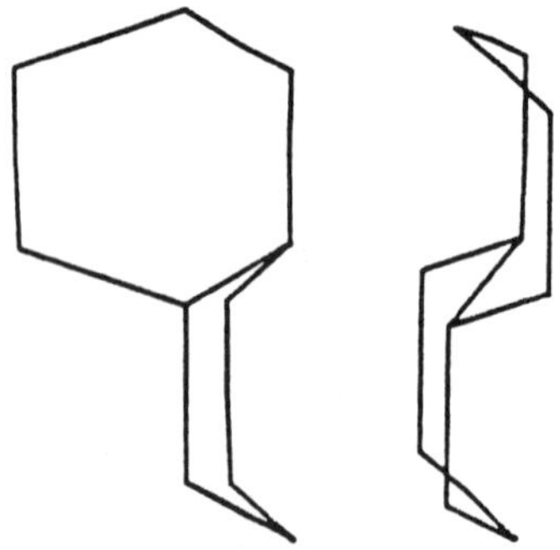

Figure 9. Ring compounds with two parallel twin planes proposed as diamond nuclei (94). *(Reproduced with permission.)*

Many workers have noted the apparent genesis of multiply-twinned crystals at the diamond/substrate interface during the earliest stages of CVD diamond growth, particularly the highly symmetric 5-fold twins (33)(44)(92)(93). Crystals imaged using HRTEM (33)(93) occasionally show the 5-fold twin at the center of the crystal. The fact that multiply-twinned crystals appear to nucleate at or near the crystal/substrate interface suggests that the nature of the crystals is determined at or shortly after nucleation (93). One model proposed that multiply-twinned nucleation occurs when a crystallite nucleates on surface steps of the {001} planes of Si and the others are generated by twinning from the initial microcrystallite (93). Another suggested that twin boundaries may be involved in forming multiply-twinned polyhedra (97).

Gas Phase Nucleation. Diamond nucleus formation may also occur in the gas phase. Kamo, (98), Frenklach (99)-(101) and Mitura (102) reported gas phase nucleation in the microwave, low pressure microwave and RF plasma environments, respectively, although the contribution of gas phase nucleation to the overall nucleation rate was not established. In hot filament and microwave plasma environments, the correlation between diamond scratching or other surface features, and enhanced nucleation (44) suggests that the majority of nucleation events occur on the surface. Alternatively, the scratches could stabilize nuclei formed in the gas phase when they condense on the substrate.

Oxyacetylene torches, high current density/pressure DC plasmas and DC plasma jets demonstrate substantially higher growth and nucleation rates on unscratched substrates than do the hot filament or microwave techniques. According to some claims (103) these techniques induced heavy and rapid nucleation without substrate scratching. However, another report detailed a strong substrate-dependence (40) of nucleation for torch growths, in which the use of a fuel-rich mixture for brief periods increased nucleation by an order-of-magnitude, apparently by depositing a highly defected, hydrogen-rich diamond-like-carbon (DLC) or amorphous carbon (AC) layer. The apparent increase in nucleation density may actually have been a decrease in the nucleation induction period. Both the DC jet and the oxyacetylene torch virtually eliminated the nucleation induction period under some conditions. Heavy nucleation on a scratched Si substrate was claimed after fifty seconds of growth with an oxyacetylene torch (104)(105). The nucleation density in torch growths has also been correlated with the position in the flame (106) and other experimental conditions.

No substrate effect was observed for plasma jet diamond growth on Si, Mo, Pt, SiC, and SiO_2 (107), nor was diamond-scratching needed to achieve high nucleation rates and continuous films. Scratching of Si substrates was also unnecessary to achieve high nucleation density with a low pressure DC plasma (108). Nucleation densities achieved were about 10^8 cm^{-2} for both cases, as compared to $<10^4$ cm^{-2} on unpolished Si in a FACVD apparatus. There was a distinct morphological difference for growth on scratched vs. unscratched samples in a cascaded arc DC plasma; the scratched samples yielded continuous films of small crystals, while the unscratched substrates grew very large, isolated crystals (47). In both cases the induction period was dramatically reduced compared to a microwave plasma or hot filament growth, while the crystal morphologies probably reflected the initial density of nucleation sites.

The reduction of the induction period with high growth rate techniques is likely a result of the high carbon fluxes and concomitant rapid carbon saturation of the substrate in the high pressure and highly-excited plasma, plasma jet and torch growth environments. The high nucleation rates might result from increased surface carbon concentration or gas-phase nucleation and cluster formation, although the latter mechanism is more speculative.

Gas phase nucleation of either diamond or graphite (sp^2-bonded) nuclei is most probable at high carbon supersaturations (19)(27). Conditions conducive to high nucleation rates, such as high carbon concentration (78) or low substrate temperature (37), also tend to generate highly defective diamond with a large concentration of planar defects and graphitic or amorphic inclusions. These conditions are undesirable for growth of high quality and/or epitaxial diamond films. Graphitic nuclei encourage formation of sp^2 carbon on the substrate, while diamond nuclei impinging on the substrate from the gas-phase should be randomly oriented in the absence of strong nucleus-substrate bonding, and disrupt the layer growth needed for high-quality homoepitaxial diamond growth (34).

Although the gas phase is probably not the primary source of nucleators on scratched substrates in hot filament and microwave plasma growth environments, nucleation is influenced by mass transfer across the gas-phase boundary layer, atomic hydrogen supersaturation near the interface, and the concentration of various precursor species. An understanding of gas flow dynamics near the sample surface is therefore critical to uniform nucleation and large area deposition.

Scratching and Seeding. Most substrates used for diamond growth are pretreated by scratching or seeding with diamond powder to enhance nucleation. Lower nucleation densities were obtained by scratching with SiC (40)(108), cBN (86), Cu or stainless steel (73), ZrB_2 (72) and Al_2O_3 (52). Diamond seeding has reportedly been accomplished by sprinkling with dry powder, dipping, spraying, spin-coating, and electrostatic development (109), and by electrophoresis (110)(111). Seeding and scratching with SiC, Si, and Al_2O_3 were also reportedly effective (109), although other workers have reported little or no nucleation with the latter two. For example, sandblasting with Al_2O_3 grit was ineffective at inducing nucleation, although it was unproven that diamond growth conditions had been attained (112).

Heavy scratching or abrading of the substrate surface with diamond grit enhances the nucleation density by up to roughly seven orders of magnitude compared with unscratched Si (10^4 to 10^{11} cm^{-2}, respectively) (49). The nucleation density is proportional to the scratching time, and the morphology changes from large isolated crystals for short scratching times to smaller, more numerous crystals with increased scratching (39)(52). The grit size of the diamond used for scratching also influences the nucleation density, with 1/4 micron grit being the most effective (85) for scratching by hand, and 40 - 50 micron the best for scratching in an ultrasonic bath using a grit suspension (49). The larger size needed for effective ultrasonic treatment compared to manual scratching may be due to the need for more massive crystals to efficiently induce scratching damage or transfer fragments to the substrate.

Transfer to the substrate of stable nuclei such as diamond fragments can enhance nucleation independent of mechanical damage, as demonstrated by the effectiveness of seeding with diamond grit (109). Scratching the substrate with diamond grit may also enhance nucleation by transferring some other residue, such as sp^3 bonded carbon material resulting from substrate-diamond grit reaction or localized heating. Alternatively, scratching may create chemically reactive substrate damage sites such as ledges, pits, and mosaic faces. The large nucleation enhancement obtained by scratching with diamond, compared with other abrading materials, and the correlation between nucleation enhancement and the lattice match of the abrading material with diamond (43) (SiC<TaC<cBN<diamond), is consistent with the transfer of fragments of the abrading material. The correlation is seen in Fig. 10, where the weight gain of Si samples in a diamond reactor (due presumably to diamond growth) is affected by the scratching material used. A close lattice match between the substrate and nucleus could stabilize nuclei in the CVD environment and shift the dynamic balance between nucleation, growth and etching in favor of nucleation (37). Crystals grown using scratching-induced nucleation and from diamond seeds are indistinguishable (44). Since virtually every substrate from Si (hard and brittle) to Cu (soft and plastic) demonstrates enhanced nucleation after diamond abrasion, it seems unlikely that a specific damage site is being generated by the scratching material. In most cases, transfer of diamond fragments or other sp^3 carbon to the substrate appears to be principally responsible for scratching-induced nucleation.

There is substantial evidence that scratching with diamond grit transfers diamond fragments and other carbon to the substrate. Diamond grit rubbed between two pieces of Si left sub-micron fragments on or in the Si (73). Figure 11 shows a HRTEM micrograph of one of the estimated 10^{11} cm^{-2} diamond fragments transferred to a Si substrate during ultrasonic treatment with diamond grit (113). However, not all of such fragments may be stable in the etching environment of the CVD reactor, particularly before the substrate becomes carbon-saturated. Yugo (49) reported an initial nucleation density of 10^{10} cm^{-2} on Si ultrasonically scratched with diamond grit, after 30 seconds of growth, and 10^9 cm^{-2} after 120 seconds of growth. The particle density of the continuous polycrystalline film resulting from further growth decreased to 10^8 cm^{-2}, possibly due to nucleus recombination or etching of the sub-critical clusters by atomic hydrogen. Nuclei formed early in the growth and before the substrate becomes carbon-saturated may be reabsorbed into the substrate, as was observed on Nb (59).

The initial size of diamond nuclei was examined by HRTEM and scanning tunneling microscopy (STM). HRTEM observations (113) of 5 nm and smaller particles with a regular crystal habit on scratched Si substrates suggest that the nuclei are at least that small. The nuclei appear to develop

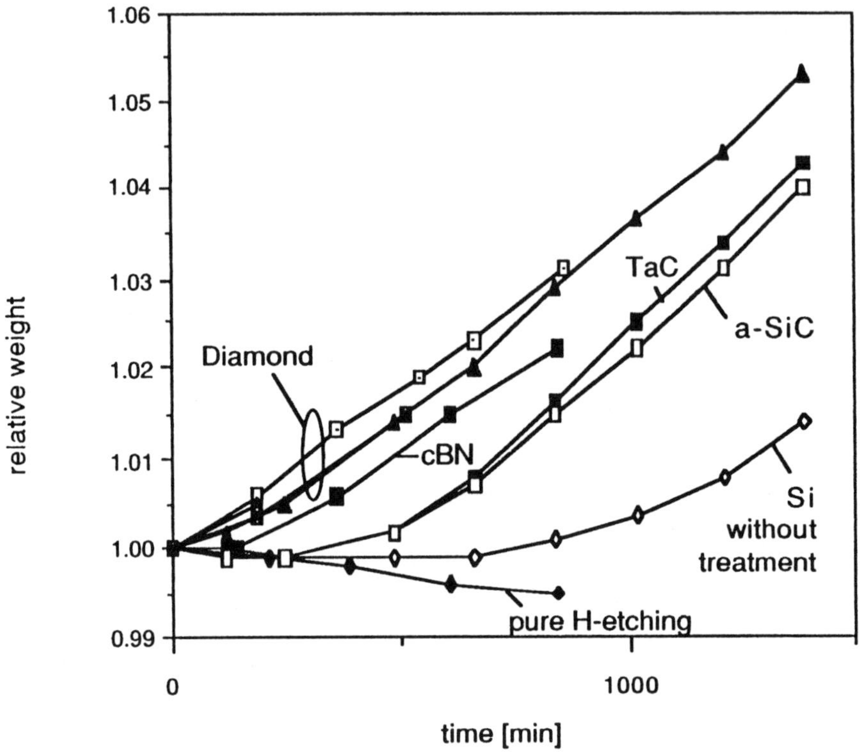

Figure 10. Relative weight gain due to diamond growth on silicon substrates scratched with different materials (as masked), or pre-etched with hydrogen, versus exposure time in a microwave plasma diamond CVD reactor. The open squares are for a Si substrate seeded with diamond grit (43). *(Reproduced with permission.)*

lattice distortions (113), possibly introduced by the scratching process, which may account for some of the defects observed in CVD diamond. Particles of 4 nm diameter on an amorphous background were observed with HRTEM for CVD diamond on polycrystalline Mo (96). STM images taken from B-doped diamond grown on scratched Si substrates reveal 100 nm patches, possibly of diamond or SiC, with a work function similar to nearby diamond crystals (114). Another STM study of nucleation on scratched Si samples also described triangular, roughly 150 nm-sided structures in the scratch tracks (115), such as shown in Fig. 12 (115). The 100 - 200 Å wide and 100 Å high ridge structures between diamond nuclei in the STM images (115) may be the source of the chain or lateral nucleation propagation. The work function of the entire surface changed after the deposition process, suggesting that the original Si surface was altered, probably by coverage with SiC. Other HRTEM images from diamond-scratched Si revealed the presence of well-faceted 10 nm diamond crystals and no particles below 3 nm diameter (49), but calculations of the critical nucleus size suggested a radius of 0.1 nm. The discrepancy was attributed to the presence of atomic hydrogen which would etch such small particles and render them unstable in the diamond CVD environment.

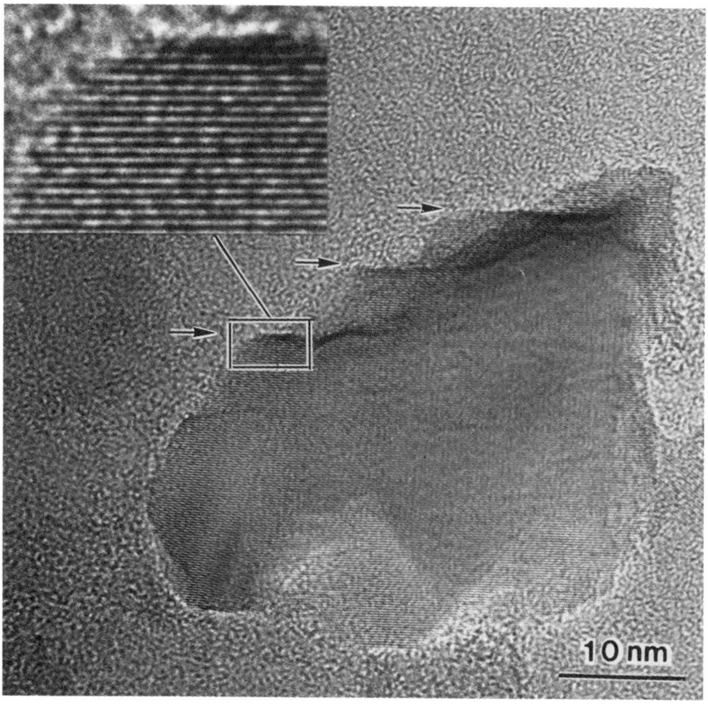

Figure 11. High resolution transmission electron micrograph (HRTEM) of an irregularly-shaped diamond particle found on Si, ultrasonically treated in a suspension of diamond grit (113). *(Reproduced with permission.)*

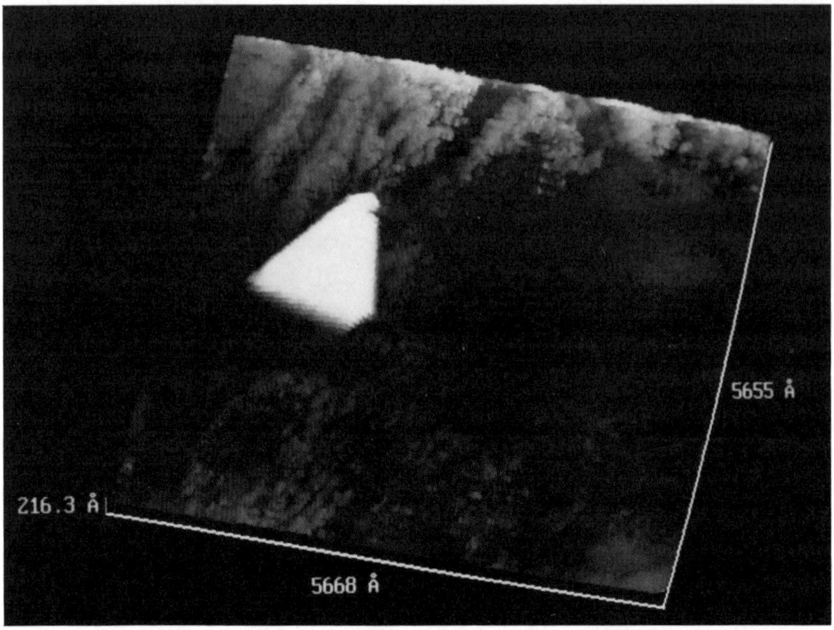

Figure 12. A Scanning Tunnelling Microscope (STM) image of a single faceted diamond nucleus located in a surface scratch track, which was produced by the initial polish of the substrate (115). *(Reproduced with permission.)*

The effects of various scratch sizes on a scratch-saturated Si surface were examined with STM, and a correlation between scratch size and nucleation density was attributed to the stabilization of critical carbon nuclei in 5 - 15 nm scratches (71). Smaller scratches had a negative influence on nucleation, possibly because they destabilized nuclei by preventing carbon diffusion. The fact that the minimum 3 nm particle size was less than the scratch width suggests that the crystals were separately nucleated, which would complicate attempts at large-scale heteroepitaxy (49).

If stable diamond fragments are indeed transferred during diamond-scratching then nucleation on scratched samples is really homoepitaxial growth on the diamond fragments. The growth rate of the initial nuclei is limited by the rate of direct impact of carbon, since carbon species that adsorb on the non-diamond substrate may diffuse into the bulk. After carbon saturation, the growth rate of the nuclei increases because it can proceed by both direct impact and surface diffusion of carbon species to the crystals

(116). This effect was illustrated by the substantial reduction in the induction time achieved by applying hydrocarbon-based oil to both carbide and non-carbide forming diamond-scratched substrates prior to growth (116), and is consistent with the accelerated formation of a carbon-saturated surface in the early stages of growth. While subsequent Auger spectroscopy did not reveal the presence of a thick overlayer of residual surface carbon on the substrate, the sub-surface region could have been saturated with carbon, stabilizing diamond nuclei resulting from the scratching. Either a carbide or simply a carbon-rich zone could act as a barrier to further diffusion of carbon into the substrate bulk. No enhancement was obtained on unscratched substrates, confirming the need for both carbon saturation and a subsequent, specific nucleation step. Similarly, enhanced nucleation was observed near diamond seed particles pressed into carbide-forming metals (117), possibly due to local carburization of the substrate through carbon dissolution and diffusion from the diamond crystals.

Diamond crystals deposited on scratched substrates appear to grow from a single nucleation site rather than over a broad area. The HRTEM image, shown in Fig. 13, shows a truncated polyhedral diamond crystal at the diamond/substrate interface. Examination of this crystal revealed line defects (traces) converging to a point at the center of the interface (67)(118). The diamond crystal and substrate appeared, to the resolution limit of the microscope, to be in direct contact only at the point of convergence. Further from the nucleation center, the Si substrate is covered with a thin, amorphous layer, presumably due to interaction between the substrate and the plasma prior to overgrowth by the diamond.

Information regarding nucleation was also obtained by oxidative plasma etching of polycrystalline films to isolate single diamond crystals (presumably etching any sp^2 carbon at a faster rate than the diamond). Under some conditions, the crystals grew as columns and tapered to a single point at the interface, consistent with a single nucleation site. Sato (81) observed that a 3 - 4% CH_4/H_2 mixture in a microwave plasma resulted in hemispherically-arrayed columns which appeared to converge slightly above the surface, suggesting that a small crystallite was the nucleator. The factors which cause crystals to grow outwardly to form polyhedra instead of vertically as columns may include the interfacial energy, and should be further examined for the development of 2-D vs. 3-D growth.

Selective diamond deposition can be achieved by masking areas of a diamond-abraded substrate surface and then Ar^+ etching (119)-(121) or irradiating with UV light to remove nucleating sites (122). By reducing the defined feature size to 1 micron^2, single diamond crystals have been generated on individual features. Electron microscopy on scratched Si substrates revealed small pits, sharp edges, strains and mosaic structures generated by the scratching. Surface roughening was thought to be the source of nucleation enhancement (119) since it was lost when the mosaic features were removed and the sharp edges rounded by the ion beam. An alternative explanation is that at least the surface of any diamond fragments or other nucleation sites on the surface was converted to graphite or amorphous carbon by the ion beam. This explanation is bolstered by the fact that removal of 50 nm of scratched Si by Ar^+ sputtering destroyed the nucleation enhancement gained from the scratching (119), while several microns of material had to be removed to achieve the same effect by chemical etching (121).

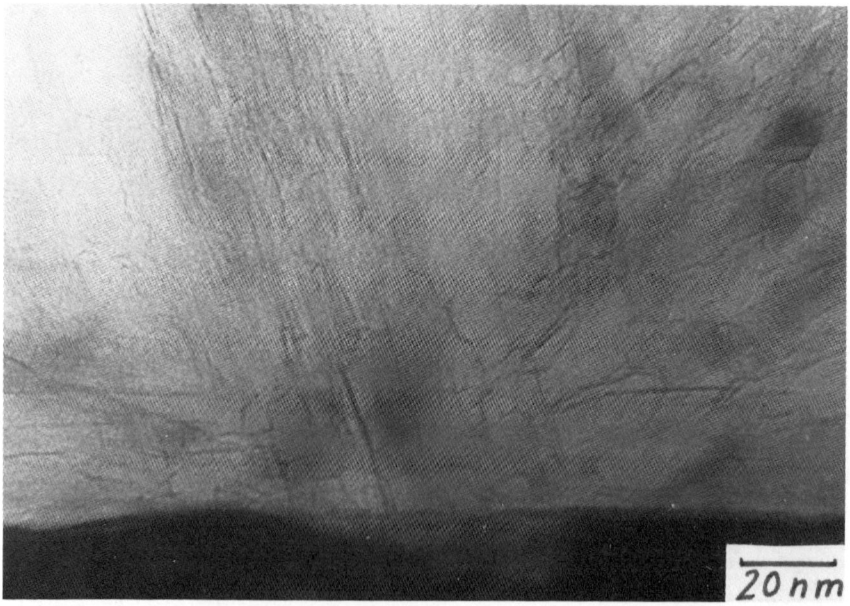

Figure 13. HRTEM image of the interface between a truncated polyhedral diamond crystal and Si substrate. Note the line defect traces converging to a central point at the interface (120). *(Reproduced with permission.)*

Standard photolithographic techniques can also be used to achieve nucleation on scratched substrates (13)(123). A diamond-scratched Si substrate was patterned with lines using photoresist, grown on for two hours, and then stripped of the resist (48), apparently leaving channels in the substrate. The area under the resist gave a lower nucleation density and larger crystals than the channels. The apparent removal of diamond fragments from the substrate surface by stripping the resist indicates the nuclei were not deeply embedded in the substrate. The nuclei grew in a linear fashion and to a characteristic size at the edge of the channels, probably due to the presence of two surfaces for attachment and nucleation via a linearly propagating defect. The differences in crystal size and morphology between the resist-coated and uncoated regions persisted even after development of a continuous film, as shown in Fig. 14. Such self-propagating behavior indicates that the ultimate size and structure of the diamond crystals comprising a film are at least partially determined by the defect density or grain size developed during the earliest stages of nucleation and growth.

Figure 14. SEM image of the microstructural development at the boundary between microlithographically patterned and non-patterned regions of Si substrate after diamond film coverage (48). *(Reproduced with permission.)*

Attempts to enhance nucleation and growth by using excimer laser excitation of scratched substrates and the gas phase species over them actually resulted in suppression of nucleation, apparently due to ablation or deactivation of the existing nucleation sites (122). Although the crystal densities in the irradiated and unirradiated areas were different, the crystals that did form were equivalent in size, indicating that the nucleation step was suppressed by the laser, while growth of any nuclei that did form was relatively unaffected.

2.2 Epitaxial Diamond Growth

We include here a section on the epitaxial growth of CVD diamond. The homo- and heteroepitaxial extension of a lattice may nominally be considered a growth issue, but begins with the nucleation of each additional layer. This section will focus on the issues relevant to the formation of epitaxial nuclei.

Heteroepitaxy. Epitaxial diamond growth on either a diamond substrate (homoepitaxy) or a non-diamond substrate (heteroepitaxy) is critical to the eventual use of diamond for electronic devices. Nucleation and growth of diamond in epitaxial registry imposes constraints on both the experimental parameters and the substrate, and requires an understanding of the underlying physics both at the substrate/deposit interface, as well as the appropriate gas phase/solid interface. We now briefly consider some of those factors applied to generic epitaxial systems, and then consider some specific aspects of diamond epitaxy.

In conventional crystal growth, atoms deposit on a substrate by either 2-dimensional layer formation (Frank-Van der Merwe), 3-dimensional particle growth (Volmer-Weber) or an intermediate mode where the growth begins as a layer but develops islands and particles on the initial layers as growth proceeds (Stranski-Krastonov) (15). The different deposition types are schematically illustrated in Fig. 15. The interfacial bonding between the substrate and the deposit, i.e. the adhesion energy, β , strongly affects the stability of the incipient nuclei, and the shape and orientation of the resulting crystals (15)-(18). Symmetric crystals having a single point of contact with the substrate have little bonding with the substrate, while truncated polyhedral crystals, i.e., crystals bisected by the crystal/substrate interface, are characteristic of the high adhesion energy conducive to 2-D growth (18). Crystals become progressively flatter with increasing adhesion energy as depicted in Fig. 16, and ultimately approach two-dimensional growth as the height/width ratio of the crystal diminishes. High adhesion energy also favors nucleation of large numbers of small grains (18).

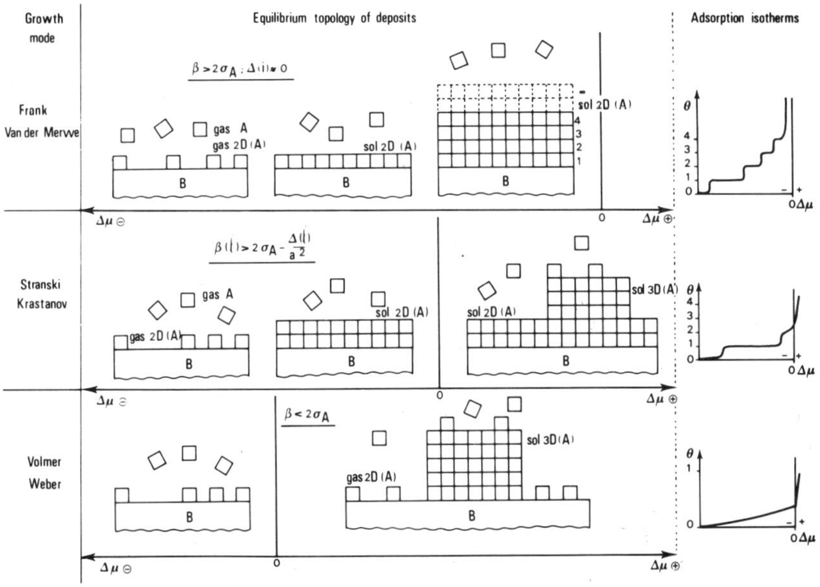

Figure 15. The three epitaxial growth modes, which depend upon the relationship between the free energy of the growing crystal surface and the adhesion energy with the surface. β is the adhesion energy, σ_a is the deposit surface free energy, and Δμ is the supersaturation of the lattice component. The three modes are two-dimensional layer growth (Frank - van der Merwe), three-dimensional island growth (Volmer - Weber), and an intermediate growth regime (Stranski - Krastonov) (18). *(Reproduced with permission.)*

Stable three-dimensional nucleation and growth requires a supersaturation of the constituent species over the crystal. True 2-dimensional growth of a single monolayer on a foreign substrate requires two conditions. The first is undersaturation, that is, the chemical potential of the deposit [$\mu = kT\ln(P/P_{inf})$] must be negative. (P and P_{inf} are the partial pressures of the lattice component at the crystal surface and at a saturated vapor phase reservoir, respectively. T is the temperature and k is the Boltzmann's constant.) The second condition for two-dimensional growth is that the adhesion energy (β) between the deposit and the substrate exceeds twice the deposit surface free energy (σ_a), i.e., $\beta > 2\sigma_a$ (18). Diamond has a very high surface free energy which limits the number of substrates that could potentially stabilize its two-dimensional growth (96). Analyses of the surface and interfacial energies of various diamond/substrate interfaces suggested that BeO and BN may be suitable candidates for heteroepitaxial growth, and further that the substrate bonding type is more important than the lattice constant match (96). Copper was also examined because it has a good lattice match with diamond. However, because copper also has a comparatively low surface energy and a low adhesion energy to diamond, that is, the excess interfacial energy between diamond and copper is high, it may have limited utility as a heteroepitaxial substrate. However, the laser heating method reported to produce epitaxial diamond on copper is evidence of the need for more study of this issue (76).

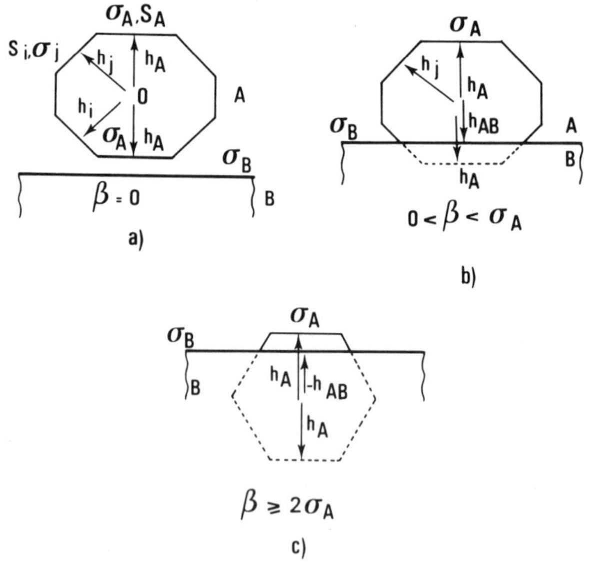

Figure 16. Equilibrium forms of a deposit A on a substrate B. (a) with zero adhesion energy, $\beta = 0$, and no contact with the substrate; (b) with intermediate adhesion energy between 0 and twice the deposit surface free energy ($0 \leq \beta \leq 2\,\sigma_a$); and (c) with adhesion energy larger than twice the deposit surface free energy ($\beta > 2\,\sigma_a$) (18). *(Reproduced with permission.)*

Truncated polyhedra are often observed (52) in diamond CVD, suggesting high crystal/substrate adhesion energies. However, some HRTEM images of truncated polyhedral crystals (67)(120) also reveal that the crystals originate from a single nucleation point on the substrate, so the influence of the initial substrate surface on the growth mode is apparently limited. The diamond interface appears to be with a coating or an interfacial layer on the substrate, rather than with the original substrate surface. Buildup of an interlayer, e.g., carbides (59) or hydrocarbons (62)(65)(124) between the diamond and substrate, during diamond CVD probably alters the growth mode from that expected on the clean substrate surface. Therefore, the development of heteroepitaxy clearly requires a detailed understanding of the true nature of the diamond-substrate interface for each substrate.

Evidence for the attainment of diamond heteroepitaxy, in which the diamond lattice is in registry with the substrate, is limited. True heteroepitaxy requires nucleation over a large area, or subsequent coalescence of separately nucleated crystals into a single domain; such growth is unlikely with the randomly-oriented multiple sites resulting from scratching. Selection of potential substrates has emphasized the lattice and thermal expansion coefficient matches between the substrate and diamond (53). Diamond growth was attempted using remote plasma enhanced CVD on polycrystalline Ni and single crystal Ni(111) (125). The former grew only amorphous carbon, while the latter grew oriented, crystalline carbon which proved to be epitaxial graphite. (However, diamond growth conditions in the remote plasma environment were not verified on other diamond abraded substrates) (125). Another report claimed epitaxial diamond on scratched, polycrystalline Ni (126). A low C/H ratio in the feed gas mixture was deemed critical in preventing formation of graphitic carbon on the substrate surface prior to diamond growth. Diamond growth was attempted in microwave and hot-filament environments on single crystal Cu, a Ni layer on single crystal Cu(111), Cu/Ni alloys, and Ni on β-SiC(100). The substrates were chosen in an attempt to lattice match the substrate to diamond (51). Although some diamond growth was achieved, it was not epitaxial. On- and off-axis β-SiC(100), α-SiC(0001), and large grain polycrystalline NbC were also tried as heteroepitaxial substrates because the expected high interfacial bond energy would encourage two dimensional rather than three dimensional growth. Nucleation densities on all these substrates with the exception of the on-axis β-SiC(100) were low, and the growth was non-epitaxial on all the substrates.

There have been reports of textured (or oriented) and even heteroepitaxial polycrystalline films on non-diamond substrates, including Si (67)(93)(127) and cubic BN powder (128). Heteroepitaxy of isolated diamond grains on Si was claimed on the basis of HRTEM and selected area diffraction (SAD) (67). It was suggested (93)(129) that textured, and possibly epitaxial, growth can result when the diamond lattice matches at least one of the major substrate planes, even if the lattice match is poor. XRD evidence of oriented and locally epitaxial diamond crystals was obtained from unscratched Si(100) exposed to a lean hydrocarbon mixture at high temperature (to encourage formation of a single-crystal SiC layer), followed by growth under standard conditions (127). These results have proved difficult to reproduce (130). In another study, preferentially oriented diamond films with [011] texturing were grown on Si substrates, where the {111} and {200} diamond planes were observed to match with the Si {022} planes (93)(129). These growths were performed after depositing a DLC layer on the substrate with a carbon-rich mixture. Growth probably occurred on this layer rather than on the substrate itself. Growth of [111]- or [110]-textured films was observed with microwave plasma deposition onto Si or microcrystalline diamond layers, respectively. Substrates exposed first to a 4% CH_4/H_2 mixture followed by a 1% CH_4/H_2 mixture developed diamond films with a [110] texture, while films grown without the initial step, presumably resulting in deposition directly on Si, had a [111]-preferred orientation (6)(38). Similar results were obtained for microwave growths on scratched Si (38). Textured diamond film growth was observed following diamond deposition onto oriented diamond seeds (131) on Si. Partial orientation of the diamond fragments which remained on the Si substrates following scratching with diamond grit also was reported (113).

Although difficult to synthesize, c-BN is a potential heteroepitaxial substrate due to its close lattice and thermal expansion match with diamond (132)(133). Morphological evidence of epitaxial diamond growth on c-BN grains was obtained (128). Nucleation was much faster on BN than on the refractory metals (Mo, W, Nb) on which the BN grains had been seeded. The delayed nucleation on the metals probably resulted from the formation of a carbide layer. Other workers also grew diamond heteroepitaxially on c-BN, based on evidence from Raman shifts, film morphology and reflection high energy electron diffraction (RHEED) (134)(135). A very high initial nucleation density of 10^{11} cm^{-2} occurred on c-BN, with growth particularly along the surface steps of the substrate (134). The nucleation on c-BN was followed by island growth and eventual formation of a network of connected islands and isolated particles and, finally, a continuous, possibly epitaxial film. The persistent high nucleation density on c-BN may reflect stabilization of the incipient diamond nuclei by both the lattice matching and high surface energy of cBN (96)(133).

Homoepitaxy. Homoepitaxial diamond CVD (136)(137) displays large variations in the growth rates on different single crystal faces, with the (100) face demonstrating the slowest growth. On polycrystalline films, the different faces also exhibit different amounts of secondary growth, and in one case, periodic renucleation, growth and coalescence (34).

Spear (26)(31) discussed different nucleation mechanisms and growth behavior on various diamond faces. Growth of a new (111) layer required the low-probability formation of a 2-4 atom initiation kernel. The adamantane molecule, shown in Fig. 17c, is also suggested as the nucleation kernel for a new (111) plane. According to the proposed acetylene-based growth mechanism (138), nucleation kernels, such as those shown in Fig. 18, are formed on a (111) face by hydrogen abstraction, followed by deactivation, reactivation and finally addition of another acetylene molecule. Likewise, the step propagates by a hydrogen abstraction from a step or ledge, followed by addition of acetylene or other acetylenic molecules, such as C_2H. Although nucleation is slow on the (111) face resulting in film morphology with predominantly (111) faces, addition proceeds readily, once initiated. During the addition process there is only a low probability of introduction of a stacking fault or twin plane, consistent with the observation of ledges or steps that spread over an entire crystal face (44)(139).

By contrast, the (100) and (110) faces require no nucleation step to begin a new layer because each carbon addition event produces a new growth site (26). Nonetheless, ledge growth characteristic of slow nucleation followed by rapid growth has also been observed on (100) faces (12)(34). At high temperature and >0.4% CH_4/H_2, the (100) face was the most common in one report (34), although others indicated the opposite trend (22). The increasing dominance of the (100) face at high temperature may indicate that the nucleation on the (111) face is no longer the rate-determining step and that the (111) faces then grow out more rapidly than the (100) or (110).

The small energy difference between formation of the boat and chair forms of the hexagonal ring of the initiation kernels, shown in Fig. 17, explains the facile formation of the twin planes commonly seen in CVD diamond. The cubic diamond lattice consists of layers of chair-shaped 6-membered rings separated by 6-membered rings also in the chair form. The stacking sequence is ABCABC... and the bonds are staggered. The lonsdaleite lattice consists of layers of chair-shaped 6-membered rings separated by boat-shaped 6-membered rings. The stacking sequence is ABABAB... and the bonds are eclipsed, resulting in a slightly higher energy.

Angus (89)(94) noted that although addition at the diamond (100) face could proceed relatively easily by single atom addition, nucleation at smooth (111) faces required much less probable 3- or 4-atom addition. However, the presence of twin planes generates the 1- or 2-atom addition sites illustrated in Fig. 19. Multiple parallel twin planes result in a reentrant site which is regenerated as crystal growth proceeds, ensuring continued nucleation and explaining the observed high incidence of multiply-twinned crystals. This is supported by SEM observations of rapid nucleation and growth of layers at features associated with intersecting, reentrant twin planes (46). The molecular analogs of parallel-twinned planes are the boat-boat bicyclodecanes and related conformers, while more complex, multiply-twinned molecules could grow from decahedra and icosahedra (89).

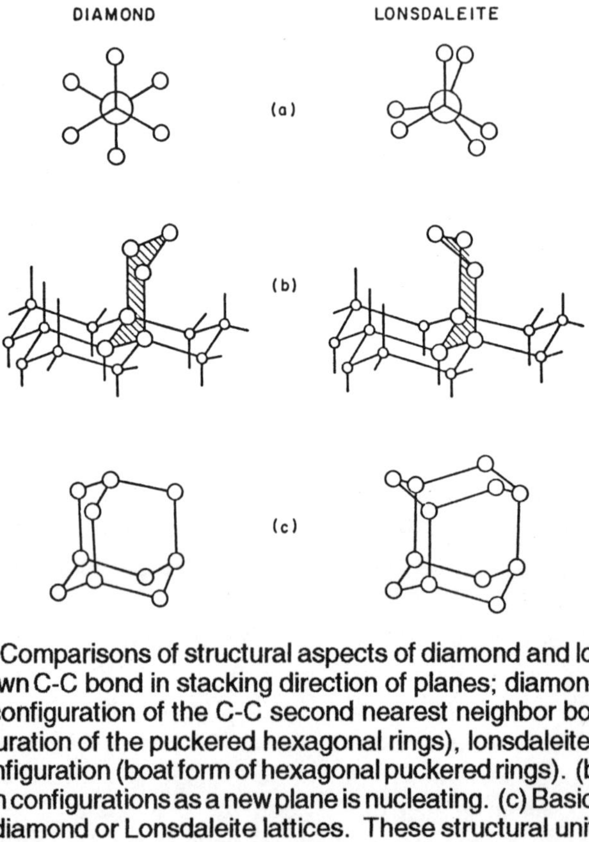

Figure 17. Comparisons of structural aspects of diamond and lonsdaleite. (a) View down C-C bond in stacking direction of planes; diamond shows a staggered configuration of the C-C second nearest neighbor bonds (chair form configuration of the puckered hexagonal rings), lonsdaleite shows an eclipsed configuration (boat form of hexagonal puckered rings). (b) possible initial carbon configurations as a new plane is nucleating. (c) Basic structural unit for the diamond or Lonsdaleite lattices. These structural units are also the nucleation kernels for new planes (31). *(Reproduced with permission.)*

2.3 Nucleation: A Summary

Nucleation cannot be described as a single unique process, but probably has several origins, with the proportions of each contribution

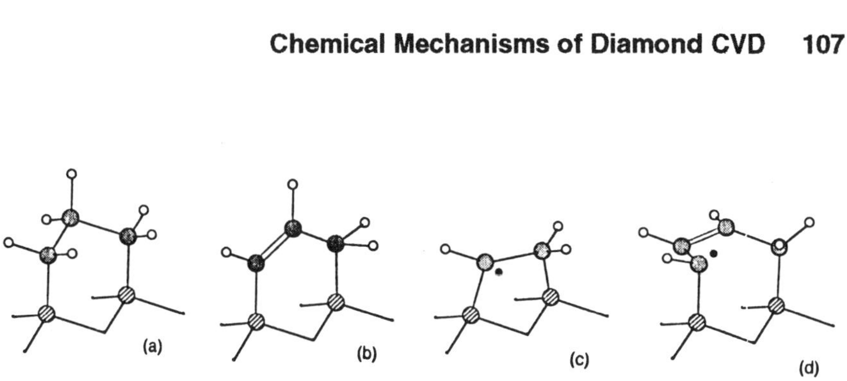

Figure 18. Schematic diagrams (a-d) of possible initiation kernels on the (111) diamond face (138). *(Reproduced with permission.)*

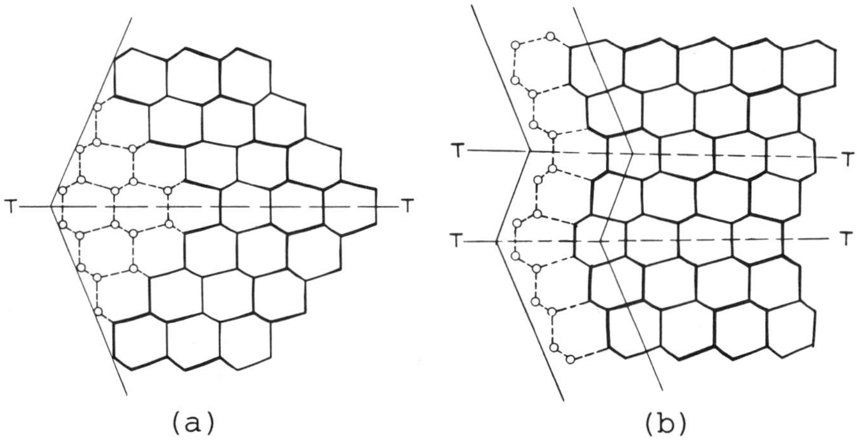

Figure 19. (a) Diamond cubic lattice: <110> projection of diamond lattice showing single twin plane (T-T) arising from single error in the stacking sequence. Broken lines and circles show addition of atoms which lead to a convex corner formed by the intersection of two {111} surfaces. (b) Diamond cubic lattice: <110> projection showing an intrinsic stacking fault from two errors (T-T, T-T) in stacking sequence. Broken lines and circles show addition of atoms. Note that a convex corner is not formed (94). *(Reproduced with permission.)*

depending on the substrate and it's preparation. However, all nucleation seems to occur by a two-step mechanism. The first step is the carbon-saturation of the substrate surface. Nucleation on unseeded or unscratched substrates depends on the attainment of a high surface carbon concentration to enhance the random formation of stable, prenuclei carbon clusters. The carbon saturation step therefore forms a stable platform for subsequent nucleation steps by preventing the diffusion of potentially nucleating carbon species into the substrate bulk and stabilizing any existing nuclei. Formation of sharp-edged features by etching probably also contributes to nucleation, although only if the features are carbon-saturated and have the correct morphology. Pre-scratching or seeding with diamond or some other suitable material leaves behind stable fragments on which subsequent diamond growth occurs. When diamond is used for scratching, such growth may be more properly characterized as isolated homoepitaxy rather than nucleation. Scratching also creates features with high-energy, multiple bonding sites, for attachment and stabilization of potential nuclei. Nucleation on most scratched substrates probably occurs both at surface defects and features, and (principally) on fragments of the scratching material. The relative contributions of homoepitaxy and "spontaneous" nucleation depend on the characteristics of the substrate, e.g., it's malleability, hardness, Debye temperature, carbon solubility, susceptibility to etching, and extent of bonding with the scratching material. Once nucleation has occurred, then the continued homoepitaxial growth of the nuclei becomes the important mechanism.

3.0 GROWTH

In this section we discuss the CVD diamond growth environment and the chemical processes leading to homoepitaxial growth, that is, the continued growth of diamond on a diamond surface after the initial nucleation process has occurred. This discussion requires an examination of both the gas phase, which determines the fluxes of species to and from the surface, and the chemistry at the gas/surface interface. Carbon addition to the bulk diamond lattice from the neutral and radical gas phase species is inherently a surface phenomenon. Thus the structure, composition, and dynamics of the diamond surface in the growth environment are key components of a detailed understanding of the CVD process. Unfortunately, in situ observations of diamond growth surfaces are nearly non-existent due to difficulty in probing the exceedingly small number of surface species. Hence, most of our understanding about the diamond growth mechanism is derived from measurements of the gaseous phase and computational models. We shall first summarize salient knowledge about diamond surface chemistry. Then we review the characterization of the gaseous environment and processes, which represent nearly all of the molecular measurements of the diamond CVD process to date. Finally, we describe the computational and theoretical models for diamond growth, which range from models of the macroscopic behavior of the system to models of the detailed microscopic dynamics.

Most diamond CVD processes employ reactants containing only carbon and hydrogen (and sometimes noble gases, e.g., He, Ar), and will be the focus of this review. The operating pressures for diamond CVD techniques range from atmospheric pressure for plasma jets and combustion torches, to 10^{-2} torr for electron cyclotron resonance (ECR) plasmas. Many processes operate at intermediate pressures of 1 to 100 torr using microwave, RF, and DC plasmas, or hot filaments. The reactants employed are usually a flow of molecular hydrogen containing 0.1 to 2.0% methane at 10 to 100 torr, with a total flow rate which depends on reactor geometry, typically 100 sccm. The substrate temperature is usually between 600 and 1100°C. Diamond growth employing oxygen-containing species has also been studied (140)-(143) and diamond CVD using halogenated species has recently been reported (143)(144).

The importance of charged species in the growth environment is difficult to evaluate. Given the range of growth chemistries, such as hot filament CVD, various types of plasmas, and combustion flames, the flux (and role) of charged species to the surface and in the gas phase is highly variable and depends on the specific CVD process. What is common to these various growth environments are many of the neutral and radical species: molecular hydrogen, methane, higher hydrocarbons, and radical species derived from methane (e.g., CH_3) and atomic hydrogen.

3.1 The Diamond Growth Surface

Much of the current understanding of the diamond growth surface is inferred from ultra-high vacuum (UHV) surface science and theoretical studies. The most studied diamond surfaces are the (111) and (100) surfaces of natural diamond crystals. These surfaces are predominantly hydrogen-terminated, meaning that the crystal lattice ends in C-H bonds which maintain the sp^3 hybridization of the carbon atoms. Other surface contaminants have been observed, such as O, N, S, F, and Cl, at less than a full monolayer coverage (145). The low coverage of non-hydrogen species results from the extremely high density of surface sites ($\sim 1.6 \times 10^{16}/cm^2$) (146), thus the inability of the surface to accommodate a full monolayer of atoms larger than hydrogen without significant strain and distortion. The inability of non-hydrogen species to completely terminate the diamond lattice may be important for the viability of other growth techniques not based on hydrogen. It is possible to desorb 50% and 100% of the surface-terminating hydrogen from the (100)(147) and (111)(148) faces, respectively, by heating in UHV to over 900°C. After desorption of the terminating species, surface reconstruction is observed for the (111) surface (>950°C)

(148) and (100) surface (>1000°C) (147). The surface reconstruction, as determined by the ratio of the half-order LEED pattern to the first order, and the desorption of hydrogen are shown versus the surface temperature in Fig. 20 (147). In the surface reconstruction, the unsatisfied valence electrons ("dangling" bonds) form pi-bonded chains and arrays among the surface carbon atoms. LEED (45)(149) and Auger carbon KVV lineshape (150)(151) studies indicate that diamond surfaces are easily damaged by electron and ion bombardment, making it difficult to obtain well ordered surfaces for study. As yet, no data exists for the structure and composition of the growth interface, but it is presumably between the extremes of a fully hydrogen terminated sp^3 bonded carbon surface and a nude sp^2 bonded carbon surface reconstruction. The precise nature of this interface should depend on the bulk temperature, and the flux of species to and from the surface.

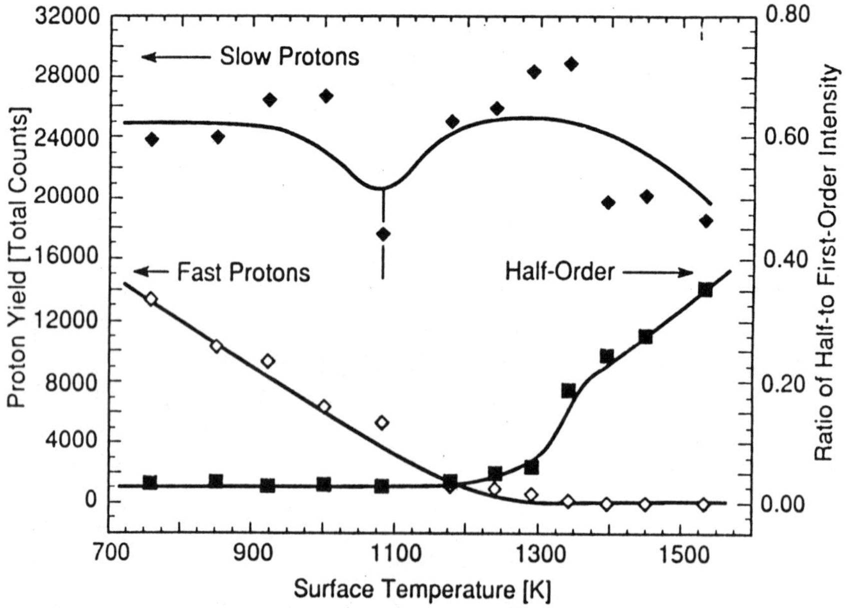

Figure 20. The proton electron stimulated desorption (ESD) time of flight (TOF) signal intensity and the ratio of the half- to first-order low energy electron diffraction (LEED) spot intensity versus anneal temperature of the diamond (100) surface. The fast proton intensity is attributed to the conversion of surface dihydride species to monohydride species. The half-order LEED intensity reflects degree of the 2x1 surface reconstruction (147). *(Reproduced with permission.)*

3.2 Characterization of the Gaseous Growth Environment

The analysis of the environment in a chemical vapor deposition process, in order to identify and understand the fundamental mechanisms, entails multiple compromises in selecting the appropriate diagnostic technique. Crucial selection criteria are *(a)* the selectivity required to identify the species of interest, *(b)* the sensitivity to detect the anticipated concentration levels, *(c)* the spatial resolution to identify the distribution of the species, *(d)* the intrusiveness, or degree to which the measurement perturbs and alters the environment, *(e)* the temporal resolution to characterize the kinetic behaviors in non-steady state systems, and *(f)* the cost of the experimental apparatus and measurements. Finally, it is important to consider the difficulty in interpreting the observations. The ease of measurement generally correlates with increased difficulty in data interpretation in terms of identifying and understanding the fundamental processes!

We divide the characterization section to differentiate in situ and ex situ analysis techniques. In situ techniques allow the direct measurement of a property (e.g., temperature) or species (reactive or stable species) within the process environment and with minimal perturbation. Ex situ techniques, in which samples of the gas or material are withdrawn for remote characterization or subsequent analysis, are perhaps simpler to implement, but are more demanding in terms of the analysis needed to extrapolate the measurements back to the actual growth conditions. The analyses reported of the diamond growth environments are primarily of the gas phase species, for which current analytical procedures have sufficient sensitivity and selectivity.

In Situ Analysis. Early in situ analysis of diamond growth environments containing H_2, CH_4, and rare gases was performed by observing the light emitted from excited species in microwave (9)(152)-(154) and DC (155) plasmas. Optical emission spectroscopy (OES) was used to detect emission from excited states of C, C_2, H, CH, CH_2, and H_2. OES provides sensitive detection of the few species with excited electronic states which fluoresce. Because of the complex nature of the excitation, quenching and radiative kinetics of each electronic state of the species, the emission intensities usually provide only qualitative information about the chemical composition of the gas. Determination of the ground state concentrations of species by OES is difficult without detailed knowledge of the excitation process, but relative changes can be deduced from actinometry (discussed later). A typical OES spectra is shown in Fig. 21 (153). The high sensitivity of OES provides for the detection of species with extremely low concentrations (e.g., the detection of CN and NH from N_2 leaks in the gas lines) (153). Because of this selective sensitivity, the use of OES data alone can provide a very misleading picture about the primary chemical reactants in the CVD process.

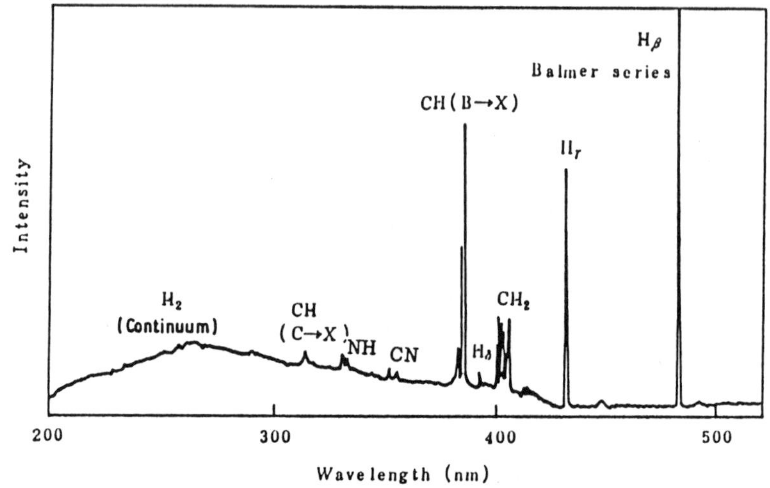

Figure 21. Optical emission spectra from the microwave plasma discharge of methane in hydrogen used for the CVD of diamond. The emission from NH and CN probably results from the leakage of air into the reaction tube (153). *(Reproduced with permission.)*

The emission intensities from various excited states of atomic hydrogen in the positive column of a DC discharge (at 200 torr) were used to derive a "temperature" of 4800 to 5300 K for excited atomic hydrogen, compared with the electron temperature of 100,000 K, measured using a Langmuir probe (155)(156). OES and the Langmuir probe were also used to examine a DC arcjet plasma (157). The different "temperatures" derived from measurement of different species indicate the non-equilibrium nature of these plasmas, and caution against the use of an equilibrium quantity such as temperature to describe these gaseous environments.

Optical emission from excited OH and CO is observed when oxygen containing reactants (141)(158)-(160) e.g., H_2O, O_2, CO, CO_2, and alcohols are added to the hydrogen-hydrocarbon-rare gas mixture used for the plasma CVD of diamond. The detection of OH in the strongly reducing H_2 environment confirms the sensitivity of OES to low concentrations and/or short lived species. OES has recently been used to probe combustion flames (161), arc (162) and arcjet (157), and microwave jet (163) plasmas, where in addition to the above observations, the quenching of emission in the stagnant gases near the surface has been used to infer the importance of kinetic processes in this boundary layer (163). In summary, the OES results reveal the presence of atomic hydrogen, hydrocarbon fragments, and many other trace species, indicating the existence of complex chemical reactions in the plasma and the non-equilibrium nature of the environment.

Plasma emission actinometry (164), a specific type of OES, employs the addition of a suitable rare gas to the plasma to assist in the relative calibration of the OES signals. Actinometry is a useful technique for situations in which the detected species and the rare gas have similar electronic energy levels, so that the emitting excited-state populations are presumed to have a similar dependance on the plasma electron energy distributions. Actinometry was used (141)(165) to infer the ground state atomic hydrogen concentrations and the results were then correlated with the quality of the deposited diamond (165). Since the addition of appreciable amounts of various rare gases to microwave CVD environments has been shown to induce significant changes in the OES spectra, as well as the quality and growth rate of the deposited diamond, care must be taken that the actinometer species does not perturb the environment (166).

The first in situ detection of the ground state species in a diamond CVD environment (167)(168) was accomplished using infrared diode laser absorption spectroscopy of the gaseous environment during diamond FACVD (169)-(173). In addition to the reactant CH_4, the detected species included the CH_3 radical, C_2H_2, and C_2H_4 (168)(172). Upper limits were set on the concentrations of many other hydrocarbons. The gaseous region between the hot filament and the deposition surface was probed using multiple reflections of the IR beam. Since absorption spectroscopy determines the integrated column density, gas phase inhomogeneities limited the analysis to the qualitative concentration estimates listed as part of Table 1.

The CH_4, C_2H_2, and C_2H_4 species were detected during microwave plasma deposition of diamond using various organic precursors in H_2 by Fourier transform infrared (FTIR) spectroscopy. CO was also detected when oxygen was present in the precursor (174). While many reactants were employed, including CH_4, C_2H_4, C_2H_2, CCl_4, CH_3OH, or C_2H_5OH, similar concentrations of CH_4, C_2H_2, and C_2H_4 were detected during diamond deposition. FTIR emission and absorption was also used to study diamond depositing oxy-acetylene torch flames (175), in which CO_2, H_2O, CO, and OH were detected. This combustion study also demonstrated a significant perturbation of the flame resulting from insertion of the substrate into the flame. For example, the CO_2 rotational temperature range narrowed from 2800 - 3300 K to 2950 - 3150 K when the growth substrate was present.

Table 1. Comparison of the measured and calculated number densities for various species observed in the FACVD of diamond at a distance of 5 mm from the substrate surface. The conditions were 1% CH_4 in H_2, filament temperature 2200°C, substrate temperature 900°C.

Species	Measured (a) (cm^{-3})	Calculated (b) (cm^{-3})
CH_4	$8 \pm 3 \times 10^{14}$	1.1×10^{14}
CH_3	$2 \pm 1.5 \times 10^{12}$	1.6×10^{13}
C_2H_2	$2 \pm 1 \times 10^{14}$	6.0×10^{14}
C_2H_4	$6 \pm 2 \times 10^{12}$	2.0×10^{12}
C_3H_4	$< 8 \times 10^{13}$	9.4×10^{10}
C_3H_6	$< 1.4 \times 10^{14}$ (c)	2.2×10^{9}
C_4H_2	$< 3 \times 10^{13}$	2.0×10^{11}
H	2×10^{15} (d) 1×10^{15} (e)	5.8×10^{14}

(a) Infrared diode laser measurements (172), except as noted.
(b) One-dimensional numerical model (213).
(c) Sum of methyl acetylene and cyclopropane.
(d) LIF measurement taken 5 mm from the filament with no substrate present (182).
(e) From the kinetic analysis of gas chromatographic data sampled at the substrate surface (194).

Resonance enhanced multi-photon ionization spectroscopy (REMPI) was used to detect both the H (169)(176)(177) and CH_3 (169)(171)(173) radicals in a FACVD reactor. A focused pulsed dye laser resonantly ionized the species of interest, and the total induced current was detected. The REMPI technique, although capable of excellent spatial resolution, requires the insertion of a potentially intrusive electrode to collect the ionization current. CH_3 radicals were detected through a resonance at 333.5 nm (178) confirming the previous IR diode laser results (167). The CH_3 radical REMPI intensity increased with both the initial reactant CH_4 concentration (171)(173) and the filament temperature (173). The spatial dependance of the CH_3 signal with respect to the filament has also been reported (173).

Ground state H atoms were detected with REMPI using three 365.5 nm photons resonant with the Lyman alpha transition at 121.8 nm (176). The detection of ground state H atoms provided for the direct measurement of one of the species believed to play a key role in the chemistry of diamond CVD. The REMPI signal, obtained at a fixed distance midway between the filament and the substrate and using molecular hydrogen as a reactant, increased with increasing filament temperature. The observed signal was proportional to the equilibrium atomic hydrogen concentration computed from the thermal dissociation of molecular hydrogen at the filament temperature. Measurements made above and below the filament indicated that diffusion is the principal H-atom transport mechanism (176). The addition of CH_4 to the H_2 resulted in lower H atom signals, compared with the use of pure H_2, as shown in Fig. 22 (176). The significant decrease in the H atom signal at temperatures less than 2200°C and CH_4 concentrations greater than 3% (176) has been attributed to reduced H_2 dissociation caused by carbon deposition on the filament surface (179).

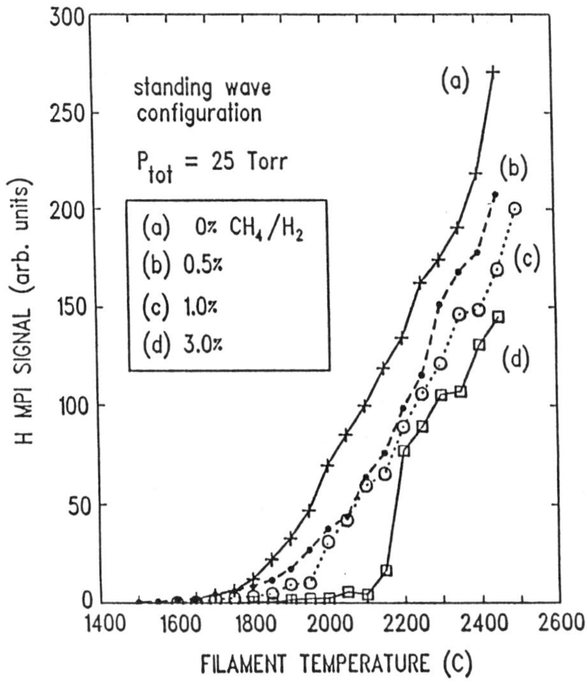

Figure 22. Dependance of the hydrogen atom multi-photon ionization (MPI) intensity on the filament temperature, T_{fil}, for standing wave excitation at 364.68 nm, using the indicated initial concentrations of methane in hydrogen (176). *(Reproduced with permission.)*

Third harmonic generation (THG) (180) was also used to detect atomic hydrogen in both the FACVD (177) and DC plasma environments (181). In the THG technique, a four wave mixing process causes the generation of coherent radiation slightly blue shifted from the Lyman alpha wavelength (121.8 nm) when light near the three-photon resonance (365.5 nm) is focused through the gas containing atomic hydrogen. The THG technique does not employ a collection electrode, so it is less intrusive than REMPI; however, it does require that the medium be partially transmissive at the third harmonic wavelength. THG holds promise for determining spatial concentration profiles of ground state H atom concentrations in diamond CVD and other environments.

The concentration of atomic hydrogen produced in H_2/CH_4 mixtures was determined using two-photon laser-induced fluorescence (LIF) (182). Diamond was not deposited during these measurements, but the absolute H atom concentrations and spatial profiles could be used as a benchmark for both calculations and other relative H atom determinations (see below). The effect of various filament materials and diameters on atomic hydrogen production was observed with LIF, but were different from those determined with the filament current-voltage characteristics (183). Species concentrations may be quite dependant on the particular reactor and the presence of a substrate. These measurements also support diffusion as the principal atomic hydrogen transport mechanism near the filament.

Other laser-based techniques used to study diamond CVD environments include LIF and coherent anti-Stokes Raman scattering (CARS). LIF, like OES, is a useful technique for the study of species with accessible fluorescent electronic states. Unlike OES, LIF can provide temperature and concentration information about ground state populations. LIF has been used to detect OH and C_2 in an acetylene torch (184), and CH in a DC arcjet (185). The rotational temperature of CH in the DC arcjet was determined (185) by LIF to be ~ 2100 K, compared with a value of 5000 K from OES measurements of electronically excited C_2. This discrepancy also demonstrates the need for care in interpretation of OES data.

The four wave mixing technique of CARS is useful in probing ground state populations with reasonable spatial resolution, but is generally applicable to only the major gas-phase components. CARS was used to measure CH_4 and C_2H_2 concentrations (186), and H_2 rotational temperature profiles (187), in an FACVD reactor. The minimal perturbation of the environment, and the ability to measure both concentrations and temperatures, are major advantages of both LIF and CARS.

The novel technique of planar laser-induced photodissociation fluorescence was applied to the analysis of a diamond-depositing oxy-acetylene flame (188). The importance of the C_2H radical, which is difficult to detect by many other techniques, was suggested. However, the assignment to C_2H of the observed laser-induced CH fluorescence needs to be confirmed.

Molecular beam mass spectrometry provides direct information about the concentrations of reactive and stable species. By sampling gases through an orifice in the deposition substrate, most reactive collisions are quenched by formation of a (collisionless) molecular beam. Distortion of the gas transport mechanisms near the sampling orifice are a concern with this technique. Molecular beam mass spectrometry sampling of diamond depositing FACVD (189), microwave plasma (189), microwave plasma jet (163), and DC plasma (190) environments has been performed and shows the presence of H, H_2, CH_3, CH_4, and C_2H_2 near the surface when either CH_4/H_2 or C_2H_2/H_2 mixtures were used as reactants. Because of the near collisionless sampling, these measurements have provided the most direct measurement of the gaseous species near the growth surface.

Ex Situ Analysis. Gas sampling and subsequent mass spectrometric analysis were performed on stable species from near the FACVD growth surface (124)(191). The species were withdrawn through an orifice flush with Si (191) and Pt (124) substrates, but the subsequent chemistry of reactive species was not quenched, as occurs with the molecular beam sampling techniques. The major species detected from CH_4/H_2 reactant mixtures were CH_4 and C_2H_2, along with smaller amounts of C_2H_4 and C_2H_6. The high sensitivity of mass spectroscopy also permitted the detection of larger hydrocarbons, after the gas was condensed and concentrated (124). The measured or computed concentrations of CH_4, C_2H_2, CH_3 and C_2H_4 were concluded to be sufficient to account for the observed diamond deposition rates. The sampling mass spectrometric technique was also used to examine the effects of oxygen addition to the CH_4/H_2 reactant mixture (192). The principal effect of the added O_2 was the formation of CO, which lowered the effective concentration of hydrocarbons. Modeling of these results also suggested the presence of the OH radical, which could play a role in removing non-diamond carbon deposits (192).

The exhaust gases of a diamond FACVD reactor were sampled and analyzed using mass spectrometry (83)(179)(193). These studies also observed the significant conversion of either CH_4 or C_2H_2 reactants into both CH_4 and C_2H_2. Major changes in the CH_4 to C_2H_2 ratio in the reactor exhaust gas were noted as a function of the filament temperature, reactant hydrocarbon concentration, and filament surface composition (83)(179). It was found that carbon deposition on the filament occurs under some conditions, and can have a significant effect on the gas phase concentrations.

Gas chromatography has been performed on gases sampled from the diamond FACVD environment (194)(195). Again, CH_4 and C_2H_2 were the major species observed from CH_4/H_2 and C_2H_2/H_2 reactant mixtures, while C_2H_6 and C_2H_4 were minor constituents. The observed C_2H_6 and C_2H_4 concentrations were sensitive to the pressure in the sampling probe, indicating that these measurements reflect the chemistry in the sampling probe, and not the concentrations in the CVD reactor (194). By assuming that CH_3 recombination in the sampling probe was the primary source of the C_2H_6 and C_2H_4, CH_3 concentrations were derived (194). Absolute H atom concentrations were also deduced, using kinetic analysis discussed later (194).

The $^{12}C/^{13}C$ ratio of CH_4 and C_2H_2 sampled from a FACVD growth environment employing isotopically tagged reagents, $^{13}CH_4$, $^{12}CH_4$, and $^{12}C_2H_2$, was provided by FTIR analysis of species condensed in a cryogenic matrix (196)(197) The carbon isotopic ratio in the deposited diamond film, derived from the shift of the Raman band, differed substantially from that determined for C_2H_2 by the gas sampling analysis, and agreed closely with that inferred for the CH_3 radical. These studies provide strong evidence for the CH_3 radical as a primary growth species.

Examination of the key carbon species for diamond deposition in a plasma assisted growth environment was performed when either CH_4 or C_2H_2 was injected into a partially dissociated hydrogen/argon gas stream, and the flux of species delivered to the surface was analyzed using computational analysis of the chemical kinetic and mass transport processes (198). Diamond was deposited in both cases, although CH_4 gave higher growth rates and better crystal quality. It was concluded that only CH_3 and CH_4 were present in sufficient concentrations to account for the observed deposition rates when CH_4 was used as a reactant. The contribution of C_2H_2 to diamond growth could not be ruled out because of the observation of diamond growth, although of poorer quality, when C_2H_2 was injected as a reactant. The localized growth of diamond near the hydrocarbon injection point suggested the importance of surface or wall losses of atomic hydrogen on the diamond growth process.

In summary, the principal species detected in diamond growth environments employing hydrogen and hydrocarbons reactants are H_2, H, CH_4, CH_3, C_2H_2, and C_2H_4. Traces of C_2H_6, C_2H and hydrocarbons with three or more carbons were also reported. In plasmas, emission is observed from C, C_2, CH, CH_2, H, H_2, and trace impurities. When oxygen or oxygen-containing organic compounds are used as reactants, CO, H_2O, and OH are also observed. The concentrations of CH_4, C_2H_2, C_2H_4, and CH_3 present in diamond growth environments are sufficient to account for the observed deposition rates. Isotopic labeling experiments suggest the CH_3 radical as a principal growth species.

3.3 Models of the Gaseous Growth Environment

Gaseous Equilibrium. The use of hot filaments, plasmas, or other activation techniques alters the composition of the gas phase, as demonstrated in the analyses described above. For example, the detection of significant concentrations of acetylene derived from the reactant methane, or methane from reactant acetylene, provides evidence for a complex sequence of gaseous reactions leading to products characteristic of high temperature environments. Because of the significant flow and species transport rates involved in most diamond CVD environments, the gaseous environment is not expected to reach thermodynamic equilibrium, either in the activation region or in the vicinity of the growth surface. However, a general picture of the gas phase composition is provided by the equilibrium computations for situations where the gas residence time in the reactor is comparable with the reaction kinetic time scales. Such thermodynamic calculations have been performed for equilibrium with solid pyrolytic carbon (199) and specifically for diamond CVD (27)(167)(168)(171)(200)-(203). Figure 23 presents the results of an equilibrium calculation for a 20 torr mixture of 0.5% CH_4 in H_2 over the temperature range of 1000 to 3000°C. The species observed in the FACVD environment (171)(172), most of which are listed in Table 1, agree qualitatively with the species predicted by the equilibrium calculation between the limits set by the substrate and filament temperatures, 900 to 2200°C. The concentrations predicted by the equilibrium model for the observed hydrocarbons, CH_4, C_2H_2, CH_3 and C_2H_4, were adequate to account for the typical diamond deposition rates, while the fluxes of CH, CH_2, C_2, and C_3 were too low to account for most diamond growth (171)(191). Quantitative differences between the thermodynamic results and the experimental observations indicate the important role of gaseous reaction kinetics and the transport processes. Further thermodynamic analysis (27) has shown that the deposition rate is proportional to the carbon supersaturation and the carbon solubility in the gas phase, the latter of which increases with pressure. The analysis was extended to explain diamond deposition by chemical vapor transport in a closed system. In that case, deposition occurs from an undersaturated solution which is in an activated, non-equilibrium state (204).

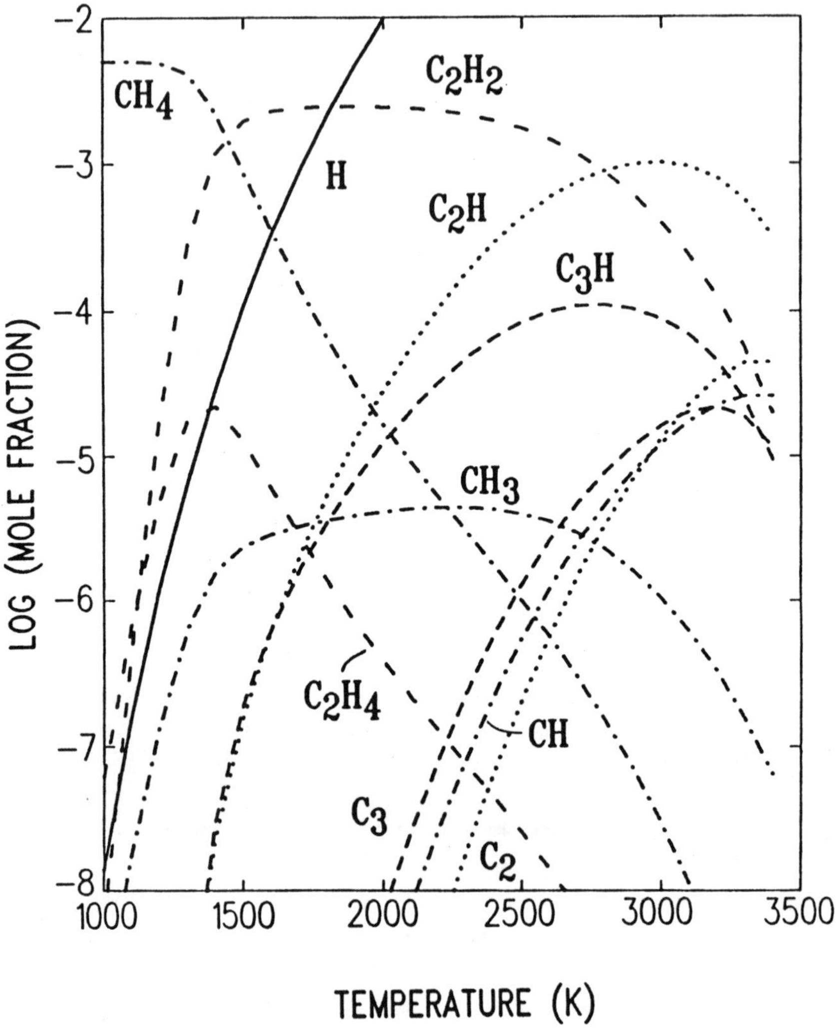

Figure 23. The calculated equilibrium concentrations of gaseous species (with mole fractions between 10^{-2} and 10^{-8}) versus temperature is shown for an initial gas mixture of 0.5% CH_4 in H_2 at 20 torr pressure. Molecular hydrogen (mole fraction >0.01) and species with mole fraction < 10^{-8} at temperatures less than 2500 K are omitted (171). *(Reproduced with permission.)*

Gaseous Kinetics. Analysis of the complex diamond CVD environment requires a comprehensive understanding of the kinetics controlling the generation of reactive species, their reaction rates, and their transport within the reactor. The radical species detected in the gas phase and near the surface indicate that the chemical lifetimes of these reactive species can exceed the time required for their transport to the surface. In lieu of the many difficult experimental observations needed to obtain a detailed knowledge of the species fluxes near the growth surface, kinetic computational models have been implemented. Although information is extracted from a number of sources, e.g., combustion chemistry, thermodynamics, experimental data, and ab initio calculations, the development of detailed kinetic models often requires simplifying assumptions when there is a lack of reaction rate data or the computations become intractable. It is therefore imperative to evaluate the region of validity of the model assumptions through comparison with experimental observations.

One of the earliest kinetic models of gaseous processes in diamond CVD was constrained to species containing only a single carbon atom, and indicated the need for high H atom concentrations (205). The insensitivity of the deposition process to the growth species was also suggested, but may have been due to the assigned values of the molecular bond energies (204). More realistic reaction rate constants were used in a number of kinetic modeling studies of FACVD chemistry. One dimensional fluid flow has typically been assumed to simplify the calculations. Such a kinetic model was used to interpret the sampling mass spectrometric detection of CH_4 and C_2H_2 in the FACVD of diamond (124)(191). This model indicated that the diamond deposition rate could be accounted for only by the calculated concentrations of C_2H_2, C_2H_4, CH_4, or CH_3. A similar kinetic model was later used to derive H and CH_3 concentrations from the detection of C_2H_4 and C_2H_6 (194). Because the reaction of $CH_3 + H_2$ to form $CH_4 + H$ was computed to be in rapid equilibrium, the measured concentrations of H_2, CH_4, and CH_3 were used to determine that of the atomic hydrogen. The resultant mole fractions of H and CH_3 are shown versus CH_4 percentage in H_2 in Fig. 24. The H atom concentration measured at the surface in this manner was within a factor of two of that determined 5 mm from the filament by LIF (182) (see Table 1).

With addition of oxygen to the FACVD environment, good agreement was obtained between kinetic calculation and mass spectral observations (192). Higher hydrocarbon reactant concentrations were possible for the growth of diamond since most of the oxygen rapidly reacted with the hydrocarbons to form CO, a stable molecule in these reducing environments.

The role of OH as a carbon etchant was also postulated (192). In a kinetic model of FACVD using methanol (206), it was concluded that the CH_3 was the principal hydrocarbon involved in diamond growth.

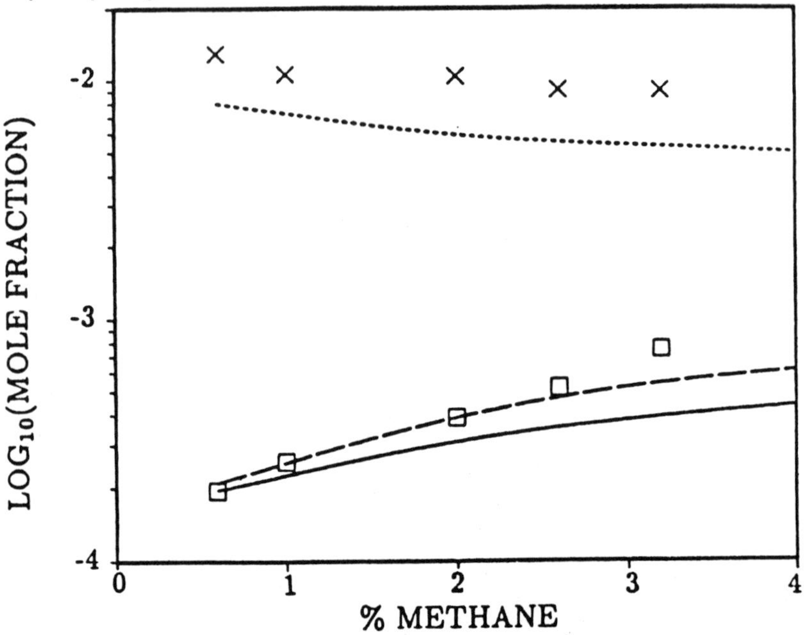

Figure 24. Comparison between measured and calculated species mole fractions as a function of initial CH_4 percentage: H - crosses (experiment) and dotted line (model); CH_3 - open squares and solid line. The dashed line is the model calculation for the mole fractions of $CH_3 + C_2H_4 + C_2H_6$ (194). *(Reproduced with permission.)*

Kinetic computations have also suggested a role for molecular hydrogen in the suppression of polycyclic aromatic species formation in the gas phase, and therefore the suppression of non-diamond carbon nucleation on the surface (28). The role of CH_4 as growth species was concluded using kinetic analysis (207) of early diamond growth experiments (208) which employed no direct gaseous activation using a plasma or a hot filament.

A kinetic model with one dimensional fluid flow was developed for diamond deposition from an oxygen-acetylene combustion flame operating at 3237 K (209). The results suggested that the surface reactions did not have a significant effect on the gas phase concentrations, but that near the cooler surface, 1250 K, the desorbing hydrocarbon was principally CH_4 which rapidly equilibrated with the high H atom concentrations to form CH_3. The observed growth rate dependance on the ratio of C_2H_2/O_2 compared with the model predictions for various species indicated the CH_3 radical was the principal carbon deposition species, as shown in Fig. 25.

Kinetic analysis was used to compute the species concentration profiles resulting from downstream injection of CH_4 or C_2H_2 into a rapid flow of atomic and molecular hydrogen from a microwave discharge (210). Correlation of the quality and quantity of the diamond deposited on silicon substrates suggested CH_3 (or CH_4) as the principal carbon deposition species, but some diamond deposition due to C_2H_2 was also observed.

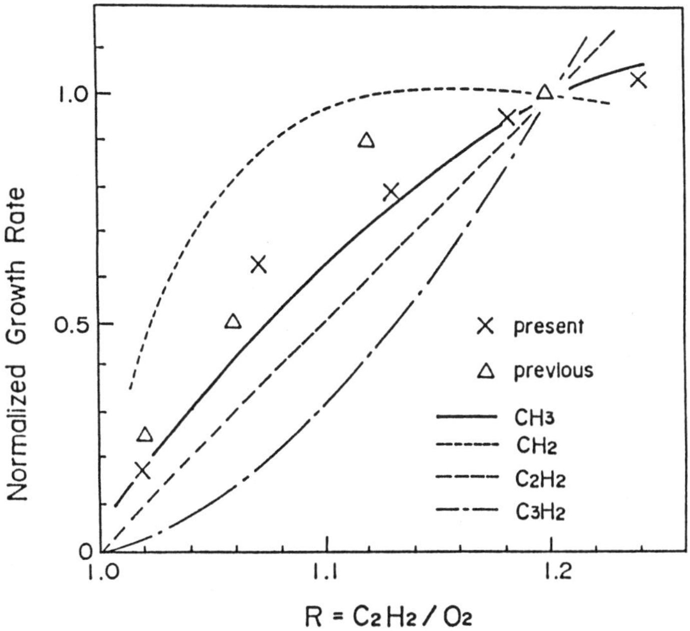

Figure 25. Measured diamond growth rates for various C_2H_2/O_2 ratios in an oxy-acetylene combustion flame (Δ and x) are shown in comparison with values calculated by assuming different growth species, as indicated (209). *(Reproduced with permission.)*

Fluid flow and heat transfer were addressed for a vertical tube FACVD reactor (211). Diamond was deposited with various substrate/filament orientations relative to the gas flow and a model was constructed for a 1% CH_4 in H_2 mixture at 30 torr and 200 sccm total flow rate. The gas composition was assumed to be in equilibrium with the filament temperature. Free and forced convection were found to be relatively unimportant in determining the film deposition rate and quality. Diffusion was the dominant mass transport mechanism between the hot filament and the substrate. The observed insensitivity of the deposition rates and material quality on the various reactor geometries was consistent with the computational results (211).

Numerical modeling of the gas transport and the kinetics was performed for an axially symmetric FACVD environment (212). The temperature, velocity and species concentration profiles were computed and compared well with the existing experimental data. The mole fractions computed for various species versus distance from the filament are shown in Fig. 26. A comparison of the results of this numerical model with various experimental data is shown in Table 1. Differences between the model and the IR diode laser absorption data are due to the line-of-sight nature of the concentration measurement. While the overall CVD environment was found to be far from equilibrium, several reactions, principally those involving H + CH_n (n = 1,2,3,4), were close to partial equilibrium (212). The concentrations of C_1 species were thus determined by the local CH_4 and H concentrations. Later work has shown that this partial equilibrium breaks down near the diamond surface, where recombination and reaction deplete the local gaseous H atom concentration (213). The only hydrocarbon species present in

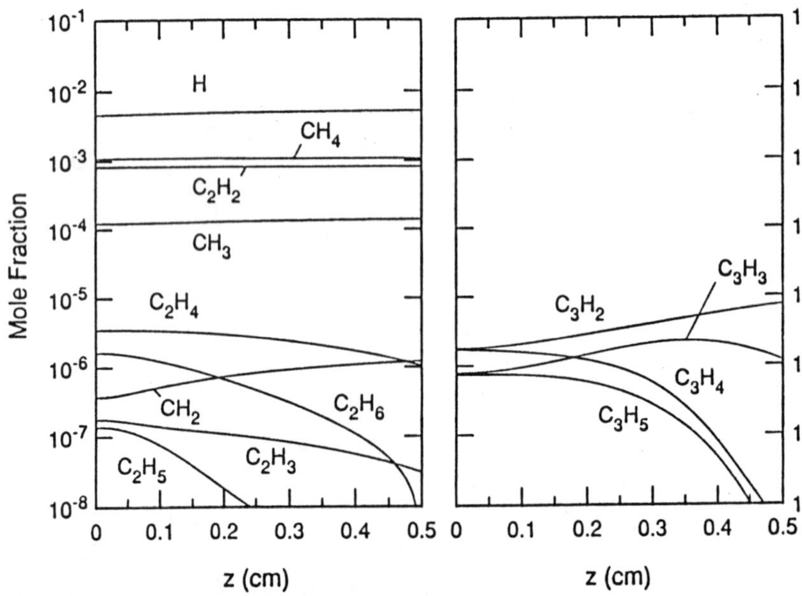

Figure 26. Mole fraction profiles for all species reaching 0.1 ppm, calculated for conditions simulating experiments of Harris (191) and coworkers (212). *(Reproduced with permission.)*

sufficient concentrations to account for the observed deposition rates were CH_4, C_2H_2, and CH_3. Gas phase mechanisms alone were found to be insufficient to account for the observed reduction (171) in H atom concentrations with increased reactant CH_4 concentration, suggesting the importance of reactions on the filament surface for the production of atomic H.

In summary, three types of models have been applied to the gaseous environment in diamond CVD. Equilibrium thermodynamic calculations have proved useful in qualitatively summarizing the complex gaseous kinetics. One dimensional kinetic models confirmed the major species present and suggested that the dominant carbon growth species for diamond deposition could be CH_4, CH_3, C_2H_2, and/or C_2H_4. Numerical models which include both fluid transport and gaseous kinetics are providing more quantitative predictions for comparison with experimental data, and suggest that diffusion is the dominant transport mechanism for the light species, such as atomic H, in diamond FACVD reactors.

3.4 Models of Diamond Growth

Dynamic Balance. The first models used to explain the growth of diamond, as opposed to graphite or other forms of carbon, were macroscopic models based on deposition from gaseous solutions simultaneously supersaturated in carbon and atomic hydrogen at the substrate growth temperature (25)(44)(201)(214)(215). The growth environment was viewed as a dynamic balance or kinetic competition between carbon deposition and etching processes for each form of solid carbon. A schematic of this simplified model is presented in Fig. 27 (171). Carbon supersaturation provided the driving force for the condensation of various forms of solid carbon including diamond, while atomic hydrogen provided a reactive environment for the removal or suppression of the non-diamond carbon by etching of solid carbons and/or converting surface sp^2 bonded sites into sp^3 moieties (171). While this macroscopic model did not identify any of the detailed gaseous or surface mechanisms leading to diamond deposition, it had utility in that it assisted in rationalizing results which are general for many growth environments. For example, higher deposition rates and concentrations of non-diamond components and crystal defects are observed with higher hydrocarbon reactant concentrations, which provide greater carbon supersaturation. The model also suggested the utility of introducing additional etchant species, such as OH or halogens, to produce higher quality diamond.

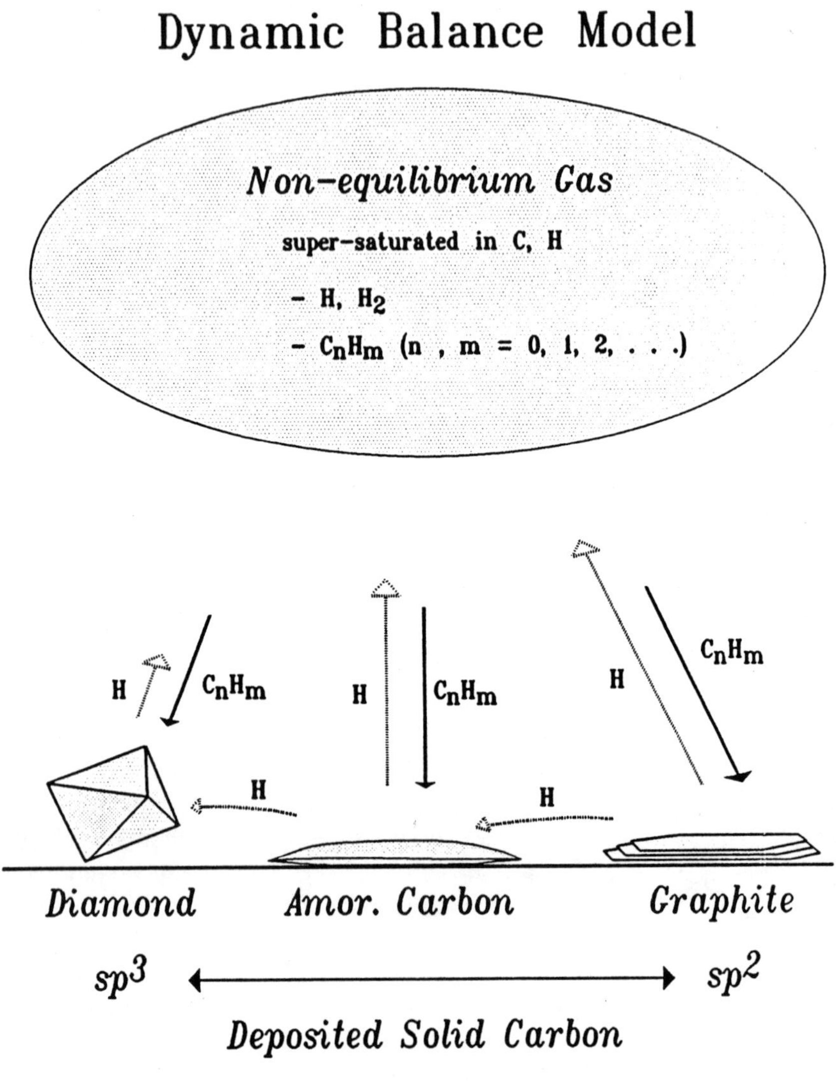

Figure 27. A schematic model of the chemical vapor deposition of diamond where the deposition of the various forms of solid carbon compete with atomic hydrogen induced reactions interconverting or etching the different surface bonding configurations (sp, sp^2, and sp^3). The quality (relative concentrations of the phases) of the diamond deposited is determined by the dynamic balance between the deposition and reaction processes (171). *(Reproduced with permission.)*

Non-Equilibrium Thermodynamics. Models derived from thermodynamics can be useful because they are simple to apply and may provide insights into the complex sequence of events comprising the deposition process. A quasi-equilibrium (QE) thermodynamic model (193)(202) has been successfully applied to diamond CVD. An equilibrium model for the solid carbon surface, and the vapor species in contact with it, is modified by the addition of a term to describe the enhanced etching of graphite by atomic hydrogen. The equilibrium carbon phase diagram is thus altered and predicts a region in the carbon phase diagram where diamond is the only stable solid phase, consistent with observations. The model was also useful in interpreting the activity of the filament materials in the FACVD of diamond. Specifically, the suppression of the gaseous atomic hydrogen observed (176) near the filament at high gaseous carbon reactant concentrations was explained by the QE model (83)(179) as due to the formation of a solid (graphitic) carbon phase on the filament surface.

Bulk diamond is thermodynamically less stable than bulk graphite, but this relationship may not be the relevant factor at interfaces such as the CVD growth surface. The structure or template of the bulk material near the interface may provide a pseudomorphic stabilization (23) for the growing diamond surface, an effect which may rationalize diamond growth, as opposed to graphite, on diamond substrates.

The local thermodynamics of the surface and near surface environments were examined as part of an application of non-equilibrium thermodynamics to a global reaction process (216)(217). While equilibrium thermodynamics is not valid for describing a system as complex as the CVD reactor, it has been shown to be valid for understanding the processes local to the growth surface (217). The growth surface is assumed to obey equilibrium thermodynamics subject to the constraint that the gas phase species concentrations in the immediate environment are determined by the flux of species to and from the region. The gas phase concentrations can be supersaturated with atomic hydrogen and/or carbon species, which can alter the stability of the diamond vs. non-diamond surfaces. The estimations (217) indicated that the hydrogen-terminated diamond surfaces have a lower energy than graphitic surfaces, and thus the growth of diamond was consistent with non-equilibrium thermodynamics. Figure 28 shows the relative fractions of graphite, diamond and Lonsdaleite versus atomic hydrogen partial pressure for a substrate at 950°C (216).

Detailed Surface Models. The first detailed surface chemical model of diamond growth was proposed for the (111) surface (218). A neutral CH_3-terminated surface, produced from the H-terminated surface by several mechanisms, was converted to a new H-terminated diamond layer by the reaction of a methyl cation, CH_3^+, with three adjacent methyl groups. The energetics of the reaction path were computed by semi-empirical quantum chemical calculations and are outlined in Fig. 29 (218). Because ionic species are not prevalent in all diamond CVD environments, e.g., FACVD, the CH_3^+ model is not sufficiently general to describe all growth systems. The CH_3^+ reaction mechanism was subsequently expanded to include the reaction of neutral methyl radicals with a positively charged site on the (111) surface (219). The energetics computed for the reaction steps in these mechanisms displayed potential barriers ranging from 5 to 27 kcal/mol (218)(219).

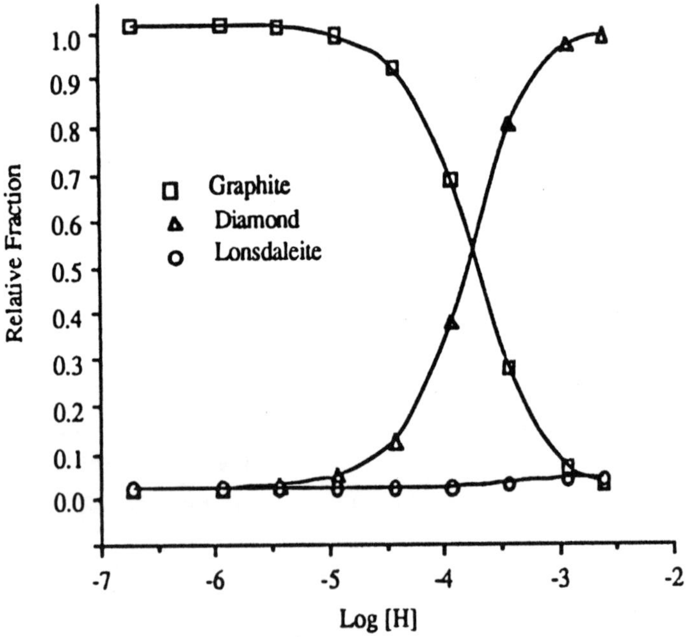

Figure 28. Variation of the predicted phase of carbon with atomic hydrogen partial pressure at a substrate temperature of 950°C. At the partial pressure of H expected in H_2 at 50 torr and 2000°C (2.09 x 10^{-3} atm.) essentially all of the solid is diamond. As the partial pressure of H approaches the equilibrium value, increasing amounts of graphite are seen. At equilibrium, only graphite is expected (216). *(Reproduced with permission.)*

Acetylene was proposed as the principal growth species for another model of growth on the diamond (111) surface. The reaction mechanism alternates between the generation of a surface radical site by H-atom abstraction and the addition of one or more C_2H_2 molecules (Fig. 30) (138). A semi-empirical quantum mechanical method was applied to computing the energetics of this mechanism (220) and indicated an exothermic reaction pathway after the initial endothermic hydrogen abstraction.

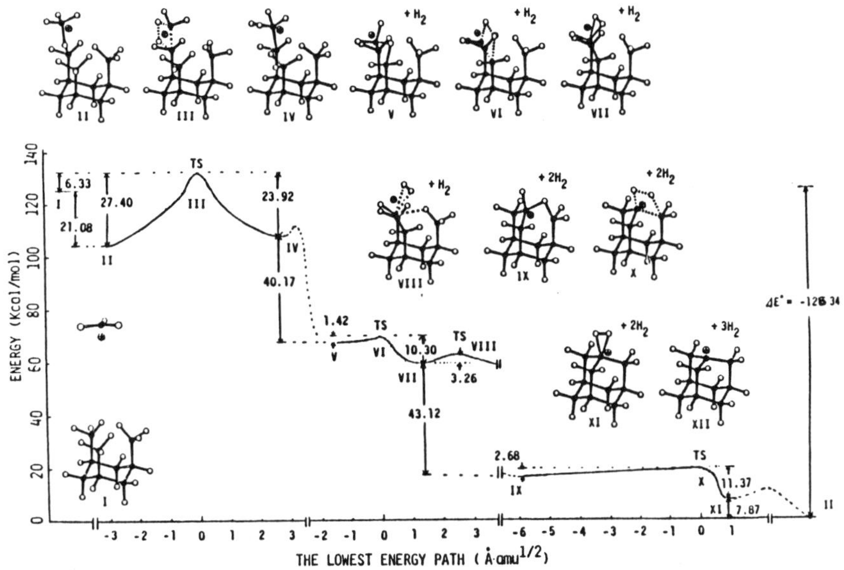

Figure 29. Detailed potential energy change following the reaction path of epitaxial growth of the diamond surface. All structures of the reaction system are shown corresponding to the transition state and the minimum point. Numerals are the energy differences in kcal/mol. Ordinate: potential energy in kcal/mol. Abscissa: The lowest energy reaction path in the mass-weighted 3N-6 dimensional Cartesian coordinate system. Solid circles represent carbon atoms, open circles, hydrogen atoms (218). *(Reproduced with permission.)*

A mechanism for diamond growth on the diamond (100) surface was proposed in which the neutral methyl radical was the main growth species (221). The reaction sequence is initiated by hydrogen atom abstraction of a surface hydrogen to form a surface radical site. Intermediate stages of the mechanism, such as methyl radical addition and diamond surface ring closure, are shown in Fig. 31. Kinetic computations of the growth rate using a nine-carbon diamond lattice model were computed to be between 0.06 and

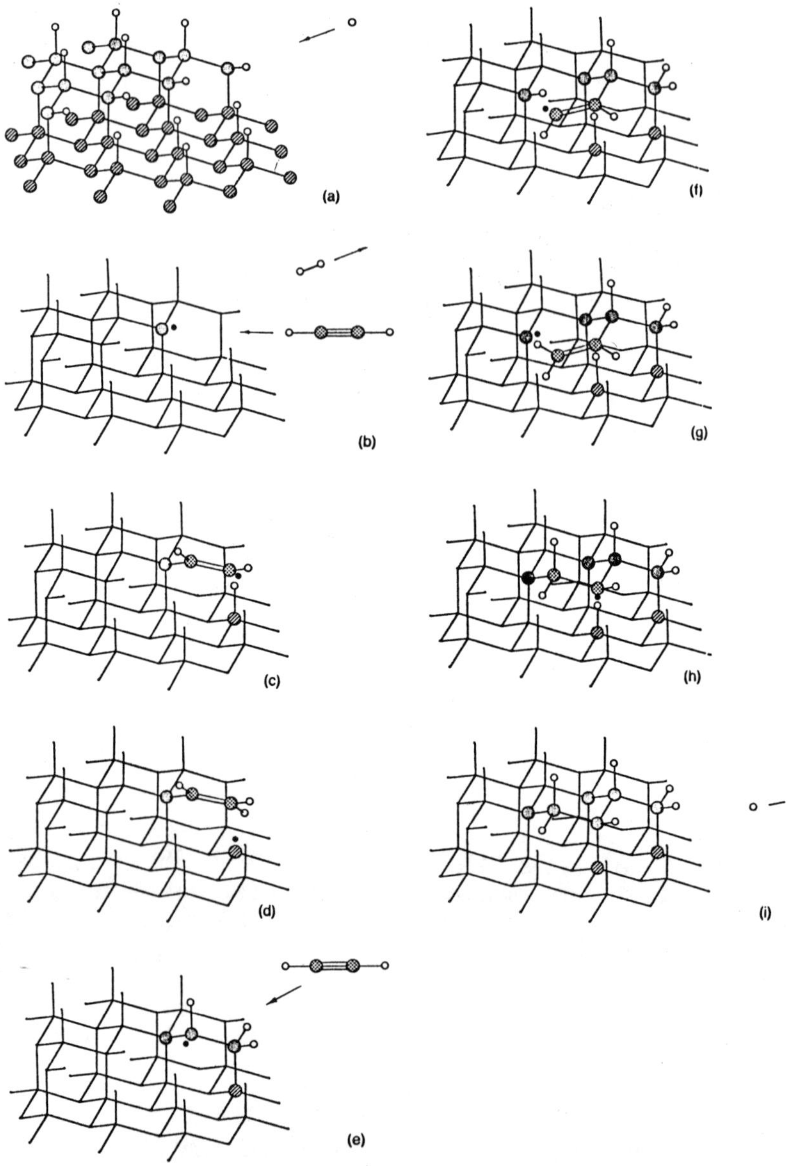

Figure 30. A schematic diagram illustrating the propagation reaction sequence for the diamond layer growth on a stepped (111) surface. The larger circles represent carbon atoms, circles with diagonal lines represent the lower layer, circles with fine mesh represent the upper layer, circles with a coarse mesh represent gaseous acetylene molecules, and the smaller open circles represent hydrogen atoms (138).

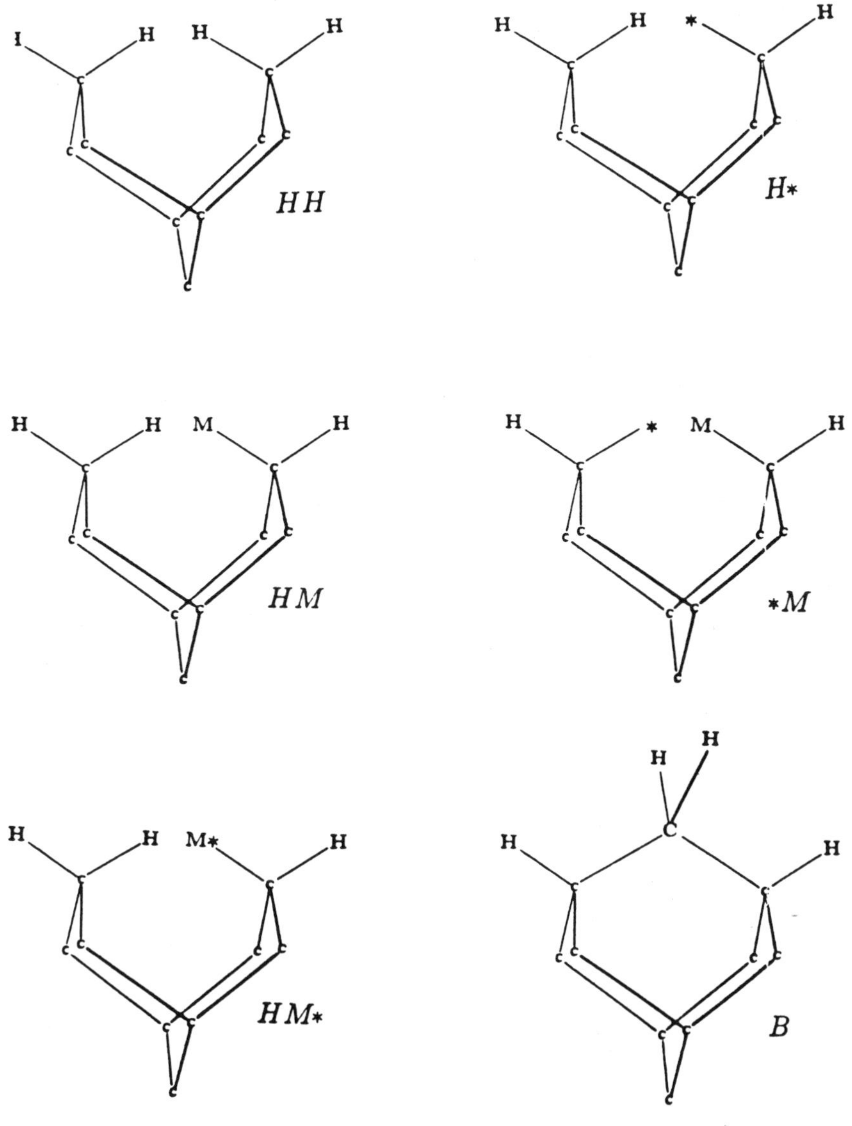

Figure 31. Surface configurations for converting a model of the diamond (100) surface, bicyclononane (HH), to a model of one added layer of growth, adamantane (B). * represents an unpaired electron (radical site); M represents a surface bonded -CH_3 species; M* represents a surface bonded -CH_2 radical (221). *(Reproduced with permission.)*

0.6 microns per hour using rate constants estimated for the model compounds and no adjustable parameters (221). These calculated growth rates are similar to those typically observed in FACVD of diamond.

An extensive kinetic model has been presented which employed both gaseous and surface kinetics to examine the details of idealized FACVD diamond deposition (222). The gaseous kinetics portion was based on data from CH_4 high temperature pyrolysis and oxidation, as well as the formation of polycyclic aromatics in C_2H_2 and C_2H_4 flames. The surface reaction rates, which are almost completely unknown, were derived from analogous gaseous reactions. Reasonable agreement was attained with a wide variety of observations including the species concentrations present in the growth environment, growth rate dependance, and material quality dependance on reagent concentration and substrate temperature. The principal conclusions are summarized in the following statements: *(a)* At low substrate temperatures, the abstraction of surface hydrogen by gaseous atomic hydrogen is the major bottleneck to diamond growth. *(b)* Growth occurs with both CH_3 radicals and C_2H_2, with C_2H_2 being the principal growth species on the (111) surface. However, the accentuated role of C_2H_2 may result from the omission of the surface desorption of C_2H_2 from the model. *(c)* Benzene condensation onto the growing surface at low substrate temperatures results in the deposition of amorphous carbon. *(d)* Molecular hydrogen and oxygen suppress the formation of polycyclic aromatics in the gas phase. *(e)* sp^2 carbon species are gasified much more rapidly by OH radicals than atomic hydrogen. *(f)* This model does not support the dynamic balance model (25)(171)(201)(215), but rather, it asserts the critical role of aromatic condensation and interconversion of sp^2 and sp^3 carbons by reaction with atomic hydrogen.

In summary, various models have been proposed for the diamond chemical vapor deposition process from hydrocarbon species in a low pressure hydrogen atmosphere. These models reaffirm the importance of atomic hydrogen in both gaseous and surface chemical processes and suggest the methyl radical, CH_3, and/or C_2H_2, C_2H_2, as the principal carbon deposition species. Detailed mechanisms for the incorporation of the hydrocarbon species into the growing diamond surface have been proposed and tested for the diamond (111) and (100) low index crystal surfaces. Less specific models employing statistical and thermodynamic concepts have been of utility in understanding the growth process over a wide range of experimental observations.

ACKNOWLEDGMENTS

We acknowledge the assistance of Dave Venezky, Helen Thorsheim, and John Russell, Jr. for discussions, careful reading of the manuscript and helpful suggestions. This work was supported in part by the Office of Naval Research, the Innovative Science and Technology Office of SDIO, Defense Advance Research Projects Agency, and the Air Force Office of Scientific Research.

REFERENCES

1. Berman, R., *The Properties of Diamond,* (J. E. Field, ed.) Academic Press, London (1979)
2. Clarke, L. J., *Surface Crystallography: An Introduction to Low Energy Electron Diffraction*, pp.184-185, John Wiley & Sons, London (1985)
3. Zilko, J. L., in: *Handbook of Thin-Film Deposition Processes and Techniques,* (K. K. Schuegraf, ed.) Noyes Publications, Park Ridge, NJ (1988)
4. Brice, J. C., in: *Properties of Gallium Arsenide,* EMIS Datareviews Series No. 2, INSPEC, London (1986)
5. Meilunas, R., Wong, M. S., Sheng, K. C. and Chang, R. P. H., *Appl. Phys. Lett.* 54(22):2204-2206 (1989)
6. Kobashi, K., Nishimura, M. K., Kumagai, K. and Kawate, Y., *Proc. 1st Electrochemical Society Intl. Symp. on Diamond and Diamond-Like Coatings,* 89(12):296-305 (1989)
7. Johnson, C. E., Hasting, M. A. S. and Weimer, W. A., *J. Mater. Res.* 5(11):2320-2325 (1990)
8. Ong, T. P. and Chang, R. P. H., *Appl. Phys. Lett.* 55(20):2063-2065 (1989)
9. Mitsuda, Y., Kojima, Y., Yoshida, T. and Akashi, K., *J. Mat. Sci.* 22:1557-1562 (1987)
10. Windischmann, H. and Epps, G. F., *J. Appl. Phys.* 68(11):5665-5673 (1990)

11. Davis, R. F., Sitar, Z., Williams, B. E., Kong, H. S., Kim, H. J., Palmour, J. W., Edmond, J. A., Ryu, J., Glass, J. T. and Carter, C. H. Jr, *Materials Science and Engineering,* B1:77-104 (1988)

12. Snail, K. A., Hanssen, L. M., *J. Crystal Growth,* 112 (4):651-659 (1991)

13. Davidson, J. L., Ellis, C., Ramesham, R., *J. Elect. Mater.* 18(6):711-715 (1989)

14. Saijo, K., Yagi, M., Shibuki, K., Takatsu, S., *Surface and Coatings Technology* 43/44:30-40 (1990)

15. Robins, J. L., *Applied Surface Science* 33/34:379-394 (1988)

16. Murarka, S. P. and Peckerar, M. C., *Electronic Materials Science and Technology,* Academic Press Inc., San Diego, CA (1989)

17. Nieman, G. C. and Klabunde, K. J., in *Thin Films From Free Atoms and Particles,* (K. J. Klabunde, ed.), Academic Press, San Diego, CA (1985)

18. Kern, R., Le Lay, G. and Metois, J. J., *Current Topics in Materials Science* 3:135-419 (1979)

19. Fedosayev, D. V., Deryagin, B. V. and Varasavskaja, I. G., *Surface and Coatings Technology* 38(1-2):1-130 (1989)

20. Gasser, R. P. H., *An Introduction to Chemisorption and Catalysis by Metals,* Oxford Science Publications, Oxford (1985)

21. Deryagin, B. V. and Fedosayev D. V., *Surface and Coatings Technology* 38(1-2):132-248 (1989)

22. Yarbrough, W. A. and Messier, R., *Science* (247):688-696 (1990)

23. Machlin, E. S., *J. Mat. Res.* 3(5):958-968 (1988)

24. Anthony, T. R., *Vacuum* 41(4-6):1356-1359 (1990)

25. Badzian, A. R. and DeVries, R. C., *Mat. Res. Bull.* 23:385-400 (1988)

26. Spear, K. E., *J. American Ceramic Society* 72(2):171-191 (1989)

27. Piekarczyk, W., Roy, R. and Messier, R., *J. Crys. Growth* 98:765-776 (1989).

28. Frenklach, M., *J. Appl. Phys.* 65(12):5144-5149 (1989)

29. Bar-Yam, Y., Moustakas T. D., *Nature* 342:786-787 (1989)

30. Jannsen, G., van Enckevort, W. J. P. and Giling, L. J., "CVD Growth of Diamond: The Multiple Role of Atomic Hydrogen," Preprint

31. Spear, K. E., Frenklach, M., *Proc. of 1st Electrochemical Soc. International Symposium on Diamond and Diamond-Like Coatings* 89(12):122-138 (1989)

32. Kikuchi, N., Komatsu, T. and Yoshimura, H., *Mat. Sci. & Eng.* A105/106:525-534 (1988)

33. Zhu, W., Badzian, A. R. and Messier, R., *J. Mat. Res.* 4(3):659-663 (1989)

34. Kobashi, K., Nishimura, K., Kawate, Y. and Horiuchi, T., *Phys. Rev. B* 38(6):4067-4084 (1988)

35. Zhu, W., Randall, C., Badzian, A. R. and Messier, R., *J. Vac. Sci Tech.A* 7(3):2315-2324 (1989)

36. Kamo, M., Chawanya, H., Tanaka, T., Sato, Y. and Setaka, N., *Mat. Sci. & Eng. A* 105/106:535-541 (1988)

37. Harker, A. B., DeNatale, J. D., *J. Mater. Res.* 5(4):818-823 (1990)

38. Celii, F. G., White, D. Jr. and Purdes A. J., *J. Appl. Phys.*, 70(10):5636-5646 (1991)

39. Singh, B., Arie, Y., Levine, A. W. and Mesker, O. R., *Appl. Phys. Lett.* 52(6):451-452 (1988)

40. Ravi, K. V., Koch, C. A., Hu, H. S. and Joshi, A., *J. Mater. Res.* 5(11):2356-2366 (1990)

41. Anthony, T., "Plasma-Assisted Deposition of New Materials", presented at the Fall Meeting of the Materials Research Society, Boston, (Nov. 30 - Dec. 5, 1987)

42. Lux, B. and Haubner, R., *Proc., 12th Int'l Plansee Seminar, Reutle, Austria,* (H. Bildstein, H. Oftner, eds.) 3(C2):615-660 (1989)

43. Bachmann, P. K., Weimer, R., Drawl, W., Liou, Y. and Messier, R., *Diamond Technology Initiative Symposium*, Paper T20, *SDIO/IST-ONR* Crystal City, VA (July 12-14, 1988)

44. Spitsyn, B. V., Bouilov, L. L. and Derjaguin, B. V., *J. Crys. Growth* 52:219-226 (1981)

45. Lander, J. J. and Morrison, J., *Surf. Sci.* 4:241-246 (1966)

46. Clausing, R. E., Heatherly, L., More, K. L. and Begun, G. M., *Surf. and Coatings Technol.* 39/40:199-210 (1989)

47. Bachmann, P. K., Lydtin, H., Wiechert, D. U., Beulens, J. J., Kroesen, G. M. W. and Schram, D. C., *Proc. of 3rd Intl. Conf. on Surf. Mod. Tech.*, TMS (1989) in press

48. Denatale, J. F., Flintoff, J. F. and Harker A. B., *J. Appl. Phys.* 68(8):4014-4019 (1990)

49. Yugo, S., Kimura, T. and Kanai, H., *Science and Technology of New Diamond*, (S. Saito, O. Fukunaga, and M. Yoshihara, eds.), pp 119-128, KTK Scientific Pub., Tokyo, (1990)

50. Farabaugh, E. N., Feldman, A., Robins, L. H. and Etz, E. S., *SPIE Proceedings - Diamond Optics*, (A. Feldman and S. Holly, eds.) 969:24-31 (1989)

51. Davis, R. F., Glass, J. T., Lucovsky, G., Bachmann, K. J., Trew, R. J. and Nemanich, R. J., *Annual Report: School of Engineering*, N.C. State University, Raleigh, NC (1988)

52. Singh, B., Mesker, O., Levine, A. W. and Arie, Y., *Growth of Polycrystalline Diamond Particles and Films by Hot-Filament Chemical Vapor Deposition*, Report CN-5300, David Sarnoff Research Center, SRI International, Princeton, NJ (1988)

53. Lee, Y. H., Bachmann, K. J., Glass, J. T., LeGrice, Y. M. and Nemanich, R. J., *Appl. Phys. Lett.* 57(18):1916-1918 (1990)

54. Ong, T. P. and Chang, R. P. H., *Appl. Phys. Lett.* 58(4):358-360 (1991)

55. Badzian, A. R., *Advances in X-Ray Analysis* (C. S. Barrett, J. V. Gilfrich, R. Jenkins, J. C. Russ, J. W. E. Richardson, Jr. and P. K. Predicki, eds.) 31:113-128, Plenum Publishing Co (1988)

56. Badzian, A. R. and Badzian, T., *Surface and Coatings Technology* 36:283-293 (1988)

57. Matsumoto, S., Sato, Y., Tsutsumi, M. and Setaka, N., *J. Mat. Sci.* 17:3106-3112 (1982)

58. Hartnett, T., Miller, R., Montanari, D., Willingham, C. and Tustison R., *J. Vac. Sci. Tech. A* 8(3):2129-2136 (1990)

59. Joffreau, P. O., Haubner, R. and Lux, B., *J. Ref. Hard Metals* 7(4):186-194 (1988)

60. Collins, R. W., Cong, Y., Kim, Y. T., Vedam, K., Liou, Y., Inspektor, A. and Messier, R., *Thin Solid Films* 181:565-578 (1989)

61. Bichler, R., Peng, J., Haubner, R. and Lux, B., *Mat. Sci. & Eng. A* 105/106:543-547 (1988)

62. Belton, D. N., Schmieg, S. J., *J. Appl. Phys.* 66(9):4223-4229 (1989)

63. Belton, D. N. and Schmieg S. J., *J. Vac. Sci. Technol. A* 8:2353-2362 (1990)

64. Rudder, R. A., Posthill, J. B., Hudson, G. C., Mantini, M. J. and Markunas R. J., *SPIE Proceedings - Diamond Optics*, (A. Feldman and S. Holly, eds.) 969:72-78 (1989)

65. Belton, D. N. and Schmieg S. J., *Surface Science* 233:131-140 (1990)

66. Belton, D. N. and Schmieg, S. J., *J. Appl. Phys.* 69(5):3032-3036 (1991)

67. Williams, B. E. and Glass, J. T., *J. Mater. Res.* 4(2):373-384 (1989)

68. Belton, D. N., Harris, S. J., Schmieg, S. J., Weiner, A. M. and Perry, T. A., *Appl. Phys. Lett.* 54(5):416-417 (1989)

69. Knight, D. S., Weimer, R., Pilione, L. and White, W. B., *Appl. Phys. Lett.*, 56(14):1320-1322 (1990)

70. Meyer, D. E., Dillon, R. O. and Wollam, J. A., *J. Vac. Sci. Tech. A* 7(3):2325-2327 (1989)

71. Yugo, S., Izumi, A., Kanai, T., Muto, T. and Kimura, T., *Proc. of the 2nd Int. Conf. on New Diamond Science and Technology, Pittsburg, PA,* pp. 385-390, Mat. Res. Soc.(1991)

72. Dennig, P. A. and Stevenson D. A., *Proc. of the 2nd Int. Conf. on New Diamond Science and Technology, Pittsburg, PA,* pp. 403-408, Mat. Res. Soc. (1991)

73. Chang, C. P., Flamm, D. L., Ibbotson, D. E. and Mucha, J. A., *J. Appl. Phys.* 63:1744-1748 (1988)

74. Pehrsson, P. E., unpublished results

75. Kirkpatrick, A. R., Ward, B. W. and Economou N. P., *J. Vac. Sci. Tech. B* 7:1947-1949 (1989)

76. Narayan, J., Godbole, V. P., White, C. W., *Science* 252:416-418 (1991)

77. Pehrsson, P. E., Hayward, I., Morrish, A. and Singer I., *International Conference on Metallurgical Coatings and Thin Films*, San Diego, CA (April 22, 1991)

78. Ravi, K. V. and Koch, C. A., *Appl. Phys. Lett.* 57(4):348-350 (1990)

79. Takata, Y., Edamatsu, K., Yokoyama, T., Seki, K., Tohnan, M., Okada, T. and Ohta, T., *Japn. J. App. Phys.* 28(7):L1282-L1285 (1989)

80. Hata, C., Kamo, M. and Sato Y., *Science and Technology of New Diamond,* (S. Saito, O. Fukunaga, and M. Yoshihara, eds.), pp 95-100, KTK Scientific Pub., Tokyo, (1990)

81. Sato, Y., Hata, C. and Kamo, M., *Science and Technology of New Diamond,* (S. Saito, O. Fukunaga, and M. Yoshihara, eds.), pp 83-88, KTK Scientific Pub., Tokyo, (1990)

82. Badzian, A. R., Badzian, T., Roy, R., Messier, R. and Spear, K. E., *Mat. Res. Bull.* 23:531-548 (1988)

83. Sommer, M. and Smith, F. W., *Mat. Res. Soc. Symp. Proc.* 162:139-144 (1989)

84. Moustakas, T. D., *Solid State Ionics* 32/33:861-868 (1989)

85. Beetz, C. P. Jr., Perry, T. A., "Hot and Cold Wall Filament-Assisted Growth of Diamond Films," *General Motors Research Publication GMR-6093*, 1-21, (November 24, 1987)

86. Sawabe, A. and Inuzuka, T., *Thin Solid Films* 137:89-99 (1986)

87. Yugo, S., Kanai, T., Kimura, T. and Ono, A., *Proc. of the International Symposium on Carbon,* 944-947, Tsukaba, Japan (1990)

88. Yugo, S., Kimura, T. and Muto T., *Vacuum* 41(4-6):1364-1367 (1990)

89. Angus, J. C., Hayman, C. C., *Science* 241:913-921 (1988)

90. Yugo, S., Kanai, T., Kimura, T. and Muto, T., *Appl. Phys. Lett.* 58(10):1036-1038 (1991)

91. Banholzer, W. and Kehl, R. J., *Surf. Coat. Tech.* 47:51:58 (1991)

92. Matsumoto, S. and Matsui, Y., *J. Mater. Sci.* 18:1785-1793 (1983)

93. Narayan, J., Srivatsa, A. R. and Ravi, K. V., *Appl. Phys. Lett.* 54(17):1659-1661 (1989)

94. Angus, J. C., Buck, F. A., Sunkara, M., Groth, T. F., Hayman, C. C. and Gat, R., *MRS Bulletin,* pp. 38-47 (October 1989)

95. Angus, J. C., Mann, J. A. Wang, Y., Sunkara, M., Washlock, P. A., Kuczmarski, M. and Li, Z., *Progress Report,* DARPA University Research Initiation Grant, Contract #N00014-89J-3167 (January 31, 1991)

96. Angus, J. C., Lambrecht, W., Pirouz, P. and Segall, B., *Annual Report,* Contract #N00014-89-1631 (September 30, 1990)

97. Williams, B. E., Kong, H. S. and Glass, J. T., *J. Mater. Res.* 5(4):801-810 (1990)

98. Kamo, M., Sato, Y., Matsumoto, S. and Setaka, N., *J. Cryst. Growth* 62:642-644 (1983)

99. Frenklach, M., Kematic, R., Huang, D., Howard, W., Spear, K. E., Phelps, A. W. and Koba, R., *J. Appl. Phys.* 66(1):395-399 (1989)

100. Huang, D., Frenklach, M., Howard, W., Kematick, R,. Spear, K. and Koba, R., *Extended Abstracts of the 19th Biennial Conf. on Carbon,* pp. 384-385, Penn. State Univ. (June, 1989)

101. Howard, W., Huang, D., Yuan, J., Frenklach, M., Spear, K. E., Koba, R. and Phelps A. W., *J. Appl. Phys.* 68(3):1247-1251 (1990)

102. Mitura, S., *J. Crys. Growth* 80:417-424 (1987)

103. Deshpandey, C. V. and Bunshah, R. F., *J. Vac. Sci. Tech. A* 7(3):2294-2302 (1989)

104. Tzeng, Y., Cutshaw, C., Phillips, R., Srivinynon, T., Ibrahim, A. and Loo, B. H., *Applied Physics Letters* 56(2):134-136 (1990)

105. Tzeng, Y., Kung, P. J., Phillips, R., Cutshaw, C., Legg, K., Solnick-Legg, H., Pope, S. and Stevenson J., Paper TP11, SDIO/IST-ONR *Diamond Technology Initiative Symposium,* Crystal City, VA (July 11-13, 1989)

106. Oakes, D. B., Butler, J. E., Snail, K. A., Carrington, W. A. and Hanssen, L. M., *J. Appl. Phys.* 69(4):2602-2610 (1991)

107. Kurihara, K., Sasaki, K., Kawarada, M. and Koshino, N., *Appl. Phys. Lett.* 52(6):437-438 (1988)

108. Suzuki, K., Sawabe, A., Yasuda, H. and Inuzuka, T., *Appl. Phys. Lett.* 50:728-729 (1987)

109. Jansen, F., Machonkin, M. A and Kuhman, D. E., *J. Vac. Sci. Tech. A* 8(5):3785-3790 (1990)

110. Guarnieri, C. R., Cuomo, J. J., Whitehair, S. J., Berry, B. S., Pritchet, W. C., Acosta, R. E., Wilson, A. D., *SPIE-Diamond Optics III*, (A. Feldman and S. Holly,eds.) 1325:326-330 (1990)

111. Valdes, J. L., Mitchel, J. W., Mucha, J. A., Seibles, L., Huggins, H., *J. Electrochemical Soc.* 138(2):635-636 (1991)

112. Meyer, D. E., Ianno, N. J., Woollam, J. A., Swartzlander, A. B. and Nelson A. J., *J. Mater. Res.* 3(6):1397-1403 (1988)

113. Iijima, S., Aikawa, Y. and Baba, K., *Appl. Phys. Lett.* 57(25):2646-2648 (1990)

114. Everson, M. P., Tamor, M. A. and Wu, C. H., *Proc. of the 2nd Int. Conf. on New Diamond Science and Technology, Pittsburg, PA*, pp. 613-618, Mat. Res. Soc. (1991)

115. Turner, K. F., LeGrice, Y. M., Stoner, B. R., Glass, J. T. and Nemanich, R. J., *J. Vac. Sci. Technol. B*, 9(2):914-919 (1991)

116. Pehrsson, P. E. and Morrish, A. A., *Proc. of the 2nd Int. Conf. on New Diamond Science and Technology, Pittsburg, PA*, pp. 397-402, Mat. Res. Soc. (1991)

117. Joffreau, P. O., Bichler, R., Haubner, R. and Lux, B., *J. Ref. Hard Metals* 7(2):92-97 (1988)

118. Kawarada, H., Ma, K. S., Suzuki, J., Ito, T., Mori, H., Fujita, H. and Hiraki, A., *Japn. J. Appl. Phys.* 26(11):L1903-L1906 (1987)

119. Hirabayashi, K., Taniguchi, Y., Takamatsu, O., Ikeda, T., Ikoma, K. and Iwasaki-Kurihara, N., *Appl. Phys. Lett.* 53(19):1815-1817 (1988)

120. Ma, J. S., Kawarada, H., Yonehara, T., Suzuki, J. I., Wei, J., Yokota, Y., Mori, H., Fujita and Hiraki, A., *Appl. Surf. Sci.* 41/42:572-579 (1989)

121. Ma, J. S., Karawada, H., Yonehara, T., Suzuki, J., Wei, J., Yokata, Y. and Hiraki, A., *Appl. Phys. Lett.* 55(11):1071-1073 (1989)

122. Celii, F. G., Nelson, H. H. and Pehrsson P. E., *J. Mater. Res.* 5(11):2337-2344 (1990)

123. Kobashi, K., Inoue, T., Tachibana, H., Kumagai, K. Miyata, K., Nishimura, K. and Nakaue A., *Vacuum* 41(4-6):1383-1386 (1990)

124. Harris, S. J., Belton, D. N., Weiner, A. M. and Schmieg, S. J., *J. Appl. Phys.* 66(11):5353-5359 (1989)

125. Rudder, R. A., Hudson, G., Markunas, R. J., Burns, R. and Posthill J., *Diamond Technology Initiative Symposium,* paper T19, SDIO/IST-ONR, Crystal City, VA (July 12-14, 1988)

126. Sato, Y., Yashima, I., Fujita, H., Ando, T. and Kamo, M., *Proc. of the 2nd Int. Conf. on New Diamond Science and Technology, Pittsburg, PA,* pp. 371-378, Mat. Res. Soc.(1991)

127. Jeng, D. G., Tuan, H. S., Salat, R. F. and Fricano, G. J., *Appl. Phys. Lett.* 56(20):1968-1970 (1990)

128. Haubner, R., *Proc. of Hard Materials Production 88,* 34:1-17, London (April 11-13, 1988)

129. Narayan, J., Srivatsa, A. R., Peters, M., Yokota, S. and Ravi, K. V., *Appl. Phys. Lett.* 53(19):1823-1825 (1988)

130. Jeng, D. G., personal communication

131. Geis, M. W., *Appl. Phys. Lett.* 55(6):550-552 (1989)

132. Pickett, W. E., *Phys. Rev. B* 38(2):1316-1322 (1988)

133. Lambrecht, W. R. L. and Segall, B., *Phys. Rev. B* 40(14):9909-9919 (1989)

134. Koizumi, S., Murakami, T., Inuzuka, T. and Suzuki, K., *Appl. Phys. Lett.* 57:563-565 (1990)

135. Yoshikawa, M., Ishida, H., Ishitani, A., Murakami, T., Loizumi, S. and Inuzuka, T., *Appl. Phys. Lett.* 57(5):428-430 (1990)

136. Geis, M. W., *Diamond Technology Initiative Symposium,* Paper T1, SDIO/IST-ONR, Crystal City, VA (July 11-13, 1989)

137. Fujimori, N., Imai, T., Nakahata, H. Shiomi, H. and Nishibayashi, Y., *Mat. Res. Soc. Symp. Proc.,* 162:109-114 (1990)

138. Frenklach, M. and Spear, K. E., *J. Mat. Res.* 3:133-140 (1988)

139. Badzian, A., Badzian, T., Wang, X. H. and Hartnett, T. M., *Proc. of the 2nd Int. Conf. on New Diamond Science and Technology, Pittsburg, PA,* pp. 549-556, Mat. Res. Soc. (1991)

140. Hirose, Y. and Terasawa, Y., *Japn. J. Appl. Phys.* 25:L519-L521 (1986)

141. Mucha, J. A., Flamm, D. L. and Ibbotson, D. E., *J. Appl. Phys.* 65:3448-3452 (1989)

142. Suzuki, J. I., Kawarada, H., Mar, K. S., Wei, J., Yokota, Y. and Hiraki, A., *Japn. J. Appl. Phys.* 28:L281-L283 (1989)

143. Rudder, R. A., Posthill, J. B. and Markunas, R. J., *Electron. Lett.* 25:1220-1221 (1989)

144. Patterson, D. E., Bai, B. J., Chu, C. J., Hauge, R. H. and Margrave, J. L., *Proc. of the 2nd Int. Conf. on New Diamond Science and Technology, Pittsburg, PA,* pp. 433-438, Mat. Res. Soc. (1991)

145. Thomas, J. M. and Evans, E. L., *Diamond Research* 3:2-8 (1975)

146. Freedman, A. and Stinespring, C. D., *Appl. Phys. Lett.* 57:1194-1196 (1990)

147. Hamza, A. V., Kubiak, G. D. and Stulen, R. H., *Surf. Sci.* 237:35-52 (1990)

148. Pate, B. B., *Surf. Sci.* 165:83-142 (1986)

149. Lurie, P. G. and Wilson, J. M., *Surf. Sci.* 65:453-475 (1977)

150. Lurie, P. G. and Wilson, J. M., *Surf. Sci.* 65:476-498 (1977)

151. Pepper, S. V., *Appl. Phys. Lett.* 33(5):344-345 (1981)

152. Matsumoto, O., Toshima, H. and Kanzaki, Y., *Thin Solid Films* 128:341-351 (1985)

153. Saito, Y., Matsuda, S. and Nogita, S., *J. Mat. Sci. Lett.* 5: 565-568 (1986)

154. Matsumoto, O. and Katagiri, T., *Thin Solid Films* 146:283-289 (1987)

155. Suzuki, K., Sawabe, A. and Inuzuka, T., *Appl. Phys. Lett.* 53:1818-1819 (1988)

156. Suzuki, K., Sawabe, A. and Inuzuka, T., *Japn. J. Appl. Phys.* 29:153-157 (1990)

157. Stalder, K. R. and Sharpless, R. L., *J. Appl. Phys.* 68:6187-6190 (1990)

158. Saito, Y., Sato, K., Tanaka, H., Fujita, K. and Matuda, S., *J. Mat. Sci.* 23:842-846 (1988)

159. Inspektor, A., Liou, Y., McKenna, T. and Messier, R., *Surface and Coatings Technol.* 39/40:211-221 (1989)

160. Saito, Y., Sato, K., Gomi, K and Miyadera, H., *J. Mat. Sci.* 25:1246-1250 (1990)

161. Yalamanchi, R. S. and Harshavardhan, K. S., *J. Appl. Phys.* 68:5941-5943 (1990)

162. Zhang, F., Zhang, Y., Yang, Y., Chen, G. and Jiang, X., *Appl. Phys. Lett.* 57:1467-1469 (1990)

163. Mitsuda, Y., Tanaka, K. and Yoshida, T., *J. Appl. Phys.* 67:3604-3608 (1990)

164. Coburn, J. W. and Chen, M., *J. Appl. Phys.* 51:3134-3136 (1980)

165. Muranaka, Y., Yamashita, H., Sato, K. and Miyadera, H., *J. Appl. Phys.* 67:6247-6254 (1990)

166. Zhu, W., Inspektor, A., Badzian, A. R., McKenna, T. and Messier, R., *J. Appl. Phys.* 68:1489-1496 (1990)

167. Celii, F. G., Pehrsson, P. E., Wang, H. -t. and Butler, J. E., *Appl. Phys. Lett.* 52:2043-2045 (1988)

168. Celii, F. G., Pehrsson, P. E., Wang, H. -t., Nelson, H. H. and Butler, J. E., *Proc. of First Int. Conf. on New Diamond Sci. and Tech.*, pp. 17-23 Tokyo, Japan (1990)

169. Celii, F. G., Pehrsson, P. E., Wang, H. -t., Nelson, H. H. and Butler, J. E., *Proc. 4th Int. Laser Sci. Conf.*, 191:747-749, Amer. Inst. of Phys. (1989)

170. Butler, J. E., Celii, F. G., Pehrsson, P. E., Wang, H. -t. and Nelson, H. H., *Mat. Res. Soc. Symp. Proc.* 131:259-260 (1989)

171. Butler, J. E. and Celii, F. G., *Electrochem. Soc. Proc.* 89-12:317-329 (1989)

172. Butler, J. E., Celii, F. G., Oakes, D. B., Hanssen, L. M., Carrington, W. A. and Snail, K. A., *High Temp. Sci.* 27:183-197 (1990)

173. Celii, F. G. and Butler, J. E., *Proc. of the 2nd Int. Conf. on New Diamond Science and Technology, Pittsburg, PA,* pp. 201-206, Mat. Res. Soc. (1991)

174. Mitomo, T., Ohta, T., Ohtsuka, K and Habu, Y., *Proc. of the 2nd Int. Conf. on New Diamond Science and Technology, Pittsburg, PA,* pp. 213-218, Mat. Res. Soc. (1991)

175. Morrison Jr., P. W., Cosgrove, J. E., Markham, J. R. and Solomon, P. R., *Carbon* 28:767-769 (1990)

176. Celii, F. G. and Butler, J. E., *Appl. Phys. Lett.* 54:1031-1033 (1989)

177. Celii, F. G., Pehrsson, P. E. and Butler, J. E., *Proc. of SPIE Conf. on Diamond Optics II,* 1146:119-123 (1989)

178. Hudgens, J. W., DiGiudeppe, T. G. and Lin, M. C., *J. Chem. Phys.* 79:571-582 (1983)

179. Sommer, M. and Smith, F. W., *J. Mater. Res.* 5:2433-2440 (1990)

180. Celii, F. G., Thorsheim, H. R., Hanratty, M. A. and Butler, J. E., *Appl. Opt.* 29:3135-3142 (1990)

181. Celii, F. G., Thorsheim, H. R., Butler, J. E., Plano, L. S. and Pinneo, J. M., *J. Appl. Phys.* 68:3814-3817 (1990)

182. Meier, U., Kohse-Hoinghaus, K., Schafer, L. and Klages, C. P., *Appl. Opt.* 29:4993-4999 (1990)

183. Jansen, F., Chen, I., Machonkin, M. A., *J. Appl. Phys.* 66(12):5749-5755 (1989)

184. Matsui, Y., Yuuki, A., Sahara, M. and Hirose, Y., *Japn. J. of Appl. Phys.* 28:1718-1724 (1989)

185. Raiche, G. A., Smith, G. P. and Jeffries, J. B., *Proc. of the 2nd Int. Conf. on New Diamond Science and Technology, Pittsburg, PA,* pp. 251-256, Mat. Res. Soc. (1991)

186. Roman, W. C., Colket, M. B. III, Hay, S. O. and Eckbreth, A. C., *Electrochemical Soc. Proc.* 89-12:330-341 (1989)

187. Hay, S. O., Roman, W. C. and Colket, M. B. III, *J. Mater. Res.* 5(11):2387-2397 (1990)

188. Cappelli, M. A. and Paul, P. H., *J. Appl. Phys.* 67:2596-2602 (1990)

189. Hsu, W. L. and Tung, D. M., *Proc. of the 2nd Int. Conf. on New Diamond Science and Technology, Pittsburg, PA,* pp. 239-244, Mat. Res. Soc. (1991)

190. Plano, L. S., Stevenson, D. A. and Carruthers, J. R., *Proc. of the 2nd Int. Conf. on New Diamond Science and Technology, Pittsburg, PA,* pp. 257-262, Mat. Res. Soc. (1991)

191. Harris, S. J., Weiner, A. M. and Perry, T. A., *Appl. Phys. Lett.* 53:1605-1607 (1988)

192. Harris, S. J. and Weiner, A. M., *Appl. Phys. Lett.* 55:2179-2182 (1989)

193. Sommer, M. and Smith, F. W., *High Temp. Sci.* 27:173-182 (1990)

194. Harris, S. J. and Weiner, A. M., *J. Appl. Phys.* 67:6520-6526 (1990)

195. Wu, C. H., Tamor, M. A., Potter, T. J. and Kaiser, E. W., *J. Appl. Phys.* 68:4825-4829 (1990)

196. Chu, C. J., Bai, B. J., D'Evelyn, M. P., Hauge, R. H. and Margrave, J. L., *Mat. Res. Soc. Symp. Proc.* 162:85-90 (1989)

197. Chu, C. J., D'Evelyn, M. P., Hauge, R. H. and Margrave, J. L., *J. Mater. Res.* 5:2405-2413 (1990)

198. Harris, S. J. and Martin, L. R., *J. Mater. Res.* 5:2313-2319 (1990)

199. Lersmacher, B., Lydtin, H., Knippenberg, W. F. and Moore, A. W., *Carbon* 5:205-217 (1967)

200. Bichler, R., Haubner, R. and Lux, B., *Proc. 6th European CVD Conf.,* pp.413-422, Jerusalem, Israel (1987)

201. Spitsyn, B., Bouilov, L. L. and Derjaguin, B. V., *Prog. Crystal Growth and Charact.* 17:79-170 (1988)

202. Sommer, M., Mui, K. and Smith, F. W., *Solid State Comm.* 69:775-778 (1989)

203. Spitsyn, B., *J. Cryst. Growth* 99:1162-1167 (1990)

204. Piekarczyk, W., Messier, R., Roy, R., Engdahl, C., *J. Cryst. Growth* 106:279-293 (1990)

205. Chen, I., *J. Appl. Phys.* 64:3742-3745 (1988)

206. Matsui, Y. and Sahara, M., *Japn. J. Appl. Phys.* 28:1023-1029 (1989)

207. Harris, S. J., *J. Appl. Phys.* 65:3044-3048 (1989)

208. Chauhan, S. P., Angus, J. C. and Gardner, N. C., *J. Appl. Phys.* 47:4746-4754 (1976)

209. Matsui, Y., Yabe, H. and Hirose, Y., *Japn. J. Appl. Phys.* 29:1552-1560 (1990)

210. Harris, S. J. and Martin, L. R., *J. Mater. Res.* 5:2313-2319 (1990)

211. DebRoy, T., Tankala, K., Yarbrough, W. A. and Messier, R., *J. Appl. Phys.* 68:2424-2432 (1990)

212. Goodwin, D. G. and Gavillet, G. G., *J. Appl. Phys.* 68:6393-6400 (1990)

213. Goodwin, D. G. and Gavillet, G. G., *Proc. of the 2nd Int. Conf. on New Diamond Science and Technology, Pittsburg, PA*, pp. 335-340, Mat. Res. Soc. (1991)

214. Spitsyn, B. V., *J. Cryst. Growth* 99:1162-1167 (1990)

215. Setaka, N., *Shokubai* 24:412 (1982)

216. Yarbrough, W. A., *Mat. Res. Soc. Symp. Proc.* 162:75-83 (1990)

217. Yarbrough, W. A., *Proc. of the 2nd Int. Conf. on New Diamond Science and Technology, Pittsburg, PA*, pp. 291-306, Mat. Res. Soc. (1991)

218. Tsuda, M., Nakajima, M., and Oikawa, S., *J. Am. Chem. Soc.* 108:5780-5783 (1986)

219. Tsuda, M., Nakajima, M. and Oikawa, S., *Japn. J. Appl. Phys.* 26(5):L527-L529 (1987)

220. Huang, D., Frenklach, M. and Maroncelli, M., *J. Phys. Chem.* 92(22):6379-6381 (1988)

221. Harris, S. J., *Appl. Phys. Lett.* 56(23):2298-2300 (1990)

222. Frenklach, M. and Wang, H., *Phys. Rev. B* 43(2):1520-1545 (1991)

4

Technology of Vapor Phase Growth of Diamond Films

Richard J. Koba

1.0 INTRODUCTION

Diamond thin films have been synthesized by at least ten different methods. Diamond film synthesis can be divided into two general categories, namely, chemical vapor deposition (CVD), and physical vapor deposition (PVD). In chemical vapor deposition, the source of carbon is an energetically excited or activated gas phase. This gas phase contains a mixture of molecules, atoms, radicals, ions, and electrons which oscillate randomly in three dimensions and will condense as diamond only upon surfaces possessing suitable characteristics, e.g., suitable chemical composition and temperature. Physical vapor deposition techniques typically involve the evaporation or sublimation of carbon from a solid source into the gas phase. The evaporated carbon is chemically and energetically altered so that it condenses as diamond on the surface of a substrate placed in its trajectory.

This chapter will primarily discuss chemical vapor deposition methods for depositing diamond thin films. Deposition of genuine diamond films by PVD methods is gaining interest in the West; however, Soviet researchers have been active in PVD of diamond since the 1970's (1). Several different CVD diamond methods have been used to deposit diamond films. These CVD methods vary widely in terms of capital equipment cost, deposition rate, film area, types of source gases used, and range of suitable substrates. However, most of these methods can deposit films with similar morphologies and Raman signatures. Additionally, any one method of film synthesis can create a wide variety of diamond films by variation of the process parameters.

It is difficult to deduce a diamond film's genesis by only analyzing the properties of the film (2).

Because of the similarity among diamond thin films made by different methods, all of the chemical vapor deposition techniques probably share certain chemical mechanisms. Various models for the chemical kinetics of CVD diamond have been proposed (3)-(5). It has not yet been determined whether there exists exactly *one* chemical kinetic theory of the CVD of diamond, or if several different theories can be valid simultaneously.

To discuss the technology of vapor phase growth, one must be aware of the general similarities between the diamond CVD techniques. By noting how each technique creates the chemical environment suitable for diamond growth, one can then compare techniques in terms of their ability to produce films with desirable properties. In the early days of diamond CVD synthesis, a certain method of deposition technology was chosen simply because a previous investigator had shown that it worked. However, because of growing awareness of the *chemical kinetics* of diamond growth, it is now possible to "comparison shop" among the demonstrated CVD methods. A researcher can intelligently select the CVD method which produces diamond films exhibiting the desired properties at the lowest possible cost.

A purpose of this chapter is to allow the reader to compare diamond CVD processes and select the method which best suits his or her needs. Some methods produce polycrystalline films at high rate over a small area. Other methods produce diamond films over a large area at a slow rate. The cost-per-carat of the diamond film also varies over a wide range and this cost is comprised of the capital equipment, raw materials, energy, labor/overhead, diamond film deposition area, and deposition rate (6). By calculating the relative costs of the various CVD processes, and by estimating the value-added of the diamond coating, scientists and engineers will be able to select the diamond film deposition method which deposits a diamond coating at the maximum performance/cost ratio.

2.0 GENERAL CONDITIONS FOR DIAMOND FILM DEPOSITION

Most methods for CVD of diamond share the following characteristics:

1. A gas phase must be activated, either by high temperature and/or by plasma excitation.
2. The gas phase must contain a carbon-containing species such as a hydrocarbon, an alcohol, carbon monoxide, or carbon dioxide.

3. There must be a sufficiently high concentration of a species which etches graphite and/or suppresses gaseous graphite precursors (e.g., polycyclic aromatic hydrocarbons). Most commonly, this species is atomic hydrogen, H. However, additional "anti-graphite" species have been proposed, including H_2, OH, O_2, atomic oxygen and F_2 (5)(7).
4. The substrate surface must be predisposed to support the nucleation and growth of diamond from the vapor phase. Practically speaking, this means that the substrate surface must not include any catalysts which promote formation of graphite. Additionally, it means that the substrate surface should be at or near the solubility limit for carbon (at the deposition temperature) in order to support the precipitation of diamond on the surface instead of the diffusion of carbon into the substrate (8).
5. There must be a driving force to transport the carbon-bearing species from the gas phase to the surface of the substrate. In most CVD processes, a temperature gradient provides a driving force for the motion of diamond-producing species via diffusion because the activated gas phase is typically much hotter than the substrate surface (9). However, convection and simple blowing may also be used to transport species from the gas phase to the substrate surface. (In PVD methods, transport of carbon species to the surface is provided by momentum supplied by the evaporation or sublimation of the carbon species.) Transport of species to the substrate may also be enhanced by electric fields.

These five phenomena are common to practically all CVD diamond methods. The absence of any one of these phenomenon typically results in the deposition (or co-deposition) of non-diamond carbon including graphite, diamond-like hydrocarbon, or diamond-like carbon. A wide variety of techniques can be used to generate these five conditions. However, most methods supply additional phenomena which may or may not enhance the stabilization of diamond relative to graphite. For example, most CVD diamond techniques create a plasma or include a hot filament which irradiates the gas phase and the substrate with intense infrared, visible, and ultraviolet radiation. Ultraviolet light may produce photochemical reactions which can affect diamond growth. Additionally, other techniques use electric fields to enhance bombardment of the substrate by either electrons or positive ions derived from the gas phase. Diamond film properties have been observed to change as a function of electron or ion bombardment (10).

As mentioned above, the chemical and physical properties of the substrate strongly influences the nucleation and growth of diamond films. Two substrates of identical dimensions which are set side-by-side in the CVD reactor can experience completely different film deposits; the film properties will be *strongly* influenced by the substrate material (5)(8). It has been observed that transition metals which are suitable solvent/catalysts for the synthesis of diamond under high-pressure, high-temperature conditions are typically poor substrates for the CVD of diamond. Since CVD of diamond occurs under conditions where graphite is thermodynamically stable, these catalytic substrates tend to stabilize the graphitic form of carbon. Additionally, these transition metals have high solubilities for carbon in their lattice making it difficult to quickly saturate their surfaces with carbon and precipitate carbon as diamond. R. H. Wentorf has listed the metals which act as solvent/catalysts for diamond. The list includes platinum, iron, chromium, manganese, cobalt, nickel, and palladium (11). Although diamond films have been successfully grown on such solvent/catalyst substrates, adhesion of these diamond films is problematic. Typically, poor adhesion is due to diamond growth upon a layer of graphite which initially nucleated on the transition metal substrate.

3.0 PYROLYSIS METHODS

3.1 Hydrocarbon Pyrolysis

Historically, hydrocarbon pyrolysis was the first method used to demonstrate the chemical vapor deposition of diamond. A patent by W. G. Eversole of Union Carbide described a cyclic process for the pyrolysis of hydrocarbons onto a diamond surface followed by a separate graphite etching step using H_2 (12). The pyrolysis of a pure hydrocarbon, (typically methane), deposited diamond plus non-diamond carbons onto a diamond substrate. This deposition step was followed by selective etching of non-diamond carbon from the deposit by molecular hydrogen, and then the cycle was repeated. Eversole's work was extended by B. V. Derjaguin, B. V. Spitsyn, D. B. Fedoseev and co-workers in the Soviet Union, who performed careful physical chemistry experiments such as thermogravimetric analysis (TGA) to definitively observe weight gain in diamond powder with complete removal of residual graphite. (Several reviews of Soviet diamond research are listed in Ref. 13). The cyclic pyrolysis method owed its success, in part, to the use of the best possible substrate for diamond CVD, namely, diamond substrates. Therefore, homoepitaxial growth diamond is the oldest form of diamond CVD. In the late 1960s, J. C. Angus and co-workers at Case Western Reserve University replicated Eversole's work and the Soviet work. They extended the science by adding boron-bearing gases to their reactor to grow boron-doped, semiconducting diamond films on diamond grit (14).

The cyclic, hydrocarbon pyrolysis method had an extremely slow diamond deposition rate ($\approx$ 1 nm/hr), and it required a diamond grit substrate (15). Historically, it is not surprising that diamond synthesis via pyrolysis was dismissed as impractical. Figure 1 illustrates a typical apparatus used for hydrocarbon pyrolysis deposition of diamond. The apparatus was simply a hot-wall furnace which surrounded a platen containing diamond powder. A gas control system was installed to alternate the flow of gases, e.g., CH_4 and H_2. In many Soviet experiments, this apparatus was used as part of a thermogravimetric analysis instrument so that weight gain due to epitaxial growth of diamond could be recorded in situ. Despite the co-deposition of graphite during the pyrolysis, the deposit plus etch process did create graphite-free diamond epitaxial films, albeit at slow rates.

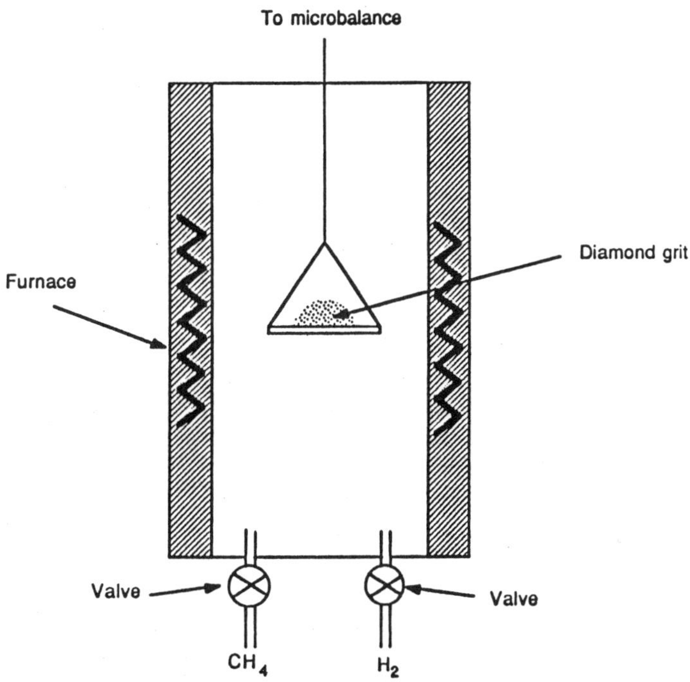

Figure 1. Schematic of a hydrocarbon pyrolysis reactor attached to a balance for thermogravimetric analysis.

Epitaxial growth of diamond via pyrolysis supports a theory that one mechanism for diamond growth is pseudomorphic stabilization (5). A diamond surface provides a crystal growth template for carbon descending from the vapor phase; the driving force for carbon to conform to the diamond template exceeds the driving force for carbon to revert to thermodynamically stable graphite.

3.2 Fluorocarbon Pyrolysis

Rosner and Strakey published a graph of various gaseous etchants for graphite which appears as Fig. 2 (16). Although atomic hydrogen is an excellent graphite etchant, it is surpassed in effectiveness by OH radicals, O_2, O°, F_2 and F°. Based on Rosner and Strakey's research, plus data on the bond strengths of combinations of C, F, and H atoms, Rudder et al. at Research Triangle Institute (RTI) theorized that epitaxial growth of diamond might proceed by the pyrolysis of fluorocarbons such as CF_4. Their theory was confirmed experimentally.

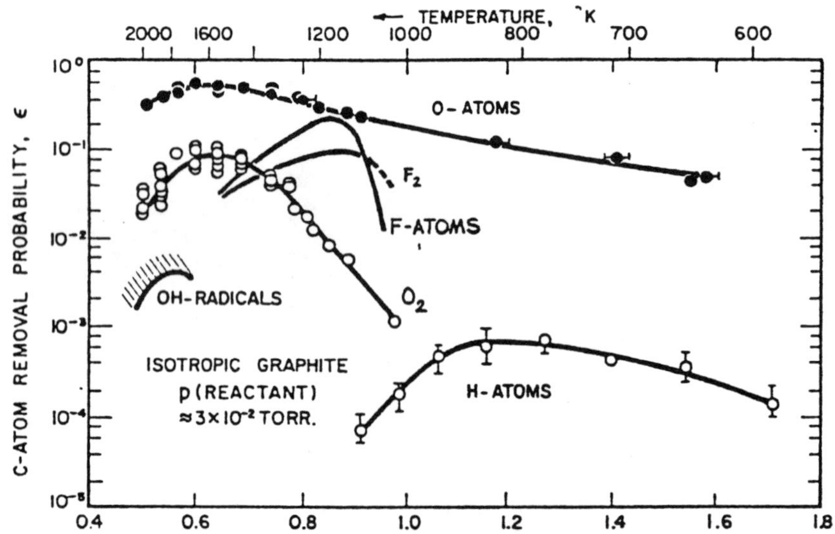

Figure 2. Carbon atom removal probabilities for the attack of an isotropic polycrystalline graphite (ρ = 1.73 g/cm³) by O, O_2, OH, F, F_2 and H (16). *(Reprinted with permission J. Phys. Chem. Copyright 1973, American Chemical Society.)*

A schematic of the RTI fluorocarbon pyrolysis system is illustrated in Fig. 3 (17). A mixture of CF_4 and F_2 diluted in He was blown onto a diamond substrate surface. The diamond substrate was heated to 875°C by a resistance heater. The presence of a diamond deposit was verified by step height measurement (via profilometry). Raman spectroscopy confirmed that the deposited film was only diamond; no co-deposition of graphite was observed. *No* plasma activation, hot filament activation, or electric field activation of the gas phase was required for epitaxial growth. The pyrolysis process occurred almost at thermodynamic equilibrium. However, it was possible to grow diamond at a rate of ≈ 0.6 µm/hr. The RTI fluorocarbon pyrolysis technique may potentially be more energetically efficient than typical CVD methods since it is unnecessary to supply energy to the gas phase (except from the hot substrate).

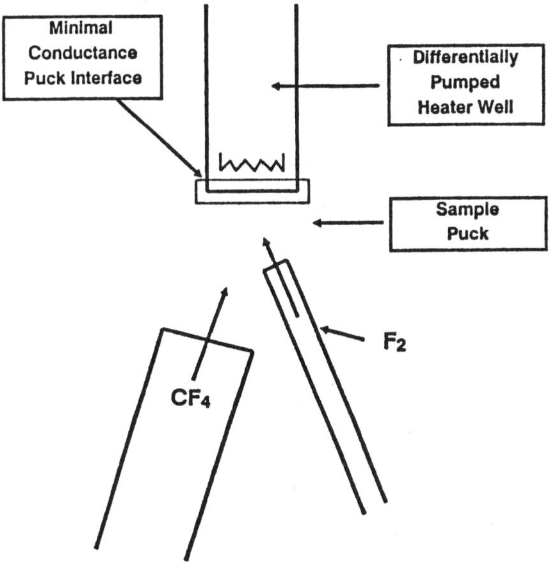

Homoepitaxial Conditions	
0.500 Torr	50 sccm He
875°C	50 sccm He 1% CF_4
Ni Foil Mask	12.5 sccm He 1% F_2

Figure 3. Schematic of fluorocarbon pyrolysis reactor and typical eptiaxial growth conditions (17). *(Reprinted by permission of R. A. Rudder.)*

The greatest limitation of fluorocarbon pyrolysis is the severe toxic and corrosive properties of CF_4 and F_2. The few engineering materials which can resist fluorine corrosion include Teflon, certain nickel alloys, and diamond. R. J. Markunas of RTI has envisioned fluorocarbon pyrolysis as being an inexpensive method to increase the thickness of diamond films on objects previously coated (by other means) with a hermetic diamond film (18). By inserting the diamond coated article in a tube furnace, also lined on the inside with CVD diamond, flow of CF_4 and F_2 over the object would serve to uniformly thicken the diamond layer. An advantage of this process is that it would have a superior ability to coat complex topographies than most CVD diamond processes or PVD diamond processes.

4.0 HEATED FILAMENT METHOD

The early 1970s, J. C. Angus suggested that the simultaneous production of atomic hydrogen (H°) during hydrocarbon pyrolysis may enhance the deposition of diamond. This suggestion was tested by Soviet researchers who generated H° by dissociating H_2 using an electric discharge or a hot filament (19). It was observed that atomic hydrogen could be easily produced by passage of H_2 over a refractory metal filament, such as tungsten, heated to temperatures between 1800 to 2300°C. When atomic hydrogen was added to the hydrocarbon, it was observed that diamond could be deposited while graphite formation was suppressed. The deliberate generation of atomic hydrogen during diamond CVD enabled *(a)* a dramatic increase in the diamond deposition rate to ≈1 µm/hr, and *(b)* the nucleation and growth of diamond on non-diamond substrates (1)(2)(5)(7)(13)(15)(19)-(22).

Figure 4 is a schematic diagram of a typical hot filament reactor system used to grow diamond films (22). Because of its simplicity, comparatively low capital cost and operating cost, filament assisted CVD has become popular in industry where it is imperative to minimize cost per gram of synthetic diamond (23). A typical apparatus includes a gas handling system which meters in a suitable gas mixture, for example, 1% CH_4 + 99% H_2. This gas mixture is passed over an incandescent filament. A wide variety of refractory materials have been used as filaments including tungsten, tantalum, and rhenium. Refractory metals which form carbides (e.g., tungsten), typically must carburize their surface before supporting the deposition of diamond films (24). Rhenium is attractive as a filament material since it does not easily form a carbide and, therefore, does not require a long carburization period before diamond deposition will commence. The hot filament is situated above the substrate which is typically also independently heated (or in certain cases, is independently cooled) to a temperature between 500 and 1000°C.

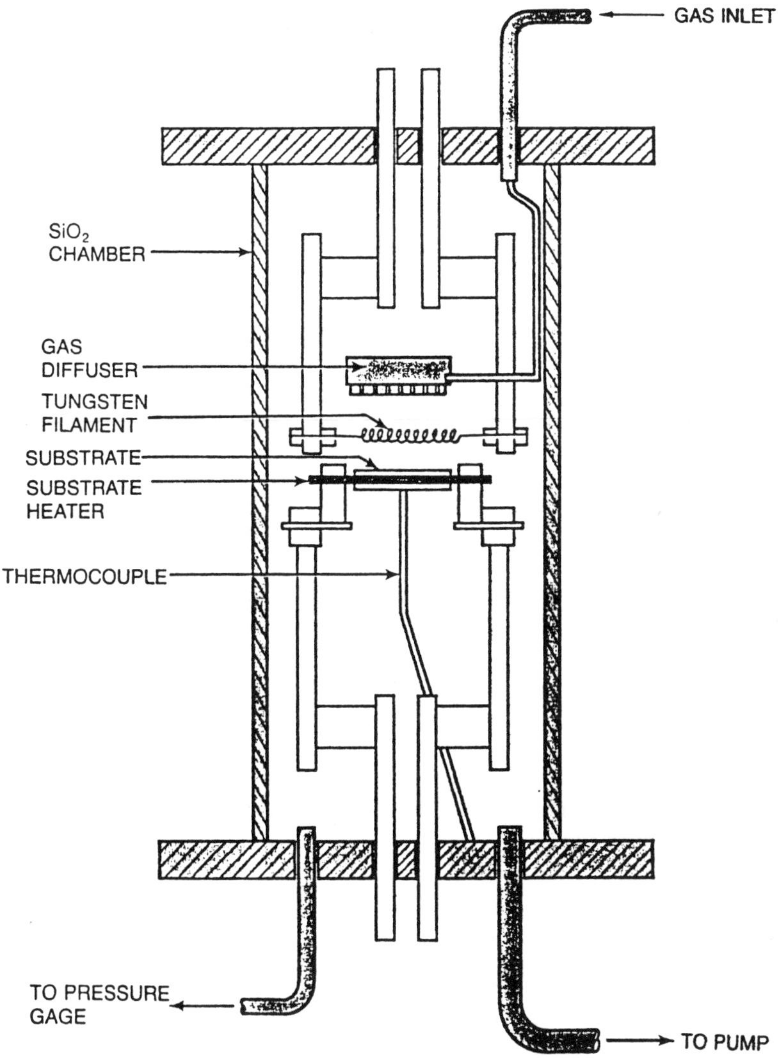

Figure 4. Schematic of a hot filament CVD reactor (22) *(Reprinted by permission of the publisher, The American Ceramic Society.)*

Because of its simplicity, heated filament CVD has been selected for extensive investigations into the gas phase chemistry of diamond growth using methods such as mass spectroscopy, laser induced fluorescence, infrared spectroscopy, and other techniques. Many of these chemical investigations are discussed by J. Butler in Ch. 3 of this book. In general, it has been learned that the hot filament transforms the input hydrocarbon (e.g., CH_4) into hydrocarbons stable at the filament temperature (e.g., C_2H_2 and CH_3). The hot filament also dissociates H_2 into H°.

A simplified theory of diamond growth via hot filament will be presented. It is believed that thermodynamic near-equilibrium is established in the gas phase at the filament surface (9). At temperatures around 2000°C, methane transforms into acetylenic species, methyl radicals and other hydrocarbons stable at high temperatures, and H_2 decomposes into H°. Atomic hydrogen and the high-temperature hydrocarbons then diffuse from the filament to the substrate surface. Although the gaseous species generated at the filament are in equilibrium at the filament temperature, the species are at super-equilibrium concentration when they arrive at the much cooler substrate (9). The reactions which generate these high temperature species (e.g., C_2H_2) at the surface of the filament proceed faster than any reactions which decompose these species during the transit time from the filament to the substrate. Consider equilibrium between methane and acetylene:

Eq. (1) $$2CH_4 = C_2H_2 + 3H_2$$

At the filament surface, the reaction is immediately driven to the right, creating acetylene. After acetylene diffuses to the substrate, thermodynamic equilibrium at a substrate temperature of 850°C calls for the formation of methane, but the reverse reaction proceeds much slower. Solid carbon precipitates on the substrate in order to reduce the super-equilibrium concentration of species such as acetylene in the gas phase. The diamond allotrope of carbon is "stabilized" by a concurrent super-equilibrium concentration of H° (and/or other "anti-graphite" species such as OH). To summarize, this simple explanation emphasizes the importance of reaction kinetics (and rates of reaction) in diamond synthesis by hot filament. Scientists continue to debate the exact identity and behavior of the gaseous species which are responsible for diamond growth (3)-(5)(22).

All of the process parameters for optimal diamond growth by hot filament CVD are highly interrelated. The relevant process parameters include: filament material, filament temperature, input gas mixture, substrate temperature, filament-to-substrate separation distance, and gas flow dynamics. In general, it has been found that the addition of oxygen-bearing species tends to expand the temperature range over which diamond deposition occurs. It is also found that the maximum allowable distance between the filament and substrate increases with decreasing pressure. Some researchers report that the flow trajectory of gases past the filament is critical in producing diamond films, while other researchers report that no intentional blowing of gases past the filament is necessary.

A typical growth condition is reported by Lee et al. of North Carolina State University (10). A tungsten filament is heated to 1930°C and is suspended 0.8 cm above the surface of the substrate. The substrate is a Si wafer polished with 0.25 µm diamond paste and is independently heated to a temperature 850°C. The feedstock gas mixture is 3% methane diluted in H_2 at a total flow of 100 sccm. The total gas pressure is 3.3 kPa (25 torr). Lee et al. discuss the effects of DC bias between the filament and the substrate, thereby creating electron or ion bombardment of the growing film. The best diamond films are formed when there is no significant electrical leakage current passing through the filament-substrate "vacuum diode."

It has been found that the hot filament assisted CVD method is able to accommodate a wide variety of carbon-bearing source gases. Successfully used source gases include methane, propane, ethane and other hydrocarbons, plus oxygen-containing hydrocarbons including acetone and ethanol. Boron-doped diamond films have been grown by the addition of boron-containing gases such as diborane and boron carbohydroxide, which is created by the bubbling Ar through an aqueous solution of boric acid dissolved in acetone (25).

The hot refractory metal need not be in the form of a filament. W. Yarbrough and R. Roy flowed a CH_4/H_2 source gas mixture through a 2100 - 2200°C tantalum *tube* pointed at silicon substrate (26). This "hot tube" method produced nanocrystalline films at substrate temperatures between 600 - 800°C.

Filament methods have found favor with U. S. industry. For example, Diamonex has disclosed that it uses the hot filament method because of its relatively low cost and ability to be scaled over large areas, e.g., up to a 30 cm diameter area. They also note that the hot filament method is not protected by any fundamental, exclusive patents since hot filament CVD of diamond was published by the Soviets in technical journals before any exclusive patent was filed (27). Hot filament CVD has also been used to deposit diamond films onto liquid metal substrates, in order to manufacture free-standing diamond membranes or plates (23).

5.0 PLASMA ENHANCED CHEMICAL VAPOR DEPOSITION

In the early 1970s, Soviet researchers found that the concentration of atomic hydrogen could be increased by use of an "electrical discharge" (1)(2)(13)(19). One source of plasma excitation employed in the Soviet Union was the DC arc discharge from a Tesla coil (28). In the West, technology to generate plasmas for materials processing was being developed for the semiconductor integrated circuit industry. Plasma enhanced chemical vapor deposition (PECVD) reactors are frequently used to deposit films such as silicon nitride, amorphous silicon, and other thin film materials. In response to VLSI design rules, plasma reactors were then developed for plasma etching and reactive ion etching of Si and GaAs substrates. The thrust of Western research into CVD diamond circa 1987 tended to gravitate toward plasma enhanced CVD methods because of the availability of equipment and the familiarity of Western researchers with its operation.

Three plasma frequency régimes will be mentioned. *Microwave* plasma CVD typically uses excitation frequencies of 2.45 GHz. *Radio frequency* (RF) plasma excitation typically employs frequencies of 13.56 MHz (or less commonly, 450 kHz). These frequencies are reserved for scientific use in the United States by the Federal Communications Commission. Non-reserved frequencies in the RF and microwave may be employed by researchers, but elaborate precautions (e.g., a Faraday cage) must be employed so that the electromagnetic signals are not broadcast outside the reaction chamber. *Direct current* (DC) plasmas can be run at low electric powers (a "cold" plasma) or at high electric powers (which create an *arc* or a *thermal* plasma).

5.1 Microwave Plasma Enhanced CVD

Microwave plasma enhanced CVD was demonstrated to grow polycrystalline diamond films on non-diamond substrates by a research group at Japan's National Institute for Research in Inorganic Materials, NIRIM. (A review of the research at NIRIM was written by S. Matsumoto, Ref. 29.) The Japanese were influenced by Soviet reports that atomic hydrogen was necessary to nucleate diamond films on non-diamond substrates. Previous work had indicated that microwave plasmas could dissociate diatomic molecules into atomic species. Microwave plasmas are different from other plasmas in that the microwave frequency can oscillate electrons. Collision of electrons with gas atoms and molecules generates high ionization fractions. Microwave plasmas are typically described as having "hot" electrons and "cool" ions and neutrals (30). NIRIM researchers correctly predicted that microwave plasma excitation of H_2 would generate super-equilibrium concentrations of atomic hydrogen.

The group at NIRIM used the microwave reactor schematically illustrated in Fig. 5 (22). Microwaves are generated by a magnetron, go through a circulator, a stub tuner, and are directed toward an applicator. The applicator is composed of an aluminum waveguide with a hole drilled through a top and bottom face. Aluminum tubes extend out from both holes; the tube extensions are water cooled. A fused silica tube then passes through the water cooled tubes and waveguide. The substrate and plasma is contained within the tube.

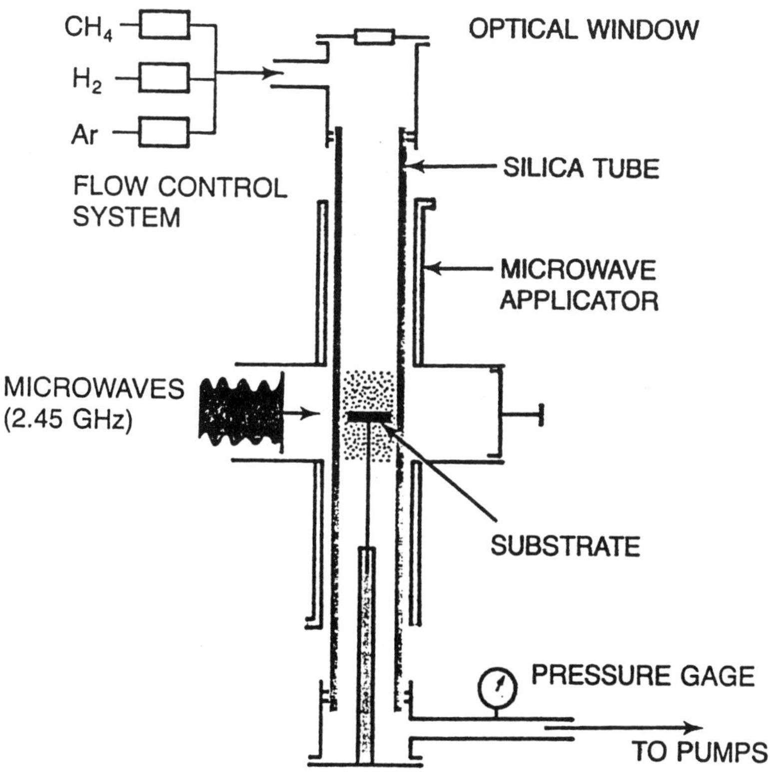

Figure 5. Schematic of a tube-style microwave plasma-enhanced CVD reactor (22). *(Reprinted by permission of the publisher, The American Ceramic Society.)*

Fused silica is used because it is vacuum compatible, capable of operating at temperatures up to ≈ 1200°C, reasonably low cost, and it does not absorb microwave radiation. The use of fused silica unfortunately has tended to produce silicon contamination in the diamond films deposited by microwave plasma enhanced CVD. Silicon monoxide vapor is generated by etching of the heated tube walls by atomic hydrogen. Hence, diamond films grown in such reactor may be contaminated by silicon. However, steps can be taken to mitigate this silicon contamination. For example, the interior of the silica tube may be coated with a thin layer of diamond-like hydrocarbon.

The substrate rests on a platen inserted from the bottom of the tube. The platen may be a susceptor which is heated by coupling to the microwave field. The platen may be independently heated a resistance heater, whose output is controlled by a thermocouple in the platen, or by an optical pyrometer pointed at the substrate. For some experiments, it may be necessary to cool the platen, usually by circulating water through the platen into a heat exchanger.

Typical deposition parameters for microwave plasma enhanced CVD are as follows: gas composition between 0.1 to 5% CH_4 diluted in H_2; total flow rate between 50 and 500 sccm; substrate temperature between 400 and 1000°C; total pressure, between 1 and 400 torr. High pressure throttle valves are now available which permit the control of total pressure independently of gas flow rates and pumping speeds, over specified limits. Electronic controllers measure pressure by a capacitance manometer and control the throttle valve to attain the set point pressure.

Because of the early popularity of the tube-style microwave PECVD reactor, it was definitively demonstrated that polycrystalline and epitaxial diamond films could be reproducibly grown by this method. However, this "tube" configuration suffered the disadvantage of a small substrate size. The largest substrate which can be inserted in a silica tube compatible with a WR284 waveguide is only ≈ 2 - 3 cm in diameter. Therefore, in 1986 microwave engineers were asked to design and build microwave plasma enhanced CVD systems capable of depositing diamond films over larger areas. U. S. companies including Wavemat, Microscience, Gerling Laboratories, and Applied Science and Technology market novel microwave applicators capable of producing a microwave plasma at least 10 cm in diameter. Innovations in hardware have also led to the development of ion guns which use microwave radiation to create an ion beam.

A schematic of a microwave system marketed by Applied Science and Technology is shown in Fig. 6 (31). Microwaves are introduced into the reaction chamber from a proprietary antenna which converts a rectangular,

WR284 microwave signal into a circular mode. The microwaves pass through a silica microwave window/vacuum barrier into a cylindrical cavity which is the PECVD process chamber. The location of the luminous plasma ball is controlled by the cavity tuning as well as the pressure, power and gas phase composition of the plasma. The vertical position of the platen changes the tuning of the cavity and affects the position and intensity of the luminous plasma ball. When argon is added to the plasma, the luminous ball will fill the chamber with a bright, magenta glow. However, under typical 1 - 5% CH_4/H_2 gas compositions, the luminous plasma contracts to a powdery white ball, 3 - 6 cm in diameter; the contraction is due to the lower ionization propensity of hydrogen as compared to argon. The luminous ball will further contract in size as the pressure in the chamber increases. The luminous ball will increase in diameter as the applied microwave power is increased.

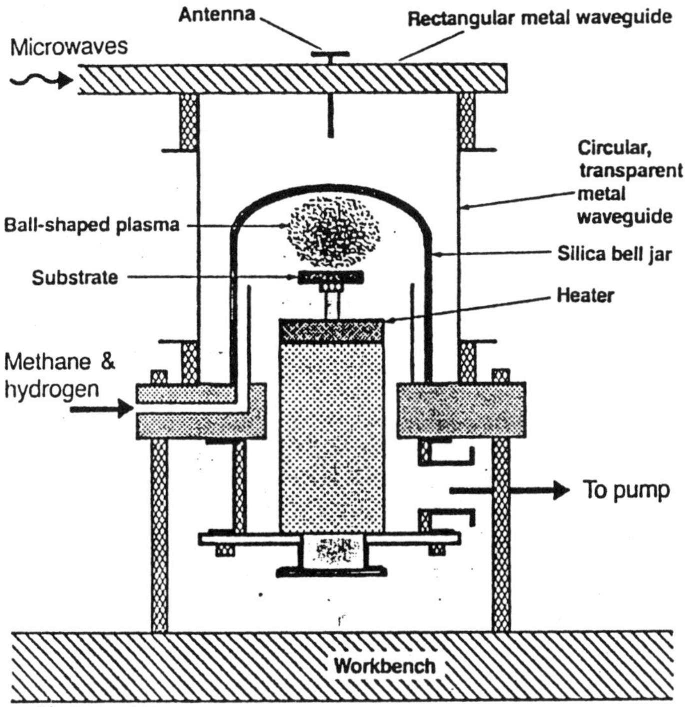

Figure 6. Schematic of a microwave plasma-enhanced CVD reactor manufactured by Applied Science and Technology, Inc. (31). *(Reprinted with permission from Chemical & Engineering News. Copyright 1989, American Chemical Society.)*

If the platen is electrically conductive, it will couple to the microwaves and become heated. A thin graphite susceptor atop a silica rod can achieve temperatures more than 800°C by the combination of microwave coupling and plasma bombardment. It is also believed that electrically conductive substrates such as silicon experience some heating due to direct coupling with the microwave energy.

Platens have been produced which can independently heat the substrate to a desired temperature up to 1000°C. Such heated platens supplement heat provided by microwave coupling and by plasma bombardment. Diamond films have been grown with the edge of the luminous plasma ≈ 2 cm above the substrate when the substrate is independently heated (32). It is important to emphasize that the luminous region of the microwave plasma visible to the unaided eye represents but a fraction of the total volume of the true plasma. Direct contact of the substrate with the luminous glow is *not* a necessary condition for diamond growth from a microwave plasma. The microwave plasma also generates intense ultraviolet radiation which may affect diamond deposition. It is speculated that optically transparent diamond films may be electrically conductive during microwave PECVD due to the photoelectric generation of electron-hole pairs.

A variation on microwave plasma enhanced CVD is the use of magnetically enhanced or electron cyclotron resonance (ECR) plasma enhanced CVD. When a magnetic field is coupled to the microwaves, plasmas can be altered in intensity and position. However, if the pressure in the chamber is held below 1.3 Pa (10 mtorr), then the mean free path of electrons in the plasma will be long enough to oscillate in electron cyclotron resonance if the appropriate magnetic field is applied. At 2.45 GHz, the magnetic field which produces the ECR effect is 875 G. However, ECR is impossible if the pressure exceeds ≈ 1.3 Pa; the process at such high pressures should be called "magnetically enhanced microwave PECVD." It has been reported that diamond films can be grown by magnetically enhanced microwave plasma CVD, as well as genuine ECR conditions (33)(34). Current ECR technology includes ECR plasma chambers capable of creating uniform plasmas over 8" in diameter by novel magnet placement and microwave field injection.

Because of the initial popularity of microwave plasma enhanced CVD, it has been used to perform various studies on the addition of oxygen-containing gas species to the deposition of diamond films. It is has been found that addition of O_2, CO, or water vapor enhances the deposition rate of diamond as well as the quality of the diamond film as measured by Raman spectroscopy (5)(31)(32)(35). The addition of O_2 to a microwave plasma permits deposition at cooler substrate temperatures (as low as 450°C). However, it has been noted that as the substrate temperature of diamond deposition decreases for a fixed film quality, the deposition rate also decreases (36).

Researchers at AT&T Bell Laboratories have proposed an updated version of the Eversole cyclic process for growing diamond films (37). J. A. Mucha, D. L. Flamm and D. E. Ibbotson grew diamond films on Si substrates by microwave PECVD from mixtures of CH_4, H_2, and O_2 diluted in He. They compared the properties and growth rate of polycrystalline diamond films grown by a conventional, "continuous chemistry (CC)" process to films grown by an "alternating chemistry (AC)" process. A typical CC process used 0.5% O_2 + 2.0% CH_4 + 97.5% H_2. The AC process was composed of 1 minute of 4% CH_4/He followed by 4 minutes of 15% H_2 + 1% O_2 + 84% He. Remarkably, the AC process produced diamond films with practically no 1530 cm^{-1} Raman feature at a net deposition rate of 9 μm/hr. This result compared favorably to the CC diamond films. The authors estimated that the deposition rate during the 4% CH_4/He half cycle was ≈ 45 μm/hr.

T. Anthony of General Electric (23) grew diamond using a hybrid microwave plasma plus hot filament reactor. A hole was cut into the bottom of an ordinary, household microwave oven. A glass belljar was inserted through the hole to set inside the microwave oven chamber. The belljar was connected to a gas flow system and vacuum pump. A hot filament growth system was positioned inside the belljar. When CH_4/H_2 mixtures were introduced into the belljar at a pressure below 1 atm, the microwaves coupled to the gas and created a microwave plasma within the belljar. Anthony emphasized the need to keep the duty cycle of the oven's magnetron near unity to maintain a stable plasma. He reported that the combination of microwave plasma and hot filament grew diamond films at a faster rate than either excitation technique alone.

5.2 RF Plasma Enhanced CVD

Generally, radio frequency power can be applied to create a plasma in two electrode configurations: capacitively coupled parallel plate and by induction. The two methods are illustrated in Fig. 7. Typically, RF plasma enhanced CVD employs a frequency of 13.56 MHz. To maximize plasma stability, an impedance matching network is wired between the RF generator and the electrode assembly in order to match the impedance of the plasma load to that of the RF power supply. An advantage of RF plasmas is that they can be easily generated over much larger areas than microwave plasmas.

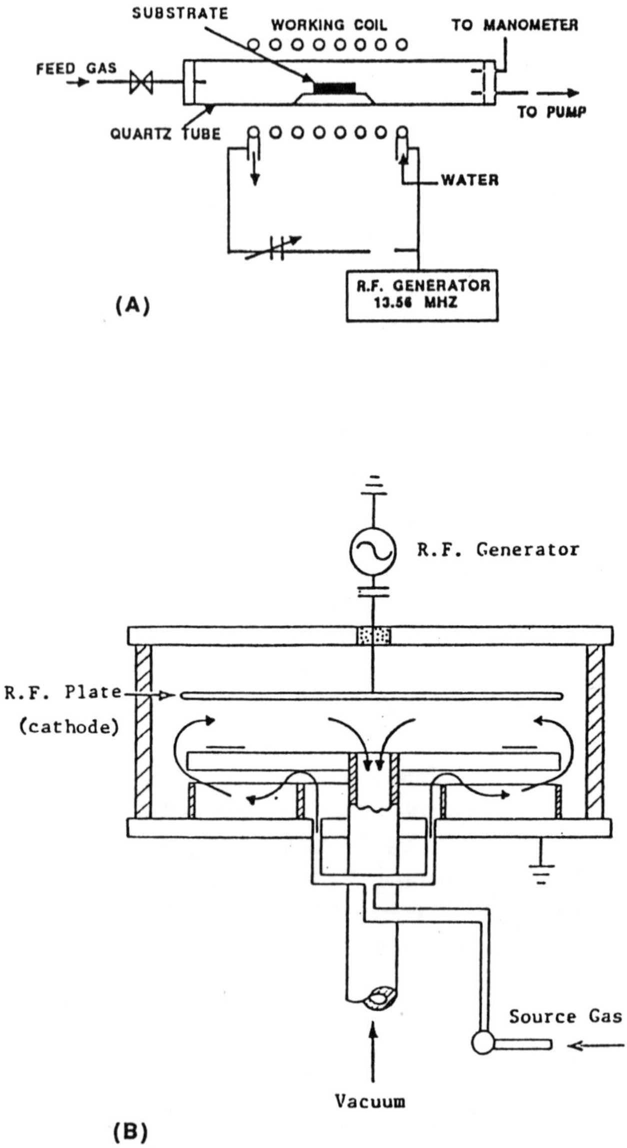

Figure 7. (A) Schematic of an inductively coupled RF plasma-enhanced CVD reactor (51). *(Reprinted by permission of the publisher, The American Institute of Physics.)* (B) Schematic of a parallel plate, capacitively-coupled RF PECVD reactor. (*Deposition Technologies for Films and Coatings*, edited by R. F. Bunshah, Noyes Publications, Park Ridge, NJ (1982) p. 372; reprinted by permission of the publisher.)

The disadvantage of RF capacitive plasmas is that the frequency of the plasma is optimal for sputtering of dielectrics, especially if the plasma contains argon. Since ion bombardment from the plasma can create diamond-like hydrocarbon films, capacitively coupled RF plasmas are commonly used to create diamond-like coatings by placing the substrate on the powered plate electrode (20)(21). Wood, et al. have reported the synthesis of genuine diamond films by parallel plate, capacitively coupled RF plasma reactor (38).

Deposition of polycrystalline diamond films by RF induction has been demonstrated by various researchers using process conditions similar to microwave plasma CVD (39). Homoepitaxial diamond films have also been grown by RF induction PECVD (17). W. C. Roman and M. B. Colket at United Technologies Research Center have used RF induction plasma to deposit polycrystalline diamond films while performing in situ gas phase laser spectroscopy (40). Their fused silica chamber permitted laser beam sampling of the diamond-growing plasma as a function of distance away from the substrate surface. They employed an advanced laser in situ spectroscopy known as coherent anti-stokes Raman scattering (CARS).

At atmospheric pressure, a thermal plasma can be created by RF induction. Matsumoto (29)(41) described the synthesis of 20 mm diameter diamond films at a growth rate of 120 µm/hr. However, the RF thermal plasma demanded 20 - 75 kW of power and an argon carrier gas flow of 20 - 30 slm.

5.3 DC Plasma Enhanced CVD

In 1951, a German publication authored by H. König and G. Helwig (42) described the deposition of hydrocarbon films from a DC plasma of benzene. The design of the parallel plate, DC plasma chamber used by König and Helwig has changed little since 1951; a typical contemporary reactor illustrated in Fig. 8 (43). In DC plasma CVD, a negative or positive DC voltage is applied to the substrate platen or to an auxiliary electrode. Intermediate electrodes may be used to guide the plasma and alter the plasma position and properties. As with RF plasmas, negatively biasing the platen to promote ion bombardment tends to produce diamond-like films. DC plasma enhanced CVD has the advantage of being able to coat large areas; the diamond deposition area is limited by the electrodes and the DC power supply. The disadvantage of DC plasma enhanced CVD is that it may self extinguish if a very thick film is grown since a biased platen will change its bias if coated by a dielectric diamond film. However, there have been reports that the deposition rate of diamond on positively biased electrodes does *not* decrease with time. This has been attributed to the generation of charge carriers in the diamond film by bombardment processes and/or absorption of ultraviolet light emitted by the DC plasma.

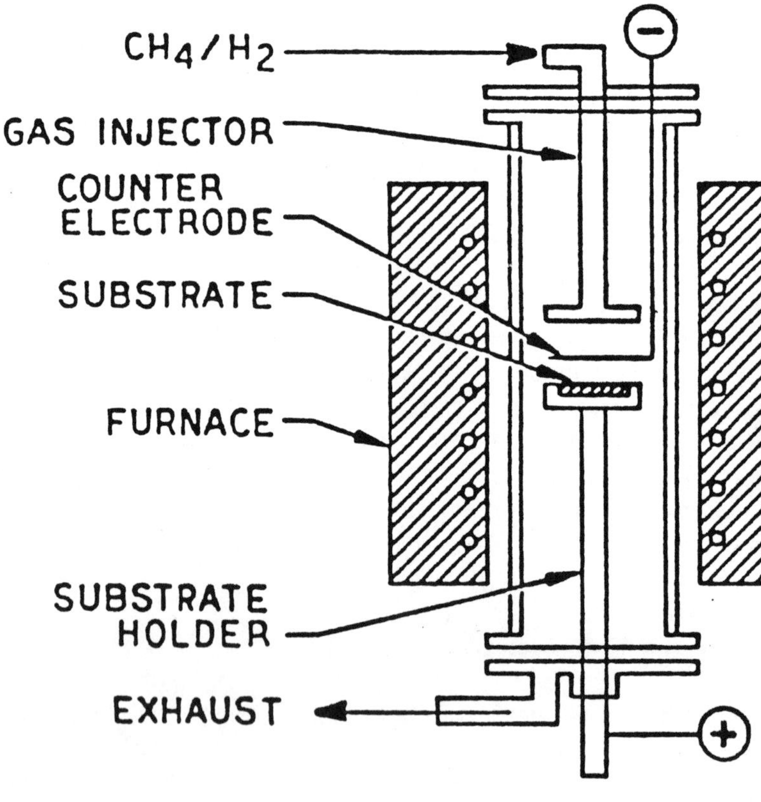

Figure 8. Schematic of a DC plasma-enhanced CVD reactor (43). *(Reprinted by permission of the publisher, The Electrochemical Society.)*

There are numerous manifestations of DC plasma CVD which generate a high energy, "thermal" plasma. Singh, et al. (44) developed a hollow cathode PECVD reactor in which a plasma plume arises from a negatively biased tube which is pointed at a grounded or positively biased substrate. Japanese researchers have been active in the development of DC arc PECVD methods which create thermal plasmas capable of growing diamond films at rates $\geq$ 20 μm/hr. Suzuki, et al. (45) power a flat plate Mo cathode to create a DC arc plasma pointed at the substrate anode. Their plasma is sustained at a pressure of 27 kPa (200 torr) by 2.4 - 4 kW/cm^2 of power. Akatsuka, et al. (46) create a DC arc between the ends of two rods positioned above a substrate. The arc is sustained by a power of 560 W. An input gas mixture of Ar + H_2 + ethanol is flowed through the arc onto the substrate at a total pressure of 27 - 41 kPa (200 - 300 torr). Akatsuka, et al. claimed a diamond film growth rate of 200 - 250 μm/hr.

Kurihara and co-workers at Fujitusu Laboratories have developed a DC plasma jet process entitled the DIA-JET process (47). DIA-JET uses a gas injection nozzle composed of a rod cathode concentric with a tube anode. Gases pass between the cathode and anode and are sprayed out from an orifice in the anode. The gas mixture is CH_4 + H_2 with a carrier gas (Ar or He); the total pressure is between 4.1 and 41 kPa (30 - 300 torr). The DC arc is sustained by 80 - 150 V and 10 - 50 A (a range of 0.8 - 7.5 kW); the substrate must be water cooled to maintain a substrate temperature between 520 - 1220°C. Typical growth rates are 80 μm/hr. The properties of diamond films made by the DIA-JET process include a Vicker's hardness of 10,000 kg/mm^2, a thermal conductivity of 8 W/cm°C and a resistivity between 10^4 and 10^5 Ω-cm. The DIA-JET process was used to synthesize freestanding diamond plates 10 x 10 x 2 mm^3. The various DC arc methods have demonstrated high deposition rates and the ability to synthesize high quality diamond, and are gaining acceptance as a commercially viable manufacturing technology for diamond film systhesis..

Fujimori, et al. (48) grew diamond films by a hybrid hot filament plus DC plasma method. Like Lee, et al. (10), a DC bias was applied between the substrate and the filament. Fujimori, et al. created a DC plasma by applying -120V to a tungsten filament heated to 2200°C. The plasma stability was probably enhanced by thermionic emission of electrons from the incandescent filament. Fujimori, et al. reported that addition of a DC plasma increased the deposition rate by a factor of three with minimal degradation of diamond film properties.

6.0 OTHER METHODS OF DIAMOND DEPOSITION

6.1 Physical Vapor Deposition Methods

Physical vapor deposition (PVD) of diamond entails the evaporation or sublimation of solid carbon into a vapor phase followed by condensation of the carbon atoms as diamond. PVD methods use a solid carbon evaporant source such as graphite, soot, amorphous carbon, etc.

Carbon may be "vaporized" by three different methods: arc evaporation, electron beam (e-beam) evaporation, or sputtering. Arc evaporation of graphite is a procedure commonly used to deposit electrically conductive carbon films onto insulating samples to enable imaging by scanning electron microscopy. Two solid carbon "pencil" rods are sharpened at the ends and pointed at each other within an evacuated belljar. An arc is sustained between the rods, causing evaporation of carbon and the consumption of the rods. "Spectroscopic purity" graphite electrodes (pencils) are commonly employed in analytical chemistry for emission spectroscopy and spark source mass spectroscopy. Electron beam evaporation of carbon is much more controllable than arc evaporation since the electron current and electron energy are controlled much more precisely using a standard e-beam evaporation gun. *Sputtering* of graphite by ion impact is extremely inefficient due to the low sputter yield of most allotropes of carbon. To maximize the sputter yield of carbon, the carbon target can be heated, the atomic mass of the sputtering gas can be increased, and the energy of the incident ions can be increased.

A Soviet emigré named Anatoly Altshuler has written a monograph entitled "Diamond Technology in the U.S.S.R." (49). Before he left the U.S.S.R., Dr. Altshuler was active in diamond research. He reports that PVD synthesis of diamond is very common in the Soviet Union. Various PVD methods have been reported capable of producing films containing diamond at substrate temperatures between 25 to 400°C. A comparison between PVD and CVD methods was discussed by Soviet researchers Spitsyn, et al. (1) in a paper published in English in 1981. They noted that PVD methods may produce films at a greater deposition rate than CVD, but that film quality is typically superior using CVD methods. For example, they conclude that homoepitaxial growth of diamond is best performed by a CVD method.

In the United States, the most notable research into PVD of diamond thin films has been performed by C. Deshpandey, R. F. Bunshah, and co-workers at UCLA. R. F. Bunshah is the inventor of the activated reactive evaporation (ARE) process in which thin films are deposited by the evaporation of a solid source into a reactive gas which is plasma excited. ARE has been successfully applied to diamond film deposition (50). The ARE methodology can provide complete decoupling of the various parameters which affect thin film growth (51). The rate of material impingement on the substrate is controlled by the evaporation rate. The interaction of the evaporant material with the gas phase is controlled by the evaporation distance, the gas pressure and the plasma power density. The substrate temperature is independently controlled by a resistance heater. The substrate bias (RF or DC) can be independently controlled by a high current power supply. The surface chemistry of the substrate can be controlled independently of the ARE plasma by bombarding the substrate using an electron gun or ion beam gun. If the input gas includes H_2, a hot filament can be positioned near the substrate to enhance the concentration of H° at the growth surface.

The ARE process for diamond is schematically illustrated in Fig. 9. A high purity carbon powder is placed in the evaporation crucible and is evaporated by an electron beam. The evaporated carbon then passes through a plasma containing a hydrocarbon and/or hydrogen. The substrate is suspended face down on an independently heated platen which can be independently biased by an RF or DC field. Typical films are nanocrystalline with x-ray diffraction and XPS spectroscopy indicating the formation of diamond. These films exhibit an ultraviolet transmission cut-off at $\lambda = 230$ nm, which is the UV cut-off of monocrystalline, type IIa diamond. It has been reported that films can be deposited at substrate temperatures below 500°C at deposition rates greater than 1 µm/hour (52).

Physical vapor deposition of diamond is probably governed by a different set of chemical kinetic mechanisms than chemical vapor deposition of diamond. In chemical kinetic models of diamond growth by CVD, hydrocarbon species such as methyl or acetylenic groups act in concert with "anti-graphite" species to deposit diamond. These chemical kinetic models probably are not applicable to PVD methods since there is no simple mechanism for the formation of acetylenic or methyl species as carbon atoms are shot from the evaporation source to the substrate. The gas pressures for PVD are typically 10^{-3} to 10^{-1} Pa, which is significantly less than the ≈ 30 kPa pressures used for CVD of diamond. A reduced gas pressure is required during PVD for the evaporated atoms to travel to the substrate with a minimal number of collisions. If perfected, PVD methods potentially could deposit diamond films at a faster rate than CVD since: *(a)* the source of carbon atoms is a *solid*, therefore, it contains a greater number of carbon atoms/volume than a gas; and *(b)* vaporized carbon atoms arising from the solid are vectored toward a substrate. The evaporated atoms obtain momentum from the electron beam heating, and the pressure of the gas phase is low enough to minimize scattering of the evaporated atoms during their flight to the substrate. PVD growth rates have the potential to exceed CVD growth rates since PVD is a *solid* → *vapor* → *solid* reaction while CVD is a *vapor* → *solid* reaction.

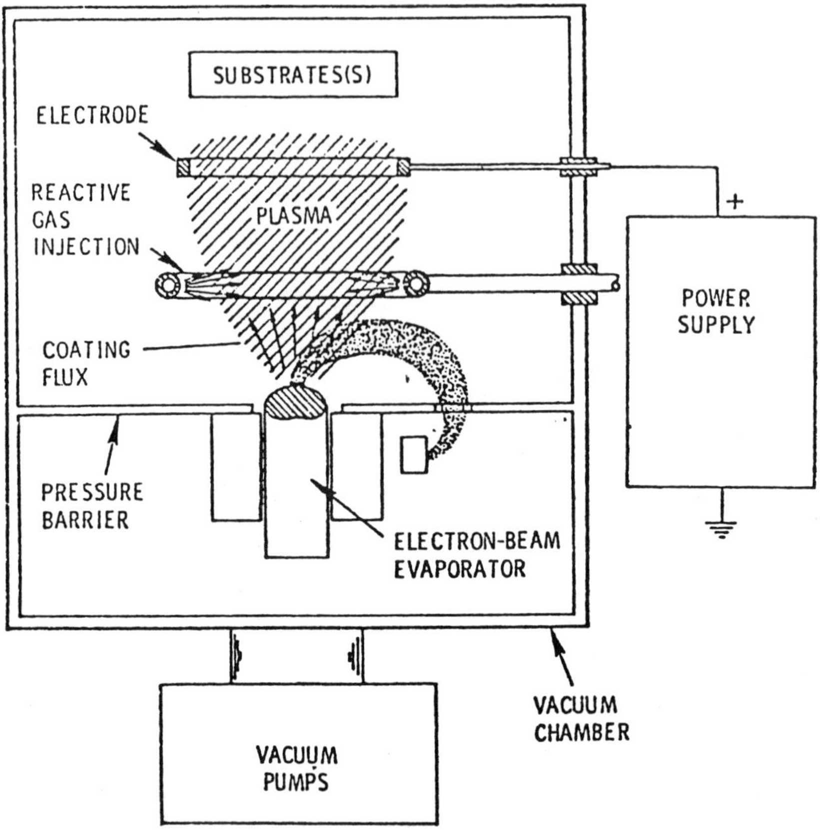

Figure 9. Schematic of an activated reactive evaporation reactor. (*Deposition Technologies for Films and Coatings*, edited by R. F. Bunshah, Noyes Publications, Park Ridge, NJ (1982) p. 126; reprinted by permission of the publisher.)

It is hoped that more Western researchers will rediscover published work in the Soviet literature on physical vapor deposition of diamond (1)(13)(49). PVD methods for diamond show great promise in applications which are not currently satisfactorily addressed by CVD methods. PVD methods would be advantageous in applications where nanocrystalline, UV-transparent diamond films are required and diamond films must be deposited at cool substrate temperatures. It is not yet known whether PVD diamond films can exhibit better adhesion to their substrates than CVD films. It is also not yet known how the performance/cost ratio PVD diamond compares to the performance/cost ratio of CVD diamond.

6.2 Combustion

In Spring of 1988, Yoichi Hirose and co-workers at the Nippon Institute of Technology announced the deposition of diamond particles by combustion (53). They placed the flame of an oxy-acetylene welders' torch onto a substrate in room air; diamond particles were deposited beneath the center of the flame. A schematic diagram of a simple combustion synthesis apparatus is illustrated in Fig. 10. An ordinary welders' torch with a brazing tip burns approximately a 1:1 mixture of C_2H_2 to O_2. Welder's grade gas purities are used, and the gas flow rates are controlled by flow meters. A 1:1 mixture is fuel-rich and is seldom employed by welders. However, if the tip of the inner luminous cone of the flame is placed in contact with a surface cooled to between 800 and 1050°C, then diamond crystals are observed to nucleate and grow where the luminous cone intersects the surface.

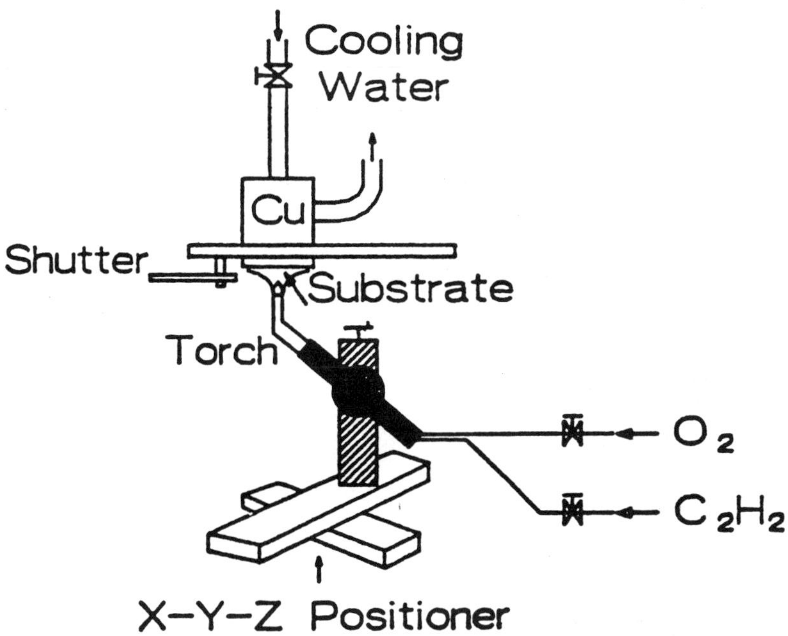

Figure 10. Schematic of a combustion synsthesis apparatus (21). *(Reprinted by permission of the publisher, The Materials Research Society.)*

As pointed out by Hirose, et al., the combustion synthesis of diamond offers many salient advantages over "conventional" CVD methods (53):

1. Diamond can be synthesized in air so that vacuum chambers are not needed.
2. The equipment is simple and inexpensive.
3. No electricity is needed for this process.
4. High growth rates can be obtained. In case of particles, the maximum growth rate is 200 µm/hr and the average is 30 - 50 µm/hr. For films with crystal habits, it is 100 µm/hr, and for DLC films, 200 - 500 µm/hr is possible. These growth rates are ten times faster than those of previous processes.
5. It is possible to deposit diamond over a large area.
6. The process seems to be useful for coating diamond film on curving substrate surfaces.

Hirose's process was quickly replicated in the United States by research groups at the Naval Research Laboratory and Penn State University (54). Since that time, researchers worldwide have been working to better understand combustion synthesis of diamond and mitigate its shortcomings. It has been concluded that combustion synthesis of diamond is a chemical vapor deposition method (5)(21)(22)(31). Some of the shortcomings of combustion synthesis with a welders' torch are as follows: *(a)* The deposited diamond films have inhomogeneous microstructure and composition. Diamond nucleates in a "bull's-eye" which is ringed by non-diamond carbon such as graphite and soot. *(b)* It is difficult to control the microstructure to reproducibly create fully dense, pore-free diamond films over a large area. *(c)* Active cooling of the substrate is required in order to maintain the substrate at ≤ 1000°C. Under such fuel-rich conditions, the adiabatic flame temperature in the inner, luminous cone exceeds 1700°C. *(d)* The welder's torch creates severe thermal gradients over the surface of the substrate, making it difficult to coat large areas without the substrate warping or fracturing due to thermal shock.

Generally, the requirements of CVD diamond film deposition are at odds with the design of brazing tip burner. A typical goal of CVD diamond deposition is to produce highly dense, homogeneous diamond films at a fast rate over a large deposition area. To achieve this goal, it is logical to conclude that thermal gradients and chemical concentration gradients over the plane of the substrate should be minimized. However, a brazing tip is designed to create a highly focused flame to locally melt the material it contacts. Creation of severe thermal gradients is enhanced using a diffusion flame whereby the oxygen and acetylene are mixed in the flame, i.e., outside the burner.

Subsequent research by Hirose and co-workers (55)(56), and by M. A. Cappelli and P. H. Paul (57) has made great strides in improving the area and film quality of diamond films deposited by combustion. It is now possible to coat large areas with a highly dense, polycrystalline diamond film. Film quality can be excellent; flame deposited films have been grown optically transparent, exhibiting excellent Raman spectra. Methods used to improve film properties and increase film area have typically involved redesign of burners, precise control of gas mixture chemistry, and placing the film deposition apparatus inside a chamber to exclude room air. As combustion synthesis science and technology continues to advance, it may eventually be used to manufacture useful diamond coatings for tribological applications (58).

Researchers at Naval Research Lab have enhanced the properties of flame-deposited diamond by adding acetone to the fuel. When acetylene was bubbled through acetone saturated with boric acid, boron doped, semiconducting diamond films were deposited (59).

M. Frenklach, et al. (60) placed a combustion flame in a microwave plasma to demonstrate the homogeneous nucleation of submicron diamond particles. They burned CH_2Cl_2 with O_2, and enveloped the burner in a "sheath" of Ar. The operating parameters included a CH_2Cl_2/O_2 flow ratio of 1.2, a pressure of 4.0 - 9.2 kPa (30 - 70 torr), and a microwave power of 300 W. Frenklach, et al. synthesized hexagonal polytypes of diamond, including the novel 6H polytype.

7.0 COMPARISON OF DEPOSITION TECHNOLOGIES

There has not yet emerged a single method of producing diamond thin films which is adequate for *all* of the proposed applications of diamond thin film technology. The key task is to identify the deposition method which will produce a diamond-containing product possessing the greatest performance/cost ratio. The performance/cost ratio of the diamond-containing product is determined by the cost/carat *and* the properties of the diamond coating. An inexpensive cost/carat diamond film is *worthless* unless it performs a useful function. (Performance requirements of a diamond film may include good adhesion, good toughness, high thermal conductivity, high UV or IR transparency, radial tension, etc.) The cost/carat of the diamond film is determined by deposition rate, deposition area, capital equipment cost, and hourly consumable/overhead cost.

An excellent discussion of the economics of microwave PECVD technology as applied to diamond coated cutting tools appears in an article written by J. P. Dismukes and K. R. Walton entitled "Manufacturing Technology Assessment for CVD Diamond Films" (6). One notable conclusion of this study is that an increase in deposition area is more beneficial than an increase in deposition rate, all other things equal. To mass produce competitively priced diamond coated cutting tools, the number of tools which can be simultaneously coated should be maximized. All present methods for diamond film deposition are batch processes. The study also implied that development of an in-line, *continuous* process for depositing diamond onto tools would reduce production costs.

What follows is a survey of applications followed by a brief discussion of current methods used to create diamond films to meet each application.

7.1 Epitaxial Diamond for Semiconductor Device Research

Epitaxial, boron-doped diamond films suitable for semiconductor research have been grown by microwave plasma, RF plasma, and by hot filament CVD methods (17)(25)(61). During epitaxial growth, it is necessary that the diamond growth ambient be extremely stable. Fluctuations or disruption in the growth process could result in the incorporation of crystallographic and/or graphitic type defects in the epitaxial layer. For semiconductor device applications, the quality of the epitaxial diamond is paramount; a fast deposition rate is of lesser importance. It is also desirable to controllably dope the epitaxial diamond with boron in situ by using gases such as diborane (B_2H_6) or boric acid dissolved in acetone. It is recommended that the diborane gas plumbing be isolated from other gas plumbing since diborane is rather "sticky". If precautions are not taken, boron-containing material will remain in a CVD system, causing significant boron doping of subsequent diamond films even if diborane gas is not flowed during growth.

7.2 Thick, Free Standing Diamond Structures

A U. S. industrial laboratory has already produced a CVD diamond "brick" of dimensions 5 x 15 x 2 cm^3 (62). In Japan, a free standing diamond dome, 5 cm in diameter by 100 µm thick, was produced by operating a hot filament CVD reactor for several months (63). For applications where thick, structural diamond parts are necessary, a method is needed where the cost/carat is minimized and film microstructure is tailored to produce 100% dense films. It is also necessary to deposit the diamond onto an easily removable ("sacrificial") substrate. If the diamond structure is a flat plate or "brick," a liquid metal substrate is ideal (23). In the case of the diamond dome, the diamond film was deposited onto a male mandrel which was rotated on a turntable beneath a hot filament (63). Structural diamond tubes can be grown by depositing diamond onto a metal rod such as Mo (64). The low thermal expansion coefficient of diamond helps the diamond tube pull away from its mandrel upon cool-down. So far, the reported methods used to create structural diamond parts have been hot filament, combustion and DC arc jets. Evidently, these three CVD methods combine a low cost/carat with a versatility of deposition geometry. For example, it would be challenging to deposit a 100 μm thick diamond coating onto a 2.5 cm diameter Mo mandrel by using a microwave PECVD reactor.

7.3 Rim Supported Diamond Windows/Membranes

Diamond windows up to 7.5 cm in diameter are now routinely being produced as x-ray windows and as x-ray lithography mask membranes. The diamond film thicknesses range from 0.5 - 5 μm. For application as mask membranes, it is especially important that the diamond window be held in radial tension on its silicon substrate (65).

A diamond x-ray lithography membrane is created by depositing a diamond film on a silicon wafer and then selectively etching away the central area of the silicon to expose the diamond. In order to engineer radial tension in the diamond film, the growth temperature should be carefully controlled, and the film microstructure must be engineered by carefully controlling the nucleation and growth of the film. In particular, the film must *not* develop a highly oriented, columnar grain structure with a rough surface. Instead, the film should be isotropic, with a mean crystallite size < 100 nm, and with a surface RMS smoothness equal to significantly less than the film thickness. Isotropic microstructures can be created by continuous re-nucleation of diamond as the film thickens. Methods demonstrated to create diamond windows/membranes include microwave plasma CVD, DC plasma CVD, and hot filament CVD.

7.4 Cutting Tools

Cutting tools present a severe challenge since the diamond layer must be hard, tough and adherent to its substrate. Diamond films will not be fully accepted by cutting tool users until their performance and reliability equals that of conventional polycrystalline diamond (PCD) tools. Because of the prevalence of cemented tungsten carbide in machine shops across the world, many researchers are working to deposit useful diamond films on cemented carbide substrates. The superb adhesion, hardness, and toughness of PCD layers bonded to cemented carbide tools resulted from many years of research and development. In a PCD, the diamond grit is bonded to cemented carbide by applying very high pressures and temperatures in a device such as a belt press (66). Duplicating the mechanical properties of conventional PCD is a challenge for CVD and PVD diamond technology. The high surface energy of diamond makes it difficult to achieve good adhesion of CVD diamond to most substrates. To achieve sufficient adhesion, the diamond layer must form a strong chemical bond to its substrate. The chemical bond must be strong enough to overcome the interfacial shear stresses and high temperatures applied during cutting.

CVD diamond layers for cutting tools may be applied either by direct deposition of diamond onto the cutting tool or by brazing a free-standing diamond plate onto the tool. Since cemented carbide tools typically contain a cobalt matrix, brazing a plate of diamond is appealing since it sidesteps the chemical incompatibility between diamond and Co, as discussed in Sec. 2.0. Unfortunately, a CVD diamond plate transfers all of the heat of cutting to the braze joint. It is difficult to develop a braze between pure diamond and cemented carbide capable of maintaining good toughness and adhesion during shear stressing at temperatures above 600°C. If a diamond thin film is deposited directly onto the tool by CVD or PVD, then an intermediate layer may be used to provide strong chemical bonding to both the diamond film and the carbide substrate.

For tribological applications, manufacturing low cost/carat diamond films is meaningless unless the diamond film possesses the requisite properties of adhesion, hardness, and toughness. It is possible that a manufacturer will demonstrate a variety of CVD or PVD methods to produce diamond films possessing similar tribological properties. Each manufacturer will then select the method over which it has trade secret protection or patent protection.

8.0 RECOMMENDATIONS AND PREDICTIONS

It is becoming clear that the technology of diamond film deposition will improve as a direct result of research into the chemical kinetics of diamond CVD. CVD methods will improve once researchers identify all of the phenomena common to the hot filament method, the plasma-enhanced methods, and the combustion method. Once these fundamentals are understood, further refinement of diamond CVD can be enacted to improve the performance/cost ratio of the diamond films. This refinement/optimization process will require a close alliance between those who develop CVD processes, (including material scientists, chemical engineers and mechanical engineers), and the end users of diamond-coated products.

It is hoped that more researchers in the West will study Soviet advances in the physical vapor deposition of diamond films. PVD of diamond may promise high deposition rates at low substrate temperatures, plus the ability to synthesize UV transparent polycrystalline diamond. PVD may turn out to be an optimal technique for coating cutting tools and optical components. Studies into the chemical kinetics of diamond growth by PVD should be initiated. Further investigations are required on the physical properties of PVD deposited diamond films to see if their properties, such as Raman spectra, IR spectra, diffraction patterns, and thermal conductivity are similar to those of diamond films made by CVD. It is entirely possible that PVD diamond films may constitute a separate class of diamond materials which can easily be distinguished from other forms of diamond by measurement of the physical and chemical properties.

As the field progresses, technical decision-makers must stay abreast of the latest published technical advances. Users of diamond film technology must decide whether to deposit diamond films in-house, or contract out coating services from other companies. The science and technology of CVD diamond is still dominated by "art" and proprietary know-how. It is believed that the CVD diamond industry will be similar to other high-technology materials industries in that it will be dominated by closely guarded trade secrets, patents, and other competitive advantages used to manufacture useful diamond films for specific products. Hence, a desire to remain competitive in diamond thin film technology requires a long term commitment of personnel, facilities and capital equipment. The technology of CVD diamond is rapidly advancing and will outrun any organization unwilling to continuously update its technology.

REFERENCES

1. Spitsyn, B. V., Bouilov, L. L. and Derjaguin, B. V., *J. Crys. Growth,* 52:219 (1981)
2. DeVries, R. C., "Synthesis of Diamond Under Metastable Conditions," General Electric Technical Information Series 86CRD247, Schenectady, NY (1987)
3. Tsuda, M., Nakajima, M. and Oikawa, S., *J. Am. Chem. Soc.,* 108:5780 (1986)
4. Frenklach, M. and Spear, K. E., *J. Mater. Res.,* 3(1):133 (1988)
5. Yarbrough, W. A. and Messier, R., *Science,* 247:688 (1990)
6. Dismukes, J. and Walton, K., in *Diamond and Diamond-like Films,* (J. Dismukes, et al., eds.), Electrochem. Soc. Proc., Vol. PV 89-12, p. 653, Pennington, NJ (1989)
7. Anthony, T. A., in *Diamond, Boron Nitride, Silicon Carbide and Related Wide Bandgap Semiconductors,* (J. T. Glass, R. Messier and N. Fujimori, eds), Mater. Res. Soc. Proc., Vol. 162, Pittsburgh, PA (1990)
8. Joffreau, P. O., Haubner, R. and Lux, B., in *Diamond and Diamond-Like Materials Synthesis,* (G. Johnson, A. Badzian and M. Geis, eds.) Mater. Res. Soc. Proc., Vol. EA-15, p.15, Pittsburgh, PA (1988)
9. Wu, C. H., Tamor, M. A., Potter, T. J. and Kaiser, E. W., in *Diamond, Boron Nitride, Silicon Carbide and Related Wide Bandgap Semiconductors,* (J. T. Glass, R. Messier and N. Fujimori, eds.), Mater. Res. Soc. Proc., Vol. 162, Pittsburgh, PA, (1990)
10. Lee, Y. H., Richard, P. D., Bachmann, K. J. and Glass, J. T., *Appl. Phys. Lett.* 56(7):620 (1990)
11. R. H. Wentorf, in *Encyclopedia of Chemical Technology,* 3rd Edition, 4:676, John Wiley and Sons, New York (1978)
12. Eversole, W. G., U. S. Patent 3,030,188 (April 17, 1962)
13. Badzian, A. R. and DeVries, R. C., *Mater. Res. Bull.* 23:385 (1988); English translations of two important Soviet books on diamond appear in *Surf. Coat. and Tech.,* 38(1-2) October (1989)
14. Poferl, D. J., Gardner, N. C. and Angus, J. C., *J. Appl. Phys.* 44(4):1428 (1973)
15. Messier, R., Badzian, A. R., Badzian, T., Spear, K. E., Bachmann, P. and Roy, R., *Thin Solid Films,* 153:1 (1987)

16. Rosner, D. E. and Strakey, J. P., *J. Phys. Chem.,* 77(5):690 (1973)
17. Rudder, R. A., Posthill, J. B. and Markunas, R. J., *Semiannual Report on Semiconductor Diamond Technology,* SDIO/ONR contract N00014-86-C-0460 (November 1989); Rudder, R. A., Posthill, J. B. and Markunas, R. J., *Electronics Letters,* 25:1220 (1989)
18. Markunas, R. J., Research Triangle Institute, personal communication (1989)
19. Angus, J. C., in *Diamond and Diamond-like Films,* (J. Dismukes, et al., eds.), Electrochem. Soc. Proc., Vol. PV 89-12, p. 1, Pennington, NJ (1989)
20. Angus, J. C. and Hayman, C. C., *Science,* 241:913 (1988)
21. Angus, J. C., Buck, F. A., Sunkara, M., Groth, T. F., Hayman, C. C. and Gat, R., *MRS Bulletin,* 14(10):38 (1989)
22. Spear, K. E., *J. Am. Cer. Soc.,* 72(2):171 (1989)
23. Anthony, T., "The Growth Rate of Filament-Assisted Plasma CVD Diamond," talk presented at the Fall 1987 Mat. Res. Soc. Meeting, Boston, MA (December 1987)
24. Kimock, F. M., Hoover, D. S., Garg, D., Dyer, P. N., Iampietro, R. L., Monk, V. A. and Tsai, W., in *Diamond and Diamond-like Films,* (J. Dismukes, et al., eds.), Electrochem. Soc. Proc., Vol. PV 89-12, p. 536, Pennington, NJ (1989)
25. Geis, M. W., in *Diamond, Boron Nitride, Silicon Carbide and Related Wide Bandgap Semiconductors,* (J. T. Glass, R. Messier and N. Fujimori, eds.) Mater. Res. Soc. Proc., Vol. 162, p. 15, Pittsburgh, PA (1990)
26. Yarbrough, W. A. and Roy, R., in *Diamond and Diamond-Like Materials Synthesis,* (G. Johnson, A. Badzian and M. Geis, eds.) Mater. Res. Soc. Proc., Vol. EA-15, p.33, Pittsburgh, PA (1988)
27. Hoover, D. S., "Scale-Up and Commercial Applications of CVD Diamond and Diamond-like Materials," paper presented at GAMI Conference on Diamond and DLC Coatings, Marco Island, FL (Oct. 15-17, 1989)
28. Woods, M., *The World and I,* 8:162 (1987)
29. Matsumoto, S., in *Diamond and Diamond-like Films,* (J. Dismukes, et al., eds.), Electrochem. Soc. Proc., Vol. PV 89-12, p. 50, Pennington, NJ (1989)

30. Decker, R., Fournier, E., Asmussen, J., Bigelow, K. and Hoggins, J., "Microwave Processing for Diamond Films," in press (1990)

31. Bachmann, P. K. and Messier, R., *Chemical & Engineering News,* 67(20):24 (May 15, 1989)

32. Liou, Y., Inspektor, A., Knight, D., Weimer, R., Pickrell, D., Badzian, A. R. and Messier, R., in *Diamond Optics II,* (A. Feldman and S. Holly, eds.) SPIE Proc., 1146:12, Bellingham, WA (1990)

33. Kawarada, H., Mar, K. S. and Hiraki, A., *Jap. J. Appl. Phys.,* 26(6):L1032 (1987)

34. Wei, J., Kawarada, H., Suzuki, J., Yanagihara, K., Numata, K. and Hiraki, A., in *Diamond and Diamond-like Films,* (J. Dismukes, et al., eds.), Electrochem. Soc. Proc., Vol. PV 89-12, p. 393, Pennington, NJ (1989)

35. Chang, C. P., Flamm, D. L., Ibbotson, D. E. and Mucha, J. A., *J. Appl. Phys.,* 63(5):1744 (1988)

36. Liou, Y., Weimer, R., Knight, D. and Messier, R., *Appl. Phys. Lett.,* 56(5):437 (1990)

37. Mucha, J. A., Flamm, D. L. and Ibbotson, D. E., *J. Appl. Phys.,* 65(9):3448 (1989)

38. Wood, P., Wydeven, T. and Tsuji, O., in *Program and Abstracts of the First International Conference on the New Diamond Science and Technology,* p. 100, New Diamond Forum, Tokyo, Japan (1988)

39. Meyer, D. E., Kustka, T. B., Dillon, R. O. and Woollam, J. A. in *Diamond Optics II,* (A. Feldman and S. Holly, eds.) SPIE Proc., 1146:21, Bellingham, WA (1990)

40. Roman, W. C. and Colket, M. B., in *Diamond and Diamond-like Films,* (J. Dismukes, et al., eds.), Electrochem. Soc. Proc., Vol. PV 89-12, p. 330, Pennington, NJ (1989)

41. Matsumoto, S., in *Diamond and Diamond-Like Materials Synthesis,* (G. Johnson, A. Badzian and M. Geis, eds.) Mater. Res. Soc. Proc., Vol. EA-15, p. 119, Pittsburgh, PA (1988)

42. König, H., and Helwig, G., *Zeitschrift für Physik,* 129:491 (1951)

43. Gardos, M. N. and Ravi, K. V., in *Diamond and Diamond-like Films,* (J. Dismukes, et al., eds.), Electrochem. Soc. Proc., Vol. PV 89-12, p. 475, Pennington, NJ (1989)

44. Singh, B., Mesker, O. R., Levine, A. W. and Arie, Y., *Appl. Phys. Lett.,* 52(20):1658 (1988)

45. Suzuki, K., Sawabe, A. and Inuzuka, T., *Appl. Phys. Lett.,* 53(19):1818 (1988); Sawabe, A., Yasuda, H., Inuzuka, T. and Suzuki, K., *Appl. Surf. Sci.,* 33/34:539 (1988); Suzuki, K., Sawabe, A. and Inuzuka, T., *Jap. J. Appl. Phys.,* 29(1):153 (1990)

46. Akatsuka, F., Hirose, Y. and Komaki, K., *Jap. J. Appl. Phys.,* 27[9]:L1600 (1988)

47. Kurihara, K., Sasaki, K., Kawarada, M. and Koshino, N., *Appl. Phys. Lett.* 52(6):437 (1988); Kawarada, M., Kurihara, K., Sasaki, K., Teshima, A. and Koshino, K., in *Diamond Optics II,* (A. Feldman and S. Holly, eds.) SPIE Proc., 1146:28, Bellingham, WA (1990)

48. Fujimori, N., Ikegaya, A., Imai, T., Fukushima, K. and Ota, N., in *Diamond and Diamond-like Films,* (J. Dismukes, et al., eds.), Electrochem. Soc. Proc., Vol. PV 89-12, p. 465, Pennington, NJ (1989)

49. Altshuler, A., *Diamond Technology in the USSR (From Synthesis to Applications),* Delphic Associates, Falls Church, VA (1990)

50. Deshpandey, C., Jou, S., Radhakrishna, M. C., Doerr, H. J. and Bunshah, R. F., in *Diamond and Diamond-like Films,* (J. Dismukes, et al., eds.), Electrochem. Soc. Proc., Vol. PV 89-12, p. 559, Pennington, NJ (1989)

51. Deshpandey, C. V. and Bunshah, R. F., *J. Vac. Sci. Tech.* A7(3):2294 (1989)

52. Bunshah, R. F., "Low Temperature Deposition of Thin Transparent Diamond Films," paper presented at GAMI Conference on Diamond and DLC Coatings, Marco Island, FL (Oct. 15-17, 1989)

53. Hirose, Y. and Kondo, N., *Program and Book of Abstracts,* p.434, Japan Applied Physics 1988 Spring meeting (March 29, 1988)

54. Hanssen, L. M., Carrington, W. A., Butler, J. E. and Snail, K. A., *Materials Letters,* 7(7,8):289 (1988); Yarbrough, W. A., Stewart, M. A. and Cooper, J. A., *Surf. and Coat. Tech.,* 39/40:241 (1989)

55. Hirose, Y., Amanuma, S., Okada, N. and Komaki, K., in *Diamond and Diamond-like Films,* (J. Dismukes, et al., eds.), Electrochem. Soc. Proc., Vol. PV 89-12, p. 80, Pennington, NJ (1989)

56. Murakawa, M., Takeuchi, S. and Hirose, Y., *Surf. and Coat. Tech.,* 39/40:235 (1989); Matsui, Y., Yuuki, A., Sahara, M. and Hirose, Y., *Jap. J. Appl. Phys.,* 28(9):1718 (1989)

57. Cappelli, M. A. and Paul, P. H., *J. Appl. Phys.,* 67(5):2596 (1990)
58. Murakawa, M., Takeuchi, S. and Hirose, Y., in *Program and Abstracts of the First International Conference on the New Diamond Science and Technology,* p. 208, New Diamond Forum, Tokyo, Japan (1988)
59. Butler, J. E., Naval Research Laboratory, personal communication, (1990)
60. Frenklach, M., Kematick, R., Huang, D., Howard, W., Spear, K. E., Phelps, A. W. and Koba, R., *J. Appl. Phys.,* 66(1):395 (1989)
61. Fujimori, N., Imai, T., Nakahata, H., Shiomi, H. and Nishibayashi, Y., in *Diamond, Boron Nitride, Silicon Carbide and Related Wide Bandgap Semiconductors,* (J. T. Glass, R. Messier and N. Fujimori, eds.) Mater. Res. Soc. Proc., Vol. 162, p. 23, Pittsburgh, PA (1990)
62. Yoder, M., in *Technology Update on Diamond Films,* MRS Extended Abstracts EA-19, p.35, Materials Research Society, Pittsburgh, PA, (1989)
63. Obata, T. and Morimoto, S., in *Diamond Optics II,* (A. Feldman and S. Holly, eds.) SPIE Proc., 1146:208, Bellingham, WA (1990); also refer to the Proceedings' Introduction, p. vi
64. Patterson, D. E., Hauge, R. H. and Margrave, J. L., in *Diamond, Boron Nitride, Silicon Carbide and Related Wide Bandgap Semiconductors,* (J. T. Glass, R. Messier and N. Fujimori, eds.) Mater. Res. Soc. Proc., Vol. 162, p.167, Pittsburgh, PA (1990)
65. Cuomo, J. J., paper presented at *Diamond Optics III,* SPIE Meeting, San Diego, CA (July 11, 1990)
66. Heath, P. J., in *Metals Handbook, Ninth Edition,* 16:105 ASM International, Metals Park, OH (1989)

5

Diamond as a Wear-Resistant Coating

Benno Lux and Roland Haubner

1.0 INTRODUCTION

Based on surprisingly simple procedures, numerous different experimental synthesis methods for producing low-pressure diamond films exist today. Various patents, many publications and a great number of international conferences bear witness to the interest these techniques have aroused (1)-(4).

The properties of diamond range from extremes in hardness and toughness to extremes in thermal conductivity and chemical inertness. Diamond is widely used as a material for first-rate tools (5). Until recently, diamond was known and available exclusively as more or less bulky crystals; however the new techniques of low-pressure synthesis now offer the novel possibilities of easily forming relatively extensive two-dimensional thin films or sheets of this material.

These thin-film and coating technologies now employed to produce diamond can lead to important improvements of industrial applications in many fields (6)-(8). If integrated into the function of a system, the lifetime and technical performance of the resulting components, such as tools, can be significantly increased. For many applications, the benefits obtained through thin films far exceed their manufacturing costs. Even if these costs are high, the resulting increase in performance of the components for their applications are justified by the highly competitive improvements in the wear-resistance of materials. Also, novel diamond products, impossible to fabricate until now, can be expected from this new technology in the near future (8).

We cannot discuss all the potential wear product uses of low-pressure diamond in this chapter. A few selected, typical products will be used to illustrate the principles and some potential possibilities for this new technology in the vast area of wear-resistant materials.

2.0 DEPOSITION AND APPPLICATIONS OF WEAR-RESISTANT MATERIALS

2.1 Improvement of Materials by Hard CVD and PVD Coatings

Selected from the many different applications, examples of improvements for metal-working tools by chemical vapor deposition (CVD) or physical vapor deposition (PVD) coatings are shown in Table 1 (7). Several of them are used today primarily as coated tools, e.g., the inserts for steel turning.

The Substrates. Substrates currently used for surface-modified metal-working tools are based mainly on cemented carbides, on high-speed or certain heat-resistant steel grades, or on ceramic materials (9) (Table 1).

Particularly the cemented carbides provide a number of basic properties essential for the successful application of CVD coatings. Without major structural changes, they withstand the many hours at high temperatures necessary during the coating procedures. They are dimensionally stable during and after this heat treatment and need no re-grinding or special treatment to equalize internal stresses. Further, during the working life of the tool they meet the requirements for properties such as hot hardness (10), elastic modulus (10), toughness, and high thermal conductivity (11). Although they cannot be as easily worked as steel, they can still be shaped reasonably well to form complex tool geometries with the necessary precision. Grinding and polishing operations are technically and economically feasible.

In contrast to cemented carbides, most steel substrates have serious limitations in some of the essential and desirable properties listed above. The severe heat treatment occurring during the deposition process leads to structural changes and detrimental loss of certain properties such as hardness and strength. Furthermore, it leads to internal stresses and dimensional deformations. Since high-speed and heat-resistant highly alloyed steel grades fit these needs best, they are most frequently used (12)(13). There are also chemical limitations with many substrates due to insufficient corrosion resistance in the aggressive chemical environment during the coating process. For these reasons often PVD coatings or low-temperature plasma-assisted CVD (PACVD) must be applied to steel substrates.

APPLICATION	SUBSTRATES	COATINGS and METHODS	
I. Chip-forming tools for			
Turning	WC/Co	TiC, Al_2O_3, HfN	CVD
Milling	WC/Co	TiC, TiN, Ti(C,N)	CVD / PVD
Drilling	HSS, WC/Co	TiN, TiC	PVD / CVD
Filing	C-steel	Cr_xC_y	pack cementation
Sawing	C-steel	Cr_xC_y	surface hardening by heat treatment. (Today mainly case and surface hardening by heat treatment, more recently also CVD and PVD)
II. Dies for non-chipping tools for			
Wire drawing	C-steels		pack cementation
Extrusion	alloyed		surface hardening
Deep drawing	steels		
III. Sharp-edge applications for			
Cutting	HSS		PVD
Punching	C-steels		surface hardening

Table 1. Examples of Hard and Wear-Resistant Coatings for Material-Working Tools (7).

Substrate/Coating Interface Interactions during Production and Use. The coating has to be chosen carefully for each application. Since it is always used together with a substrate, the basic properties of the substrate are obviously also of great importance and must be suitable for the specific application. Thus a wide range of interactions must be taken into consideration in a coated material system. Figure 1 illustrates this schematically for a metal-working tool (14).

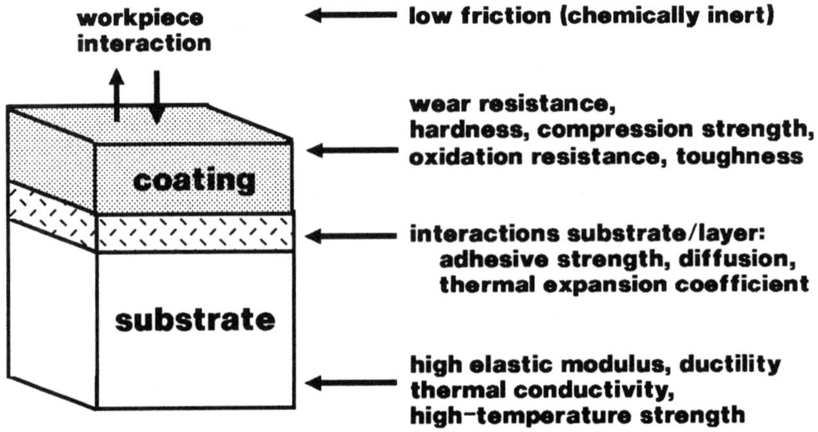

Figure 1. Interface interactions between a metal-working tool and the enviroment/coating/substrate (14).

Trace elements originating from the substrates may considerably disturb the regular formation and adhesion of the CVD layers. For example, direct deposition of an Al_2O_3 coating onto cemented carbide (WC/Co) is difficult as the Co binder promotes Al_2O_3 whisker formation (15).

The working environment/coating interaction is responsible for the lifetime and efficiency of the coated tool itself, which depends mainly on the chemical properties of the coating material (14) such as hardness and wear resistance. These, in turn, are linked with properties such as chemical bonding of the applied coating material and its mechano-chemical reactivity (friction, galling) with the workpiece at high pressures and high temperatures (16).

Among the coating types currently popular for industrial applications in the coated tool business we can distinguish:

(a) single-phase layers which are applied as one layer consisting of a single phase compound such as Al_2O_3, TiC, TiN, HfN, Ti(C,N), and (Ti,Zr)C (14)-(19).

(b) multilayers, consisting of two or more different single layers deposited sequentially. Such *composite coatings* on a tool show improved performance as compared to the respective single layer. Multilayers are predominantly used for most industrial cutting tool applications (17)(20)-(23).

Comparison between CVD and PVD. Physical vapor deposition (PVD) typically needs pressure ranges below one torr, as it uses ions, electrons and/or molecular beams. Chemical vapor deposition (CVD) with reasonable growth rates works best above this pressure range. Since CVD is based on chemical reactions, its deposition rates increase with pressure, temperature and the concentrations of the gaseous reactants. Although industrial CVD is frequently operated at atmospheric pressure, for special purposes CVD can also be applied at rather low pressures (24).

Gas compositions can be similar for PVD and CVD. The addition of neutral gases, such as argon, is also possible for both methods, but is less frequently applied for CVD.

The essential difference between PVD and CVD lies in the gas activation. In the case of CVD, the gases are activated by temperature only. Thermodynamics and kinetics determine the deposition. In PVD the reactants come from a heated source or from a bombarded target. The atoms and molecular beams can, in addition, be activated by a plasma and can be accelerated by bias voltages. High growth rates need a high mass flow between target/source and the substrate. Gas activation by the plasma permits formation of compounds at lower temperatures than with CVD. Typical differences in the experimental deposition parameters are shown in Table 2 (20).

The technological aspects of the coating methods depend strongly on temperature and chemical limitations of the substrates as well as on the economy of the particular method. For many substrates, the temperatures applied during the CVD process are too high and PVD or PACVD methods must be used. Often specific final surface qualities are required such as:

— controlled grain sizes in the layer
— well-defined surface roughness
— polishing possibilities

Industrial Examples of CVD and PVD Coatings: Most hard materials used today for CVD coatings are covalently or metallically-bonded intermetallic compounds such as carbides, nitrides, borides [e.g., TiC (9), SiC (25), TaC (26), TiN (9), HfN (27), Si_3N_4 (28)] or ionicly bonded oxides such as Al_2O_3 (17)(28), SiAlON (29), AlON (30) and mixtures thereof. As mentioned above, in many coating applications only one polycrystalline layer (TiC, Ti(C,N), TiN,...) is applied.

	PVD		CVD
A. DEPOSITION PARAMETERS			
pressure	lower	1 torr	higher
temperature of the substrate	lower	500°C	higher
deposition rate	high		low
layer crystal structure	amorphous to fine		fine to facetted
adhesion	good		excellent
porosity/cracks	less		sometimes
B. PRODUCTION PARAMETERS			
size of charges	small		large
throwing power	low		high
equipment costs per unit	high		low
costs per piece	high		low

Table 2. Comparison between PVD and CVD for cutting tool coatings (20).

The CVD growth process leads normally to rather coarse, columnar structures while PVD deposits are more finely grained (30)-(32). For fine-grained CVD coatings two or more layers are frequently used today (28)(33)(34). Compared to single-layer coatings, in such multilayer coatings each individual layer has a much finer structure since every layer requires new nucleation (20).

A general disadvantage of thick coatings is the "rounding" of sharp edges (35).

Wear-resistant coatings are usually produced by PVD when a low substrate temperature is required, as for example for coating steel-based drills, as well as decorative coatings used for fashion jewelry, watch cases, pens, lighters, etc. (36)(37).

For example, typical materials for producing watch cases are die-cast zinc, brass, stainless steels, hard metals, ceramics and even polymers. Simple, precise and economic fabrication techniques for the substrate material, such as drop forging, automatic machining or injection casting, etc., are applied. Required surface conditions before and after the coating application are obtained by grinding, polishing, lapping, sand blasting, tumbling, etc. As the surface must be corrosion- and scratch-resistant a broad range of different coatings are necessary (36).

Here are some typical examples:

(a) Obtaining a natural artificial gold color on a watch case demands a well-defined surface roughness of the TiN layers. Imitation of the various gold tones also depends strongly on impurities such as O and C in the TiN (38).

(b) The best gold colors are obtained by covering objects with an extremely thin layer of real gold in addition to the TiN. Since the TiN layer is very hard, scratches usually do not penetrate it and damage in the gold layer remains so small that the yellow TiN underground makes it invisible to the naked eye.

(c) Intense black and grey colors are produced by a-C:H films containing 10 - 20% TiC (39). These layers are obtained by a sputter-CVD process at gas pressures of about 10^{-2} torr. The PVD of Ti(O,C,N) results in gray layers with different intensities (36).

2.2 PCD and PCBN, The Superhard High-Pressure Products

Polycrystalline diamond (PCD) and *polycrystalline cubic boron nitride* (PCBN) products currently have a well-established market (40)(41). Their properties and performances under severe working conditions are generally considered to be outstanding and highly competitive in comparison to most ordinary wear-resistant materials. They are also the ultimate standard to be reached or—if possible—even to be exceeded by the new low-pressure diamond products now being developed.

Both PCD and PCBN (41) are sintered products, prepared from diamond or superhard BN powder mixed with appropriate amounts of a binder, and then compacted at high temperatures and ultrahigh pressures which are within the thermodynamic stability range of the respective superhard phases. Frequently the PCD or PCBN layers are bonded directly to a cemented carbide substrate of appropriate thickness during the sintering operation (Fig. 2) (41).

In terms of applications, PCD tools are used primarily in cutting, chipless forming, and wear part applications of abrasive workpiece materials. For example:

— wood-based products (41)
— non-ferrous alloys: Al-Si (42), Cu,
— green ceramic and cemented carbides
— abrasive reinforced plastics (43-51), man-made resin-bonded wood particle boards, plastic-laminated chip boards
— stone and rock drills (52)-(54)

— wire-drawing dies (55)(56)
— other general applications of wear-resistant surface layers (57)-(59).

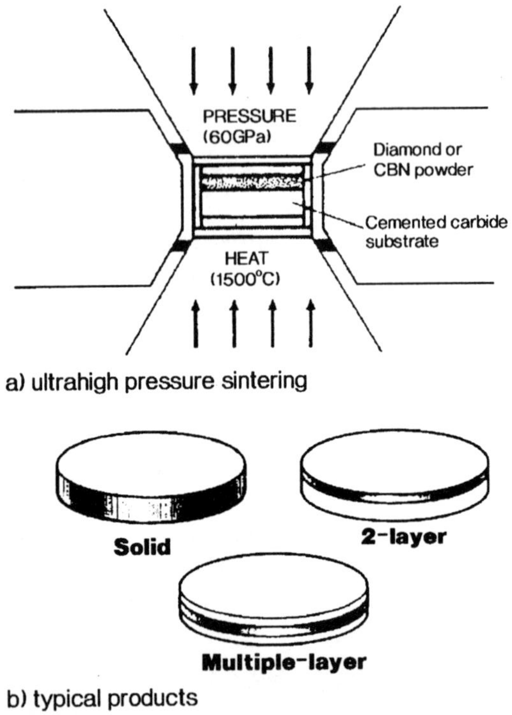

Figure 2. Polycrystalline diamond or CBN compacts (41).

Figure 3 gives a survey of cutting parameters for some typical uses (40)(60).

In contrast, PCBN tools (41)(61)-(63) are preferably used for machining (turning, milling, boring, honing, etc.) and chipless forming of ferrous metals, such as Ni-based superalloys (Inconel, Nimonic), gray and hard cast irons, hardened ferrous steels (> 45 Rc), and special hard facing materials.

Although simple three-dimensional shapes could be produced, the extreme pressures and temperatures required have restricted the common PCD and PCBN products to flat disc formats. These discs are in the form of a solid blank, a monolayer of ultrahard material supported and bonded to a solid cemented carbide, or a multiple-layered blank. Figure 4 shows the increase in the maximum producible size of these products over the last twenty years. From these large blanks, smaller sizes can be cut for subsequent mounting onto a tool holder or shank before grinding the final cutting edge geometry required for the finished tool.

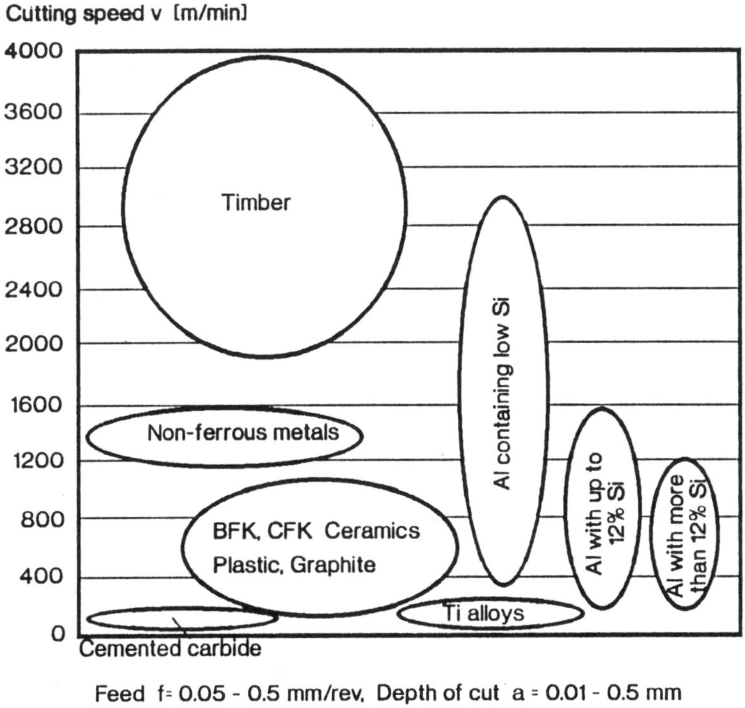

Figure 3. Survey of cutting parameters for some typical uses of PCD tools (40).

Tools frequently need complex, three-dimensional shapes so that, for many applications, the flat blanks are not applicable as such. Special electrically conducting PCD grades have had to be developed in order to manufacture such complex tool shapes by electro-erosion wire cutting (41). Such items are, however, difficult and expensive to produce.

Different workpiece materials also need different grades of ultrahard tools. Diamond particle sizes ranging from 2 to 25 microns and different superhard material/binder ratios are applied (41). A higher cobalt binder content, for instance, increases the edge quality of the grade so that it is less susceptible to edge chipping during tool impacts.

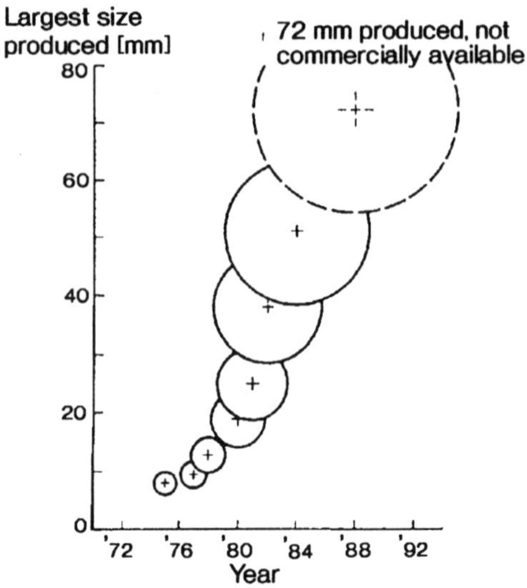

Figure 4. Increase in the maximum producible size of PCD products and the over the last 20 years (41).

Serious restrictions for certain wear applications exist:

- — The ultrahigh-pressure technology leads to relatively expensive products.
- — Fabrication of 3-D shapes is difficult and generally linked with time-consuming, complex, high-cost operations, as can be deduced from the need to use electro-erosion machining.
- — Due to the high-pressure technology applied, the products have definite upper size and shape limits, which cannot be exceeded without prohibitive cost increases.

The new low-pressure diamond technology should help to overcome at least some of the above limitations of the ultrahigh-pressure products.

2.3 Low-Pressure Diamond Layers as a New Challenge

Properties Requirements of Low-pressure Diamond for Wear Applications. Superhard, wear-resistant protective coatings for tools and anti-wear parts as well as decorative, scratch- or corrosion-resistant surfaces need and utilize mainly the following outstanding properties of diamond:

— superhardness (high Young's modulus, rigidity, etc.)
— high degree of chemical inertness (anti-sticking surface, etc.)
— high thermal conductivity

These properties are relatively insensitive to lattice defects. Most applications tolerate polycrystallinity and even a certain amount of sp^2 bonds (64) without major detrimental effects on the anti-wear performances of the diamond. This greatly simplifies the development of industrial wear applications as compared to other uses (65).

Tomlinson and Wedlake (66) consider that a binderless PCD product with the irregular fracture surface of a ballas-type material would be ideal for grinding and polishing operations. The natural ballas-type diamond crystals are frequently spherolites and free from foreign inclusions (67)(68).

High quality polycrystalline low-pressure diamonds are faceted, pure and binderless as are natural ballas diamonds. In artificial low-pressure products, however, the surface crystal facets, the crystal orientation, their internal defect structures, sp^2/sp^3 bond ratios, grain sizes, etc., can be controlled within certain limits by the synthesis parameters. Furthermore, these diamonds are available not only as bulky crystals but also in large two-dimensional shapes of defined thicknesses.

If used as an anti-wear surface layer and if not loaded to fracture, low-pressure diamond also permits the full utilization of diamond's extreme hardness, high modulus, thermal conductivity and other properties. Diamond's high cleavage tendency must be controlled by selection of appropriate structural parameters, mainly the grain size.

However, if for specific applications the characteristics of a surface must be extended beyond the natural properties of the pure diamond, a different set of properties can be created by a *multilayer* approach. It permits the development of tailor-made surfaces, with special properties better suited for meeting the challenge of a specific application. This technique proved to be successful and is extensively used today to improve in situ coated conventional CVD cutting tools (13)(69). During the in situ CVD coating of tools, thin low-pressure diamond films and conventional hard materials, such as TiC, TiN, Al_2O_3, can be combined to form a multilayer coating (70)(71). The multilayer approach is not limited to the in situ fabrication. It can also be applied to improve flat or curved freestanding diamond sheets (72)(73). Novel *multilayer sheet products* analogous to the in situ coated products are within reach.

Low-Pressure Techniques. Two different techniques exist today for applying low-pressure diamond as an anti-wear surface layer. Typical practical examples are shown in Fig. 5 (74).

(a) In situ CVD coating (Fig. 6) (75)(76) is the analogy to the classical CVD process described in Sec. 2.2. The deposition temperatures are below 1100°C, and the coating of 3-D shapes and large surface areas should be possible with low-pressure methods at relatively low costs.

(b) Bonding of freestanding diamond sheets (Fig. 7) (75)(76) is quite similar to the process applied in the classical PCD tool fabrication (see Sec. 2.4), and is an interesting alternative to the direct in situ coating. A diamond sheet is fabricated separately and then bonded to the substrate.

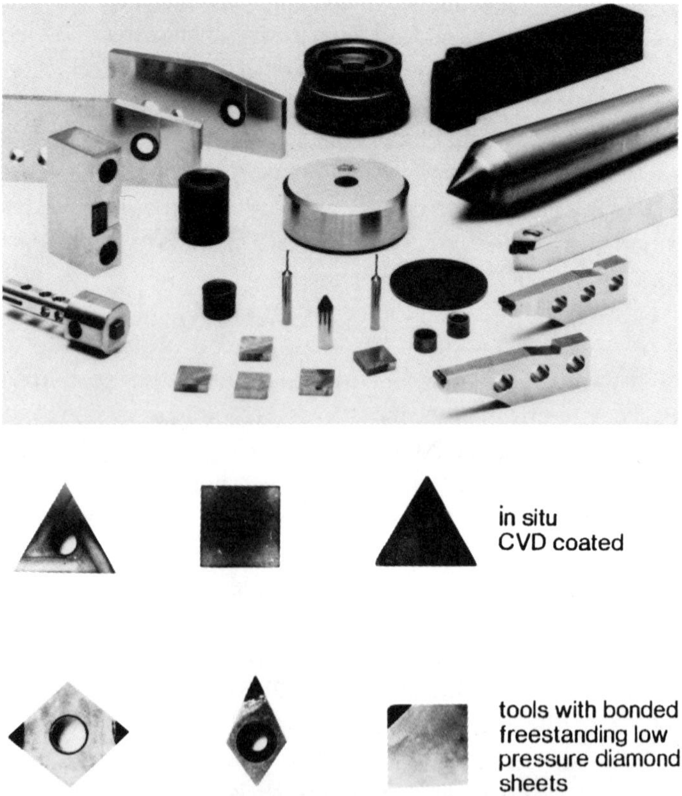

Figure 5. Typical examples of CVD diamond coatings and bonded, free-standing layered products. *(Courtesy of Asahi Diamond Industrial Co.)* (74).

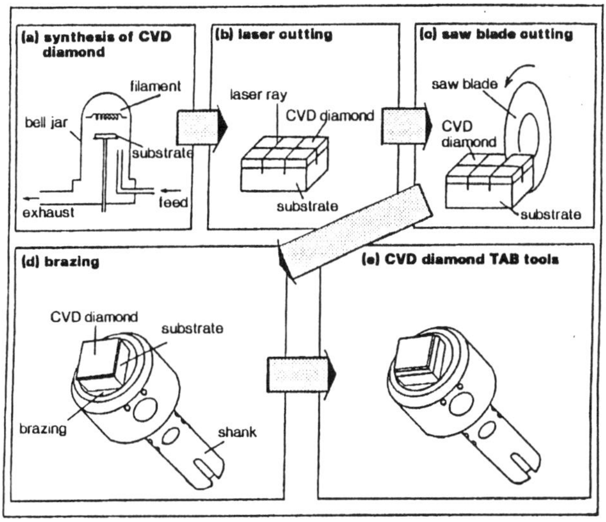

Figure 6. Production process of an in situ CVD diamond coated TAB tool (76).

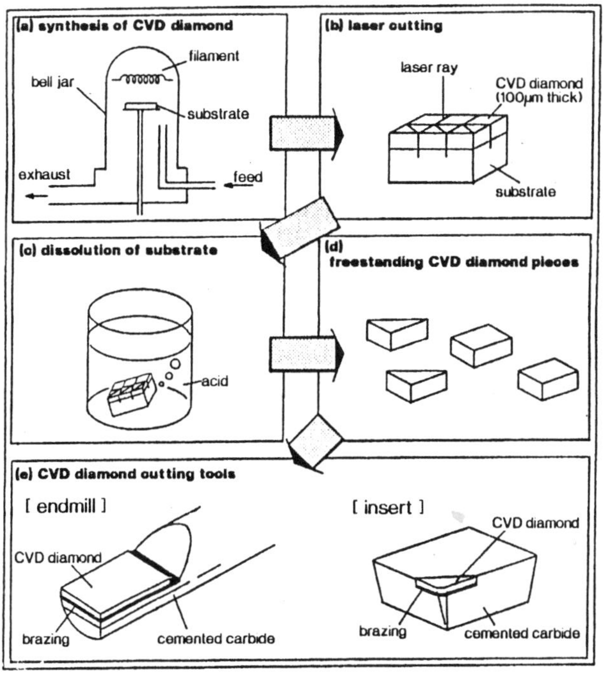

Figure 7. Production process of freestanding CVD diamond sheets brazed to cutting tools (76).

This process is carried out in three steps. *(a)* A relatively thick diamond layer is deposited by a low-pressure method on a dummy substrate. Figure 8 shows a small, laser-cut, freestanding diamond sheet after removal from its dummy substrate (77). Such flat sheets are currently produced in sizes of several hundred square centimeters. Thus, at reasonable costs, much larger sizes are feasible as compared to the conventional high-pressure technique. *(b)* The sheet is bonded to a tool surface as illustrated in Fig. 9. This is done after the tool has been shaped close to its final geometry (74). *(c)* The layer is ground to its final tolerances. Figure 9 shows the details of the microstructure of the bond to a cutting tool insert. Good bonding can be achieved. Such tools are already commercially available (see Sec. 3.1.) (74).

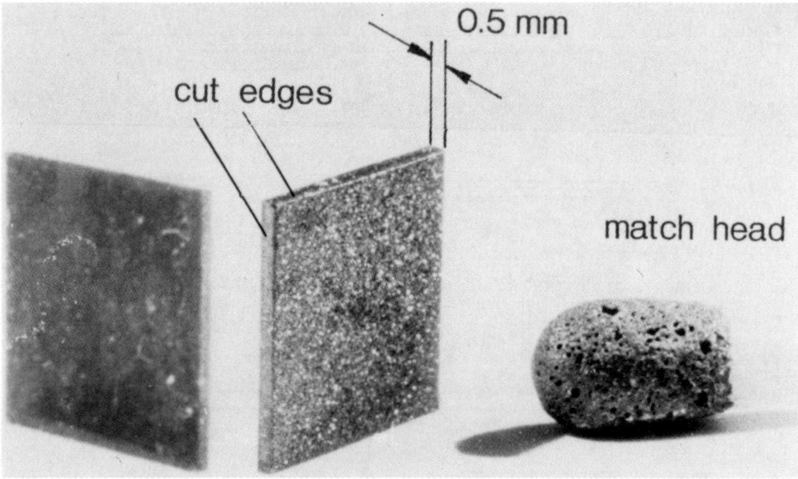

Figure 8. Laser-cut, freestanding diamond sheet after removal from its dummy substrate (77).

Compared to high-pressure techniques, the low-pressure approach offers more flexibility, allowing good performance while remaining a simpler and less expensive technology, appropriate for novel lower-priced products intended for mass production. New applications and broader marketing possibilities should result from it, and in many cases, the lower cost and the better performance of the low-pressure technology as compared to the ultrahigh-pressure technology will be key factors.

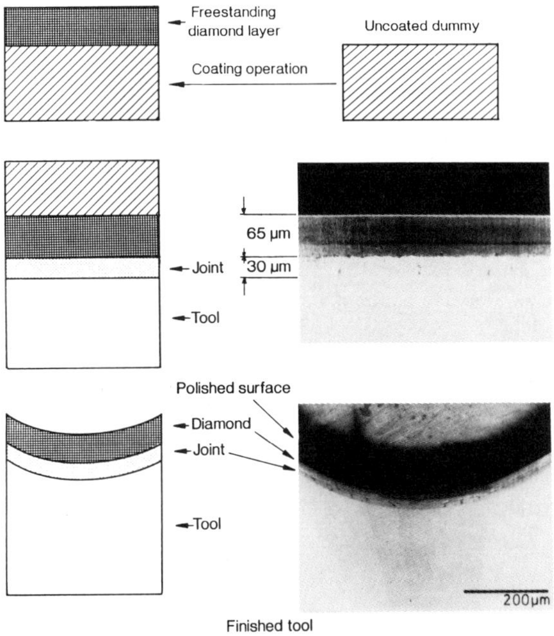

Figure 9. Bonding of the freestanding diamond sheet on a tool surface. *(Courtesy of Asahi Diamond Industrial Co.)* (74).

3.0 POTENTIAL PRODUCTS AND INDUSTRIAL APPLICATIONS

Based on known principles, research and development efforts currently being undertaken by a number of important companies have already led to new products.

3.1 In Situ CVD Diamond Deposition (CVD Coating Analogy)

Coating of PCD Tools. Coating PCD tools with thin diamond films can significantly improve their cutting performance and lifetime, as shown by Otake et al. (Fig. 10) (78)-(80). Under comparable cutting conditions, normal, uncoated PCD tools suffered from severe flank wear, while the tools covered with a low-pressure diamond coating did not wear at all at either the flank face or the rake face. Such improvements can be explained by the covering of the soft binder phase at the tool surface with the hard diamond coating (see Sec. 2.0).

Figure 10. Low-pressure diamond coated PCD tool (fracture and surface appearance) (78).

Microdrills. Microdrills made of cemented carbide were among the early examples of CVD diamond coatings applied to a tool (Fig. 11) (81). This example of a tiny precision tool also shows that the roughness of the "as-grown" faceted diamond surface can be a problem for its application (Fig. 12) (78). Grinding and polishing prior to its use may be necessary.

Twist-drills. Twist-drills with diamond coatings can be used to perforate difficult materials with which are normally associated with high drill wear, such as non-ferrous alloys (Al-Si), green ceramics, cemented carbides and reinforced plastics. Figure 13 shows the cutting edges of cemented carbide twist-drills before and after testing as well as the comparison between uncoated and diamond-coated inserts (60).

TAB (Tape Automated Bonding) Stamps TAB's are used in microelectronics to connect chips to their wires and circuits. In this application the high thermal conductivity of the diamond as well as the high wear resistivity are utilized (Fig. 14a,b). Recently, there has been a large increase in the demand for flat packaging (IC cards, electric calculators, liquid-crystal displays, etc.). To increase the through-put of these products and to decrease their costs during production, the TAB process is now gradually being accepted by many electronic goods suppliers. Figure 14c (75)(76)(82) shows the schematic drawing of continuous surface mounting of IC's using the TAB process. During the TAB process, the tools are heated to a temperature between 500°C and 600°C and pressed down by a load.

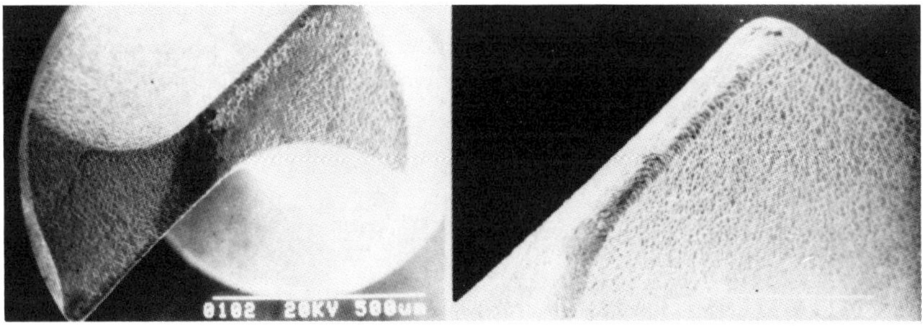

Figure 11. CVD diamond coated microdrill. *(Courtesy of Asahi Diamond Industrial Co.)* (74).

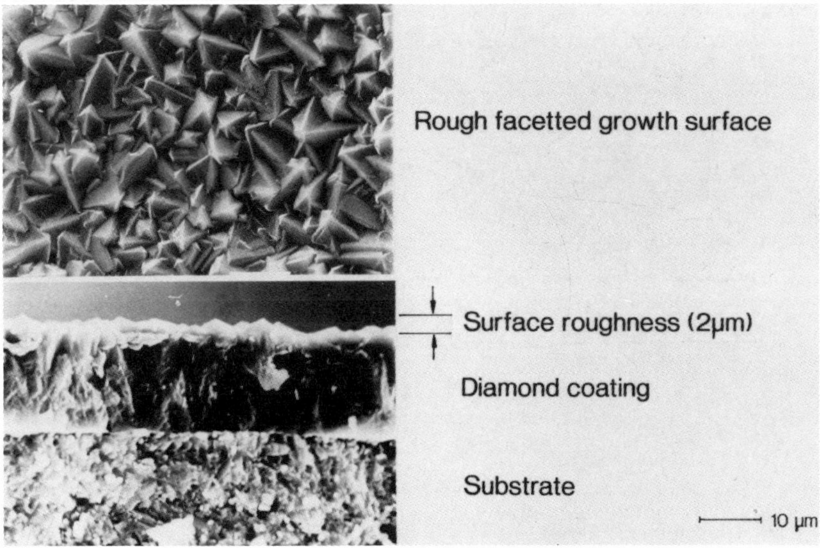

Figure 12. Roughness of an as-grown CVD diamond coating (78).

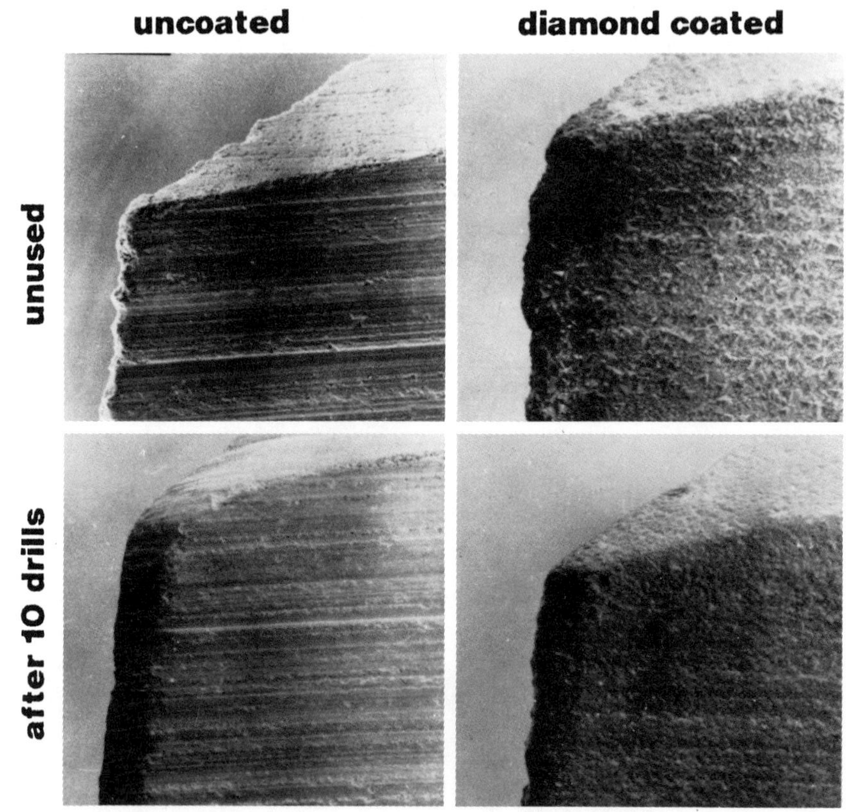

Figure 13. Cutting edges of cemented carbide twist-drills before and after testing and the comparison between uncoated and diamond coated surface of the drill (60).

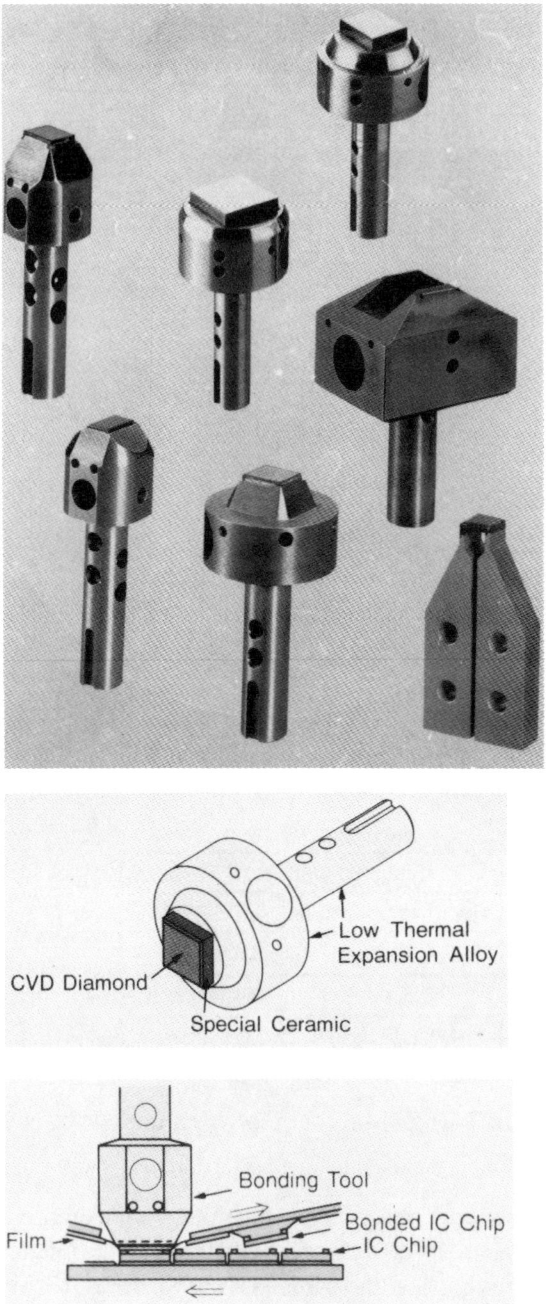

Figure 14. Freestanding diamond layers used for TAB bonding in microelectronics to connect the chips and circuits (82).

The materials for these new low-pressure diamond TAB tools are special ceramics coated with a 50 μm thick hot-filament CVD diamond layer. The CVD diamond coating is first cut into sections by a laser (Fig. 6). Then the ceramic is cut along the laser grooves by a diamond circular saw blade. The special ceramic is brazed on a shank body of low thermal expansion alloy to form the final TAB bonding stamp. The diamond surface is polished by a lap grinding wheel until its surface roughness reaches 0.1 μm or less. Finally the edges of the tool head are chamfered.

Two kinds of tests to estimate the flatness and wear resistance of the CVD diamond were carried out. Marketed PCD qualities (diamond-Si-SiC compound with 85 vol% diamond and diamond-Co composite with 90 vol% diamond) and a synthetic high-pressure diamond single crystal (sumicrystal) were compared with the CVD diamond. In comparative performance tests, in situ CVD-coated TAB tools exhibited surface roughness and surface flatness equivalent to those of high-pressure diamond single crystals.

Flatness. In a preliminary test, the surface flatness is measured after polishing at room temperature and compared with the flatness at 550°C. The CVD diamond TAB tool and the single-crystalline diamond TAB tool showed excellent flatness compared with that of the PCD TAB tool as shown in Table 3. To reach the same performance as that of the CVD diamond tools, the two PCD TAB tools had to be reground three times (Table 4).

	Flatness (μm/7mm)	
Tool Materials	Room Temperature	550°C
CVD Diamond	0.6 (Convex)	0.5 (Convex)
Sumicrystal	0.2 (Convex)	0.4 (Convex)
PCD1	1.0 (Concave)	5.0 (Convex)

Table 3. Surface flatness of TAB tools at room temperature and at 550°C (75,76).

Wear Resistance. The wear resistance of the three surface materials used was examined via abrasion using a cleaning whetstone. The influence of cleaning time on surface roughness and flatness is shown in Fig. 15. From these results, it is clear that the cleaning damage to CVD diamond is less than that of PCD, and that the CVD diamond TAB tool showed almost the same resistance as the single crystalline diamond tools (75)(76). Besides the economic advantage, the low-pressure diamond layers further offer the possibility of producing much larger tools than can be done with single-crystalline high-pressure diamond.

Tool materials	CVD diamond	Sumicrystal	PCD1	PCD2
Tool life (x10^{3}shots)	600	2,000	200	80
Max regrinds	---	3	3	3
Total tool life (x10^{3}shots)	600	8,000	800	320
Tool price (CVD unit base)	1	2.5	1	1
Regrinding price (PCD unit base)	---	2.5	1	1
Bonding cost (CVD unit base)	1	1	3.2	2.9

Table 4. Results of tool life tests and bonding cost analysis (75)(76).

conditions

tool size	8x9 mm
temperature	540°C
load	3 kg
movement speed	20 mm/sec
whetstone	WA(Al_2O_3)

movement pattern

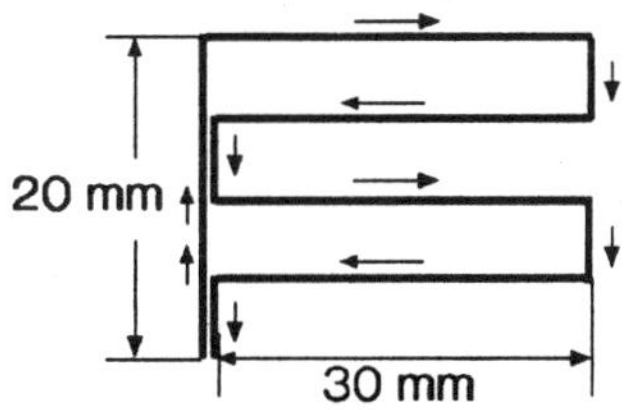

results

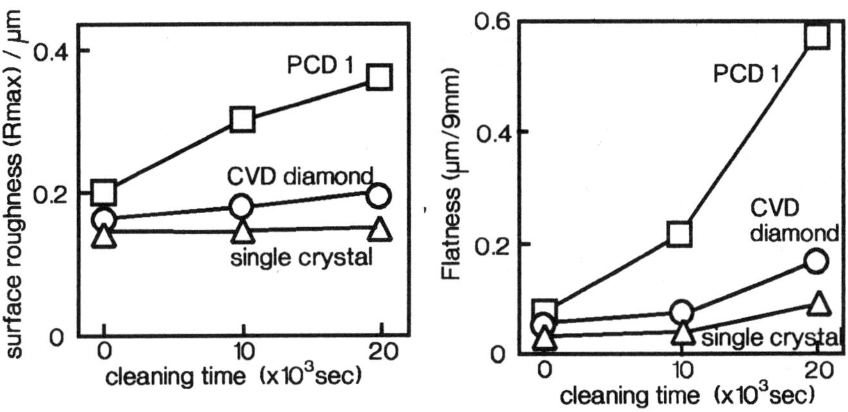

Figure 15. Influence of cleaning time of a TAB tool on surface roughness and flatness (75)(76).

Tool Cost Assessment. In terms of the cost per bond, including regrindings, the costs of both the CVD diamond TAB tools and the single-crystalline diamond TAB tools are considerably lower than those of the PCD TAB tools. The single crystalline diamond TAB tools, which have the longest tool life, have as many as 8,000,000 individual bonds. This long lifetime makes the high initial costs acceptable. Such tools are most suitable for mass production of products that may vary only slightly in design. The CVD diamond TAB tools and the PCD TAB tools show almost the same lifetimes of several hundred thousand shots. The initial costs are also similar. They are best suited for medium-variety and medium-quantity production or large-variety and small-quantity production (75)(76).

Metal-cutting Tool Inserts. Tool inserts for turning and milling of hypereutectic Al-Si alloys are also now a preferred application examined recently by a number of research groups (8)(75)(83)-(94). Figure 16 shows the typical features of a coated and a worn-out uncoated cutting edge as well as a comparison of the surface quality obtained on the Al-Si workpiece (89). (See the summary of cutting parameters, performances, tool lifetimes, etc., given in Sec. 3.3.)

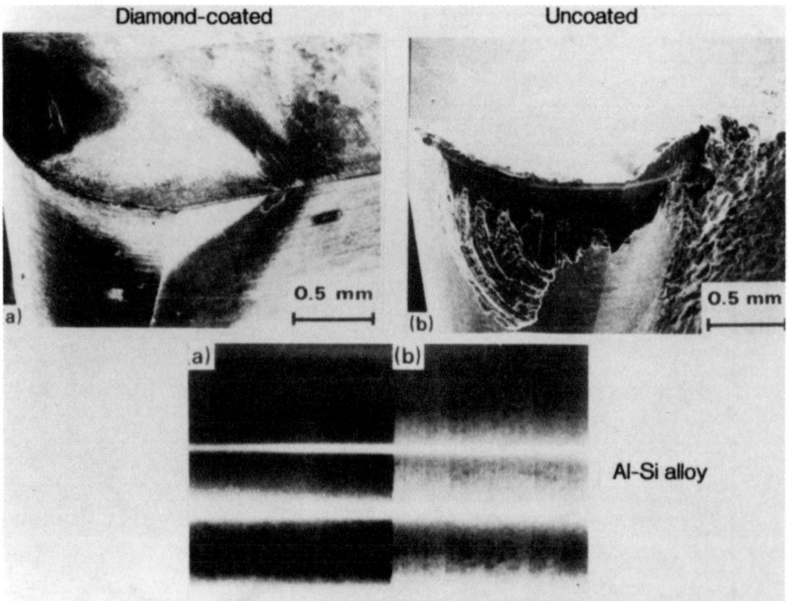

Figure 16. Typical features of a coated and an uncoated cutting edge before and after cutting an Al-Si workpiece and surface appearance of the work piece (Cutting conditions: Al-11% Si alloy, cutting speed 500 m/min, feed 0.1 mm/rev, depth of cut 1 mm, cutting time 5 min) (89).

3.2 "Bonded" Freestanding Diamond Sheets (PCD Analogy)

Industrial metal-cutting tools now commercially available are small endmills and some indexible inserts.

Endmills. Figure 7 shows a typical endmill with a bonded freestanding diamond layer (76)(95).

Metal-cutting Tool Inserts. Tool inserts produced by the freestanding layer technique are of great interest because of their relatively simple production method (Fig. 7). Freestanding layers have no Co or other binders in the diamond sheet. This is an additional advantage over PCD products.

3.3 Cutting Performance of Low-Pressure Diamond Tools with Al-Si and Other Workpieces

The different cutting parameters applied during the tool tests by various authors make a comparison of literature data difficult. For the same reason the comparison with commercial PCD tools is not easy. If the differences in cutting speeds are too high, different wear mechanisms can occur, making meaningful extrapolations practically impossible (96).

Small-diameter endmill cutting performance during machining of copper-coated printed circuit boards (GFRP) reveals the clear superiority of bonded freestanding CVD layers over cemented carbides (75)(76). The cutting conditions and results of a single-flute and double-flute CVD-coated endmill are shown in Fig. 17. The flank wear width of the CVD diamond endmill did not significantly increase with cutting time. On the other hand, using the carbide (K10) endmill, the flank wear width increased rapidly. Using the carbide endmill, a great amount of burring was found on the finished surface, corresponding to an increase in wear. Almost no burring was experienced when using the CVD diamond endmill. In the machining process of these products, cemented carbides have been usual, but now, with the application of CVD diamond, tool life can be improved by more than ten times, resulting in reduced tool changing time. Figure 17 shows the low flank wear observed with a diamond surface as compared to that occurring with conventional cemented carbide endmills (75)(76).

Continuous cut turning performances on Al-Si alloys with different Si contents are shown in the Fig. 18. A brief summary (total chip volume per cutting edge and the chip volume with time) of published cutting test results available now is given in Table 5.

Figure 19 illustrates the excellent adhesion and wear behavior of a cutting edge after 150 min cutting time on the very abrasive Al-20% Si alloy at a cutting speed of 200 m/min (91)(92).

Single-flute endmill

Conditions

Work Piece	GFRP Covered with Copper Foil
Tool	Ø 1.5 (1 Flute)
V	94 m/min
f	800 mm/tooth
d(Axial) d(Radial)	0.4 mm 1.0 mm
Cutting Fluid	Dry

Results

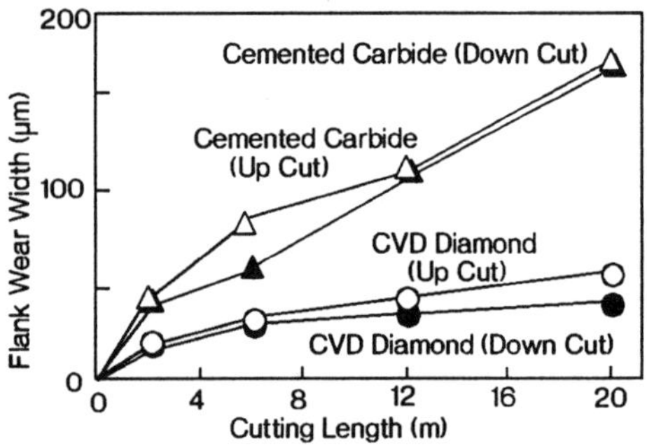

Double-flute endmill

Conditions

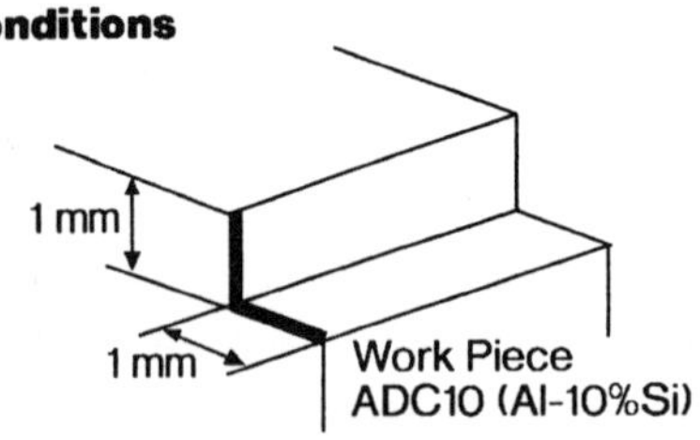

Tool	Ø 3 (2 Flutes)
V	100 m/min
f	0.02 mm/tooth
Cutting Direction	Up Cut
Cutting Fluid	Water Soluble

Results

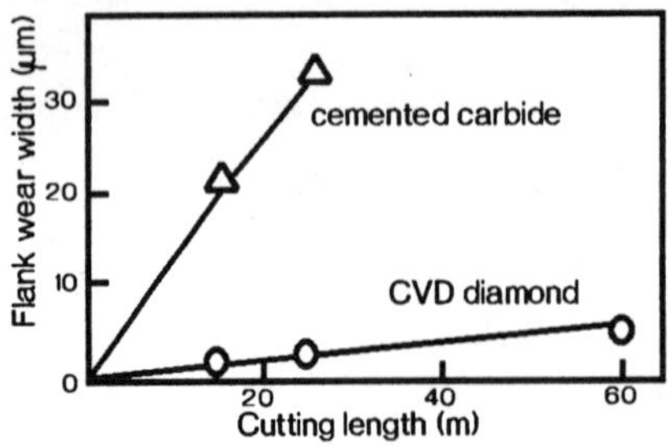

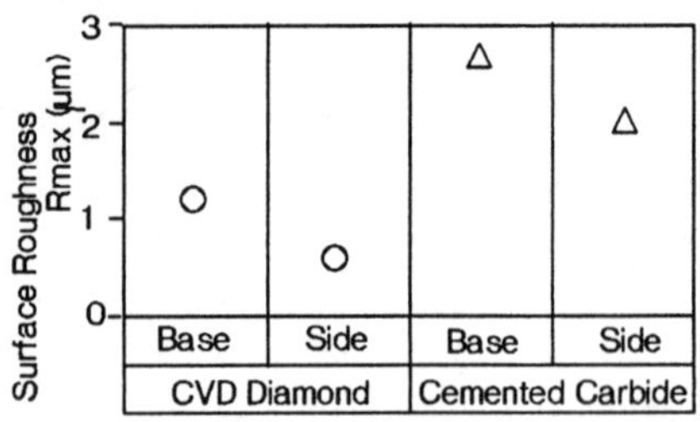

Figure 17. Wear resistance of single-flute and of double-flute CVD diamond endmills (75)(76).

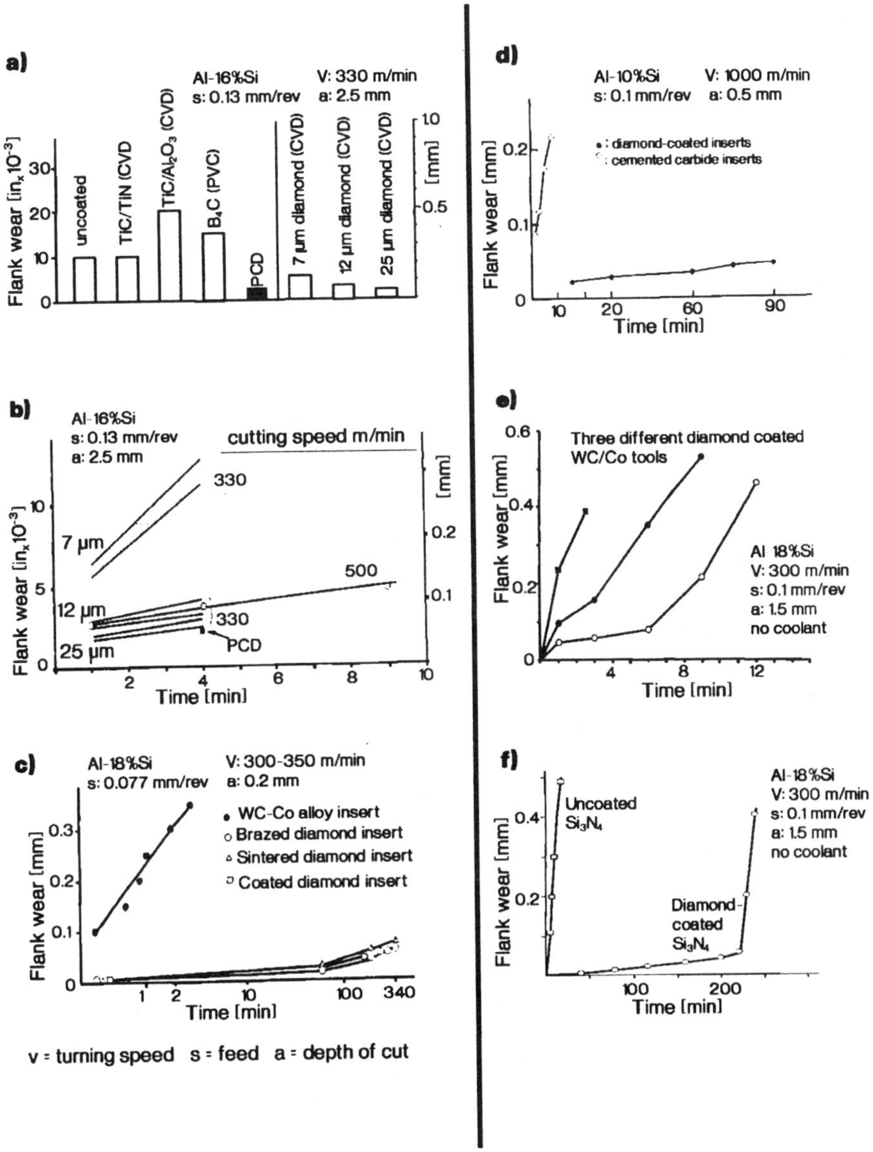

Figure 18. Typical results from turning tests using Al-Si alloys: a) and b) TU-Vienna and Teledyne Firth Sterling (35); c) and d) TUNGALOY (91)(92): e) and f) SANDVIK (84)-(86).

SUBSTRATE Composition	DEPOSIT L [μm]	M	CUTTING CONDITIONS MATERIAL [Al-Si%]	v [m/min]	f [mm/rev]	d [mm]	max. CHIP VOL. t [min]	per TIME [mm^3/min]	VOL. [mm^3]	Ref.
TURNING TESTS										
deposited diamond films:										
WC/Co 4.0%	4-5	MW	18 Si	200	0.1	0.5	5^e	10 $*10^3$	0.05$*10^6$	83
WC/Co	?	MW	18 Si	300	0.1-0.3	1.5	9	135 $*10^3$	1.2 $*10^6$	84
Si_3N_4(CC 690)	?	MW	18 Si	300	0.1	1.5	230	45 $*10^3$	10.4 $*10^6$	85
SiAlON	5	MW	18 Si	300	0.2	1.0	-	60 $*10^3$	-	86
SiAlON	?	MW	9 Si 3 Cu	-600	-0.35	1.0	63^e	210 $*10^3$	13.2 $*10^6$	87
WC/Co 0, 2, 4%	¯4	HF	18 Si	200	0.1	0.5	25^e	10 $*10^3$	0.3 $*10^6$	88
WC/Co 3.5,5.5%	?	HF	17 Si	310	0.4	1.5	30	186 $*10^3$	5.6 $*10^6$	7
WC/Co 6.0%	¯8	HF	11 Si	500	0.1	1.0	-	50 $*10^3$	-	89
-"-	¯8	HF	18 Si	250	0.1	1.5	-	37 $*10^3$	-	89
WC/Co	12-15	F&T	18 Si	-350	0.08	-0.7	-	19 $*10^3$	-	90
WC/Co ?	?	?	20 Si	200	0.1	0.5	200	10 $*10^3$	2.0 $*10^6$	91
WC/Co ?	?	?	10 Si	1000	0.1	0.5	»90	50 $*10^3$	»4.5 $*10^6$	92
WC/Co ?	?	?	?	800	0.3	-0.3	-	72 $*10^3$	-	93
brazed diamond films:										
WC/Co (H05)	-200	HF	18 Si	-350	0.077	0.2	»340	5.4 $*10^3$	»1.8 $*10^6$	94
MILLING TESTS										
SiAlON	?	MW	17 5	-350	0.2	1.5	-			87
WC/Co ?	?	?	12 Si	1000	0.1*	1.0	30			93
WC/Co ?	?	?	10 Si	100	0.02*	-	-			95
WC-Co ?	?	?	GFRP	94	80cm/min	-	-			95

* [mm/tooth]
e .... extrapolated value

L ... diamond layer thickness
M ... diamond deposition methode
F&T ... Hot Filament & Combustion Flame
HF ... Hot Filament CVD
MW ... microwave CVD

v ... cutting speed
f ... feed rate
d ... cutting depth
t ... cutting time - tool life criterion: 0.2 mm flank wear width
max .. calculated from maximum value

chip volume per time (turning) = v * f * d [mm^3/min]
total chip volume (turning) = v * f * d * t [mm^3]

Table 5. Survey of published cutting test results (total chip volume per cutting edge and the chip volume/time).

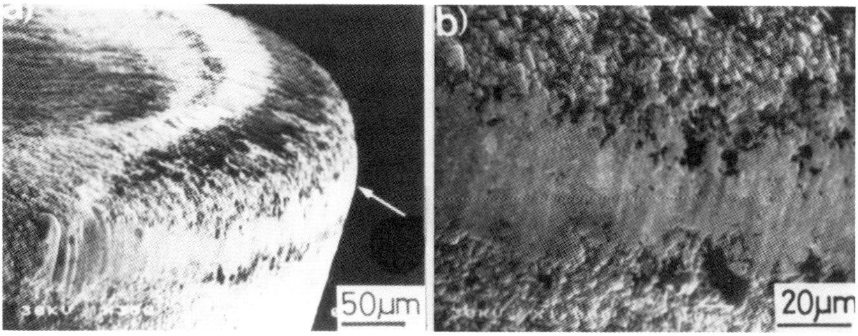

Figure 19. Cutting edge of diamond-coated insert after cutting an Al-20% Si alloy for 150 min (cutting speed: 200 m/min). The area labelled (b) shows area indicated by arrow in (a) (91)(92).

Figures 20 and 21 show the results of turning a round Al-12% Si alloy (AC3A) and Al-17% Si (A390) with PCD inserts (DA150: diamond-10% Co) and CVD diamond inserts (75)(76).

Figure 22 shows that under *interrupted cutting* conditions the CVD diamond behaves excellently as compared with cemented carbide and PCD (DA 150: diamond-10% Co and DA 200: diamond-15% Co) (75)(76).

Face milling tests on Al-12% Si alloy using CVD diamond inserts and PCD inserts (diamond-10% Co composite) show identical performances (Fig. 23) (75)(76). Face milling in factory machining tests shows that diamond-coated inserts can machine up to 250% more die-cast Al alloy cylinder heads than normal cemented carbide inserts (Fig. 24a) (97). For face milling ceramic workpieces (calcium silicate stones) the performance was excellent (Fig. 24b) (97). Face milling applications for die-cast Al automotive parts, copper alloys, graphite boards and wear-resistant grinding wheels are shown in Fig. 25 (97).

With respect to industrial standards, performance levels of in situ diamond-coated inserts are presently between those of the PCD and the cemented carbide tools. In certain cases, however, they even can be superior to PCD surfaces, as has been demonstrated by several of the above applications.

The test results mentioned here are only excerpts from current research projects. Their common goal is to achieve the ultimate performances. Most likely many of the results were obtained with specimens coated in yet unsophisticated experimental laboratory units (typically less than 20 specimens) or in small prototype reactors (up to 200 specimens). Also, further optimization of the substrate surface chemistry and surface preparations (98), as well as improvements in the production process itself are necessary. Additional progress towards optimized coatings with respect to their structure, grain size, thickness and surface roughness with minimized internal and interfacial stresses giving the best possible adhesion, can be expected. This should lead to further important improvements in performance.

3.4 Estimations of Production Costs for Diamond-Coated Cemented Carbide Cutting Tool Inserts

A rough estimate of the in situ coating costs in a very small laboratory unit in comparison with commercial mass product prices of PCD inserts is given in Table 6 (8). Extrapolations to large-scale production units (Fig. 26) suggest a rather competitive situation (99).

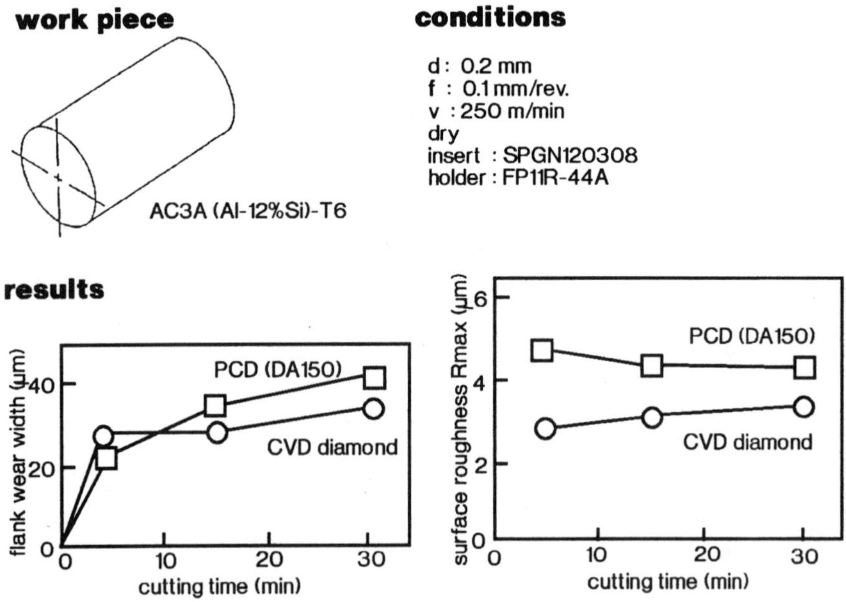

Figure 20. Comparison of flank wear width and surface roughness during dry continuous-cut turning (75)(76).

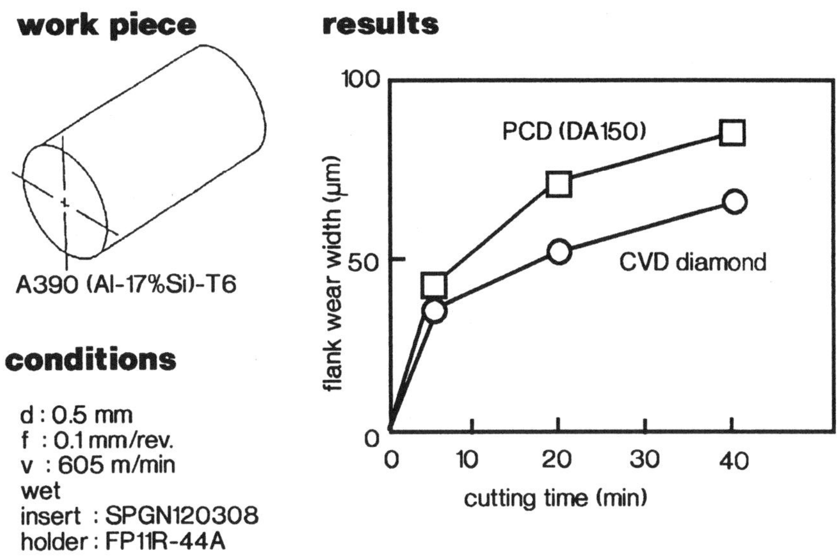

Figure 21. Comparison of the performance of PCD and CVD diamond layers during wet continuous-cut turning (75)(76).

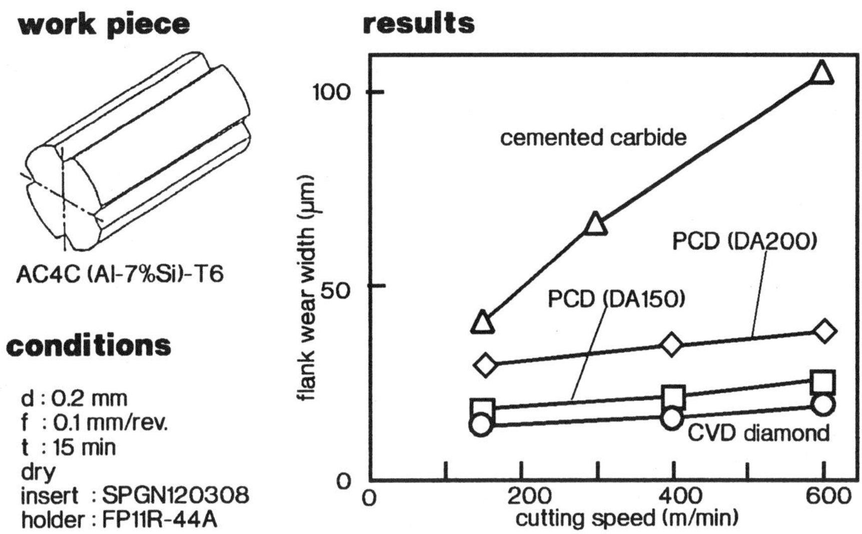

Figure 22. Comparison of the performance of cemented carbide and different diamond qualities during interrupted-cut turning (75)(76).

conditions

work piece :
AC8A(Al-122%Si)-T6

d : 0.5 mm
f : 0.15 mm/tooth
v : 1.000 m/min
t : 60 min
dry
blade : DABR15
body : DAM160R

results

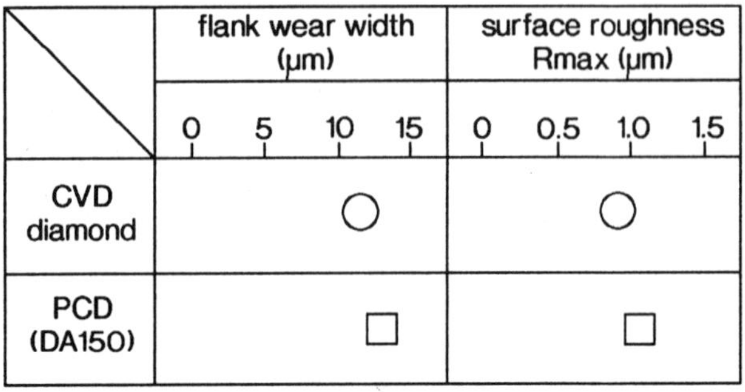

Figure 23. Results of face milling test on AC8A plates (75)(76).

a)

Face milling of **cylinder head**
(for automobile engine)

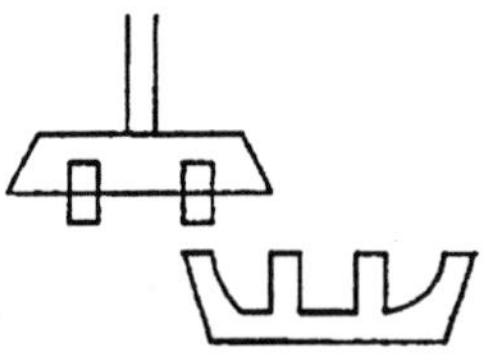

Conditions
Work : Die-cast aluminum alloy
Tool : SEA 42C10GR
Speed: 845 m/min
Feed : 0.1-0.2 mm/tooth
Depth of cut: 2 mm

Results
Cemented carbide insert (K10)
200 pieces / corner
Diamond coated insert
750 pieces / corner

b)

Face milling of **ceramic board**
(for buildings)

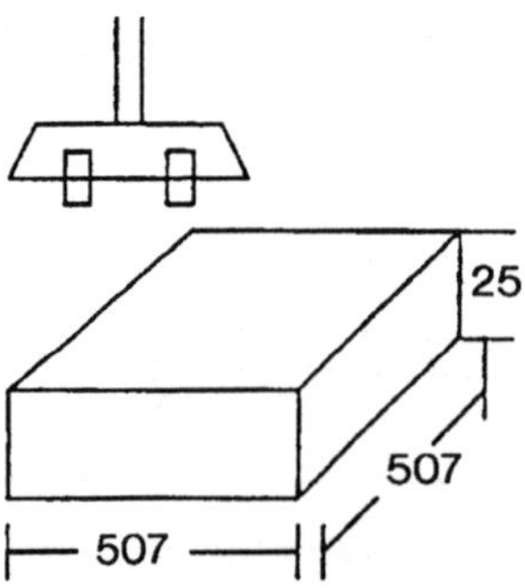

Conditions
Work : Calcium Silicate ($CaSiO_4$)
Tool : SEEN 53A FFN1
Speed: 410 m/min
Depth of cut: 10 mm

Results
Cemented carbide insert (K10)
2,000 pieces / corner
Diamond-coated insert
10,000 pieces / corner

Figure 24. Face milling in factory machining tests (97).

Wheel for automobile 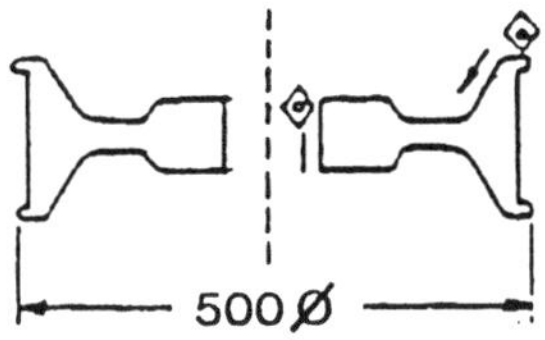	Conditions Work : Die-cast aluminum alloy Tool : DNPA 432 Speed: 1800 m/min Feed : 0.3 mm/rev Depth of cut: 0.2 - 5 mm Results Cemented carbide insert (K10) **96 pieces / corner** Diamond coated insert **280 pieces / corner**
Parts for automobile 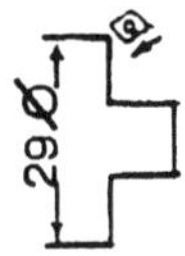	Conditions Work : Die-cast aluminum alloy Tool : SPP321 Speed: 1800 m/min Results Cemented carbide insert (K10) **10,000 pieces / corner** Diamond coated insert **50,000 pieces / corner**
Bearing of sintered copper alloy 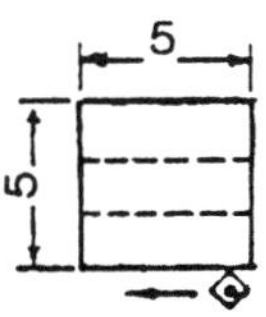	Conditions Work : Sintered copper alloy (Cu-10%Sn) Tool : SPP422 Speed: 55 m/min (3200 rpm) Feed : 0.05 mm/rev Depth of cut: 0.2 mm Results Sintered diamond insert **20,000 pieces / corner** Diamond coated insert **200,000 pieces / corner**
Face milling of graphite board 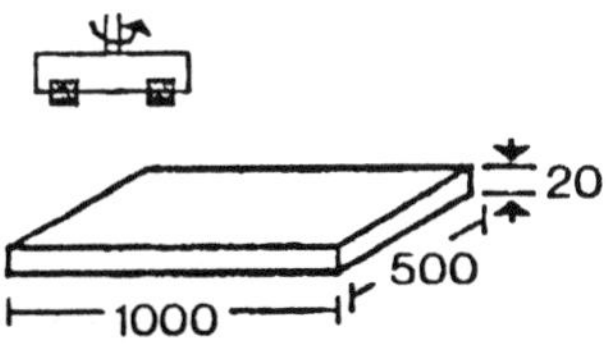	Conditions Work : Graphite Tool : SNC43B2G Speed: 200-300 m/min Depth of cut: 3 mm Results Cemented carbide insert (K10) **40 pieces / corner** Sintered diamond insert **210 pieces / corner** Diamond-coated insert **210 pieces / corner**
Wear-resistant parts 	Results Cemented carbide parts (K10) **3 days / exchange** Diamond-coated parts **14 days / exchange**

Figure 25. Industrial performance tests (97).

1. ESTIMATION of the DIAMOND COATING COSTS

Costs in US$ for:	Costs per hour	per deposition run 5 h (5 μm)	16 h (15 μm)
- Materials	0.40	2.00	6.40
- Energy	0.80	4.00	12.80
- Amortization of equipment	0.80	4.00	12.80
- Labor	50.00	32.00	32.00
Total	---	42.00	64.00

2. MARKET PRICES for PCD and uncoated CEMENTED CARBIDE INSERTS:

Average commercial prices for

(1) PCD insert with one cutting edge:	US$ 80 to 160
(2) uncoated cemented carbide insert, with 3 to 4 cutting edges:	US$ 12 to 20

3. TOTAL PRODUCTION COSTS for a CVD DIAMOND COATED INSERT:

Average production costs of	
(1) the uncoated insert:	US$ < 16
(2) the diamond coating:	US$ 64
Total cost of the diamond coated insert:	US$ < 80
Cost per cutting edge (for an insert with four coated edges):	US$ < 20

Table 6. Estimated costs for a CVD diamond coating operation, carried out in a laboratory production unit (8).

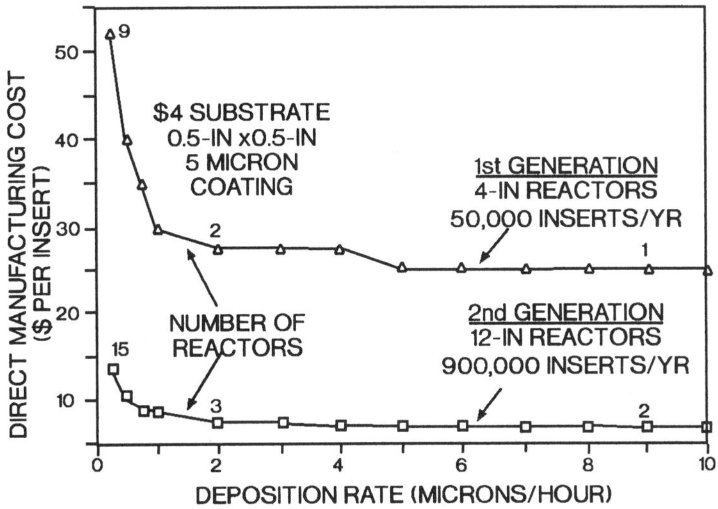

Figure 26. Cost calculation for large-scale industrial CVD diamond production equipments (99).

4.0 PROBLEMS RELATED TO THE FABRICATION AND ADHESION OF LOW-PRESSURE DIAMOND PRODUCTS

4.1 Influence of Substrate Chemistry

A decade ago spontaneous diamond nucleation without significant epitaxial relationship with the substrate became possible (100)(101). Heterogeneous nucleation has since been observed on many solid and liquid substrates (102) of very different chemical natures. However there is still no theory to give detailed practical guidance (98).

Diamond nucleation on highly perfect substrate surfaces, for example Si wafers, is sluggish and starts only after a long initial time period, which is apparently needed to form an extremely thin SiC intermediate layer (103).

Nucleation is also retarded if the carbon, arriving at the substrate surface, is removed too rapidly by diffusion into the substrate (104)(105). Diamond nucleation can start only when the minimum carbon concentration at the surface necessary for the formation of stable diamond nuclei is exceeded. Thus the nucleation and the nucleation density strongly depend on the balance between the flux of the carbon atoms arriving at the substrate surface and the carbon bulk/surface diffusion (104). Chemical reactions occurring between substrate and reaction gas (carbon, atomic and molecular hydrogen) can be another important influence of the substrate materials on nucleation (106).

4.2 Nucleation Enhancement

"Scratching" ceramics and metallic substrates with diamond, SiC, B_4C, or other hard particles is generally enhances the diamond nucleation density (105). On substrates with difficult spontaneous diamond nucleation, e.g., the surface of Si wafers (107)-(109), this produces spectacular effects.

Without such nucleation enhancement, it can be a major problem to obtain coatings thinner than a micron since with most methods, such as normal microwave or hot-filament activation, the natural nucleation density on many substrates is not high enough. This, in combination with the faceted diamond growth habit, also leads to the important surface roughness of the as-grown diamond layers, which is undesirable for many applications (Fig. 12). In such cases the costly operations of grinding and polishing the tool surfaces after coating (74)(110) can become necessary (Fig. 27). Nucleation enhancement by scratching can lead to finer grained, thinner diamond layers, with smooth as-grown surfaces (111).

Special synthesis procedures and modified equipment, such as activation with a tubular hot filament (112) in combination with an appropriate choice of other growth parameters, have led to so-called "nanocrystalline diamond", which is an ultrafine-grained, thin and very smooth diamond film. Both, modified methods or special surface treatments, leading to extremely high diamond nucleations are alternate ways of overcoming the nucleation difficulties encountered on certain substrates.

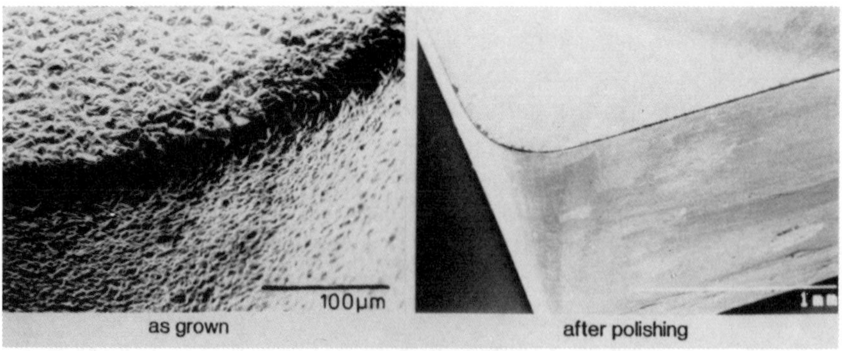

Figure 27. CVD diamond coated cutting edge before and after polishing (74).

4.3 Adhesion

Adhesion is strongly associated with interfacial stresses, depending mainly on the different expansion coefficients of the layer and the substrate. Interface stresses and stresses within the layers were determined recently (86)(113)(114).

The substrate surface temperatures during deposition, and thus the interfacial temperatures during the diamond formation, are different from the bulk temperatures. This results in a thermal gradient at the diamond/substrate interface during layer growth. After the coated tool is cooled, this gradient can lead to tension or compression at the interface (115), depending on the expansion coefficient of the substrate. Surprisingly, x-ray measurements (114) revealed tension in diamond layers on SiAlON substrates, where—based on room temperature expansion coefficients—compression would be expected (106)(115). Figure 28 shows the expansion coefficients of various substrates and the range for which neither tension nor compression can be expected for a substrate surface growth temperature of 1050°C (106)(115). The experimentally observed adhesions on different substrates shown in Fig. 28 support the assumption, known i. e. for Al_2O_3 coatings, that tension, rather than compression, is needed for good adhesion of coatings. In agreement with the theoretical predictions, it was shown recently that lowering the surface temperature during the diamond deposition on cemented carbide tools, which have a slightly higher expansion coefficient than diamond, can significantly improve the adhesion of the diamond coating (35).

Different substrate interactions with the gas phase can also influence the adhesion (105)(106). Formation of gaseous species or solid intermediate layers not only influences the nucleation but also the adhesion of diamond layers. SiO_2, which forms an intermediate SiC layer during diamond deposition, exhibits the expansion mismatch not at the SiC/diamond, but at the SiO_2/SiC interface (105).

A reasonable match of the expansion coefficients and the mastering of substrate reactions alone are, however, not sufficient to meet the high degree of adhesion required to satisfy the requirements for most wear-coating applications (35)(86)(116). At least some local chemical bonding between the diamond coating and the substrate surface also seems to be necessary. Only through chemical bonding of the Al_2O_3, TiC or TiN CVD coatings with the substrate surface atoms, can cutting tools perform as they do. Although reproducibility still has to be improved, diamond-coated substrates are able to achieve excellent adhesion performances (see Sec. 3.3).

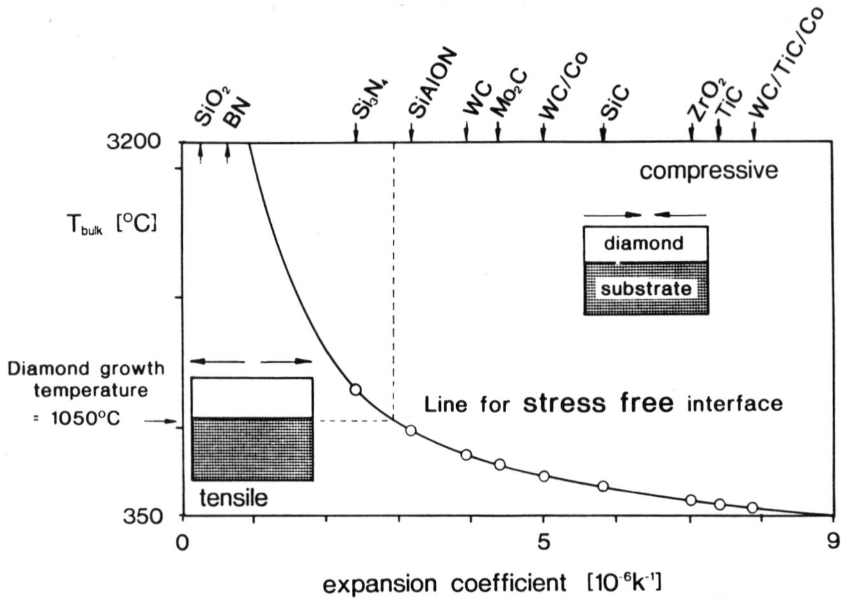

Figure 28. Influence of temperature and expansion coefficients on interfacial stress at a constant substrate surface temperature of 1050°C (78)(106).

5.0 TECHNICAL AND ECONOMICAL PROBLEMS LINKED WITH SYNTHESIS METHODS, INDUSTRIAL PRODUCTION UNITS AND FABRICATION

5.1 Diamond Synthesis Methods

During the last few years a large number of different new methods for low-pressure diamond deposition have been developed. Generally sufficient atomic hydrogen is needed during the diamond CVD synthesis and can be obtained in two ways (8):

— Activation by high temperature (TA-CVD = thermally activated chemical vapor deposition). Examples include: *(a)* hot filament or other hot solid surfaces (117)(118), *(b)* arc discharge (119) and arc plasma jets (120), and *(c)* chemically induced hot gas flames (121)-(123)

— Activation by electric or electromagnetic gas discharge (PA-CVD = plasma-activated chemical vapor deposition). Some examples are: *(a)* microwave (124)(125), *(b)* radio frequency gas discharge (126), *(c)* DC and AC glow discharges (127), and *(d)* plasma jets induced by RF (128)(129), microwave (130) and DC (131)

Combinations of the two principles such as hot filament plus microwave (132)(133), and DC discharge plus BIAS (Electron-Assisted CVD = EA-CVD) (134)(135) have also been successful and can lead to improvements.

Table 7 gives a short survey of the parameters used for the most common synthesis methods.

CVD Method	HOT FILAMENT	EA-CVD	DC-DISCHARGE	ARC DIS.JET	RF INDUCTION	MICRO-WAVE	COMB. FLAME
Electrical power [kW]	0.3	?	1.6	9.0	60.0	1.5[1]	-
Chemical power [kW]	-	-	-	-	-	-	0.33
Gas Input [% CH_4]	0.96	0.5-2.0	2.0	0.75	1.0-10.0	2.5	-
H_2 Flow [sccm]	478	5-50	98.0	1500	0-12000	800	-
CH_4 Flow [sccm]	4.64	0.25-1.0	2.0	60	100-1200	20.0	
C_2H_2 Flow [sccm]	-	-	-	-	-	-	975
O_2 Flow [sccm]	-	-	-	-	-	-	1025
Ar Flow [sccm]	-	-	-	6500	8000-35000	-	-
Deposition rate [um/h]	1.2	4.0	250	400	100	2.2	100
Deposition area [cm^2]	1.6	0.25	0.25	3.1	3.1	15	1
C supplied [mg/h]	138.9	7.5-30.0	60.0	1800	3000-36000	600	58000
C deposited [mg/h]	0.676	0.352	22	436	65-164	12	9.86
CCE* [%]	0.5	1.2-4.7	37	24	0.2-5.0	2	0.017
Elec. sp.yield[2] [kWh/g]	355	?	73	21	370-920	267[3]	-
Chem. yield[4] [kWh/g]	-	-	-	-	-	-	33
Pressure [mbar]	105	53	266	250	1000	100	1000
Surface temp. [°C]	700[5]	700-900	950	1000	700-1200	1000	-1200
Substrate	SiAlON	Si, SiC, Mo, WC	Si, SiC, Al_2O_3, Mo W, Diam.	Mo	Mo	SiAlON	Si, Mo, TiC, Ta, BN, Cu
Reference	[117]	[136]	[137]	[138]	[128]	[105]	[123]

* CCE = Carbon Conversion Efficiency

1 ... 1.5 kW Microwave + 1.7 kW supporting furnace

2 ... Electrical specific yield, related to electrical power input

3 ... related to 3.2 kW (1.5 kW Microwave + 1.7 kW supporting furnace)

4 ... Chemical specific yield, related to chemical power input (incomplete combustion of C_2H_2 to CO and H_2O)

5 ... bulk temperature

Table 7. Survey of parameters applied for the most common methods of diamond deposition.

5.2 Industrial Production Units

Mass production of in situ and freestanding layers for tools and wear applications requires scaling up of the reactors. Most important technical factors for the choice of the synthesis method for industrial production are the reliable control of the deposition process in order to obtain a constant and reproducible high-quality diamond as well as the economy of the process. Size and shape of the products will largely determine the production units. The following discussion considers mainly the problems of small sized mass products such as indexible inserts for cutting tools and similar items. In principle either a method with extremely high local growth rates (i.e., small deposition area), or a method which has a sufficiently large deposition surface or volume, with only a moderate local growth rate, can be chosen.

For batch-type in situ coating production reactors and low growth-rate diamond methods, the deposition area must be large enough to ensure a reasonably high throughput. If the diamond growth rates are very high, shorter deposition times become possible so that even continuous production units might be feasible.

However, for reasons of control and reproducibility, the lower growth rate methods seem preferable, at least for the beginning. They also allow a better control of the diamond quality. Currently, hot-filament and microwave gas activation methods fulfill these requirements best. Also, a large body of data is available for these two methods, as most basic research on diamond nucleation and growth has been performed with them. Therefore they are now considered the preferred choice for scaling up units.

High growth rates are obtained with jet methods, such as the plasma arc jets (139) or the surprisingly simple oxygen/acetylene flame, both of which can work at atmospheric pressure (120)(121). Deposits of high-quality diamonds were obtained on small areas with both methods (121)(122)(139). A small surface covered by a jet or flame can be a serious handicap however as the outer regions of the jet flame generally produce a low-quality sp^2-rich diamond. By moving the deposition area over large substrates in order to obtain complete surface coverage, this outer zone is also relocated and leads to an uncontrolled mixed quality of deposited diamond (140). Over prolonged production periods, the high-speed processes might also be more difficult to control than the low-speed ones.

The interesting synthesis possibility via halogenides was recently demonstrated (141)-(143) but is not yet well enough described for conclusions to be drawn. However these methods could, if successful, have a great potential for inexpensive diamond grit.

Freestanding layers could be produced by both batch-type and continuous units. Most of the above considerations for in situ diamond production are also valid for freestanding layers, with the extremely important exception that for freestanding layers the optimal substrate for nucleation and growth can be freely chosen. No undesirable chemical interaction, no expansion mismatch, etc., must be solved by difficult-to-find compromises. "Ideal substrates" can be used since interface bonding requirements between dummy and diamond are much less critical than for the in situ coating and thus much easier to meet.

Freestanding sheet materials are simply cut to the required shape and size needed for the final superhard surface before bonding (Fig. 7). The intermediate product needed is a freestanding diamond-coated dummy composite (Fig. 9) (8).

Economical, continuous high-speed production of diamond-coated dummies can be envisaged with diamond synthesis methods having high growth rates if several deposition sources are applied consecutively. Cooling of a continuous substrate could pose problems if high-speed continuous diamond deposition is performed. A moving highly polished "ribbon substrate" coated in continuous production could avoid this problem and also permit the fabrication of extremely thin, fine-grained and smooth as-grown surfaces. Such "dummy ribbons" could furthermore allow easy handling during automatic bonding operations of mass products and should lower their cost. A typical practical example would be the diamond-improved teeth of wood-cutting saw blades. Also, the earlier mentioned difficulties in obtaining layers thinner than a micron could be overcome by a high nucleation density. Taking into account the enormous differences in hardness between diamond and almost all the materials to be cut, even very thin layers should suffice for many applications if high-quality diamond can be so perfectly bonded that no losses by premature fracturing or flaking off of larger diamond patches occur. The use of thinner films could also facilitate adhesion of the coatings by reducing mechanical stress at the interface (86).

For thick sheets, highly polished dummy surfaces can produce a smooth replica diamond surface which can be used directly if the interface is turned inside out and used as the working surface (Fig. 9). In connection with precision bonding it could minimize or possibly avoid the costs of the final grinding of the tool.

5.3 Comparison of the Two Concepts for Applying Low-Pressure Diamond

Both techniques, the in situ coating and the bonding of freestanding diamond sheets, have specific benefits and drawbacks. The pros and cons of the in situ coating approach are listed:

a) Pros

(i) A simple one-step operation

(ii) Applicable simultaneously to a great number of substrates

(iii) Relatively high throwing power has the potential for evenly covering 3-D geometrical shapes, including the inner surfaces of hollow products such as tubings, dies, nozzles, or even porous structures (6)

(iv) Multilayers can be used (see Sec. 2.5)

b) Cons

(i) Layer formation, adhesion, etc., are frequently jeopardized by the chemical nature of certain substrates:

- Diamond growth can be disturbed due to diffusion of certain elements to the growing diamond/gas interface (144), e.g., the detrimental influence of Co from WC/Co hard metal substrates (117)
- Diamond adhesion can be adversely affected by modifications of the substrate surface due to excessive carbon diffusion into the substrate (104).
- Carbon or atomic hydrogen reactions with the substrate surface (106)(105).

(ii) Adhesion difficulties with substrates having high expansion coefficients can occur (106).

(iii) Rough as-grown surfaces need expensive grinding and polishing.

(iv) The prolonged heat treatment of the substrates at deposition temperature during the coating process can lead to undesirable changes in the substrate bulk properties. This is particularly detrimental for certain steel substrates (106) and must also be considered for cemented carbide substrates (145)(146).

(v) Synthesis methods having high growth rates are difficult to use as they usually also have high energy outputs which require complicated cooling of the substrates.

The pros and cons of the freestanding layer bonding approach are:

a) Pros

(i) The separation of layer formation and the layer adhesion by an independent and adaptable bonding operation can obviously avoid a great number of important problems encountered during the in situ CVD diamond coating. The dummy substrate can be freely chosen to best fulfill all the requirements of nucleation and growth.

(ii) Compared with the conventional PCD products, they are binder-free. Moreover, the freestanding sheets can be produced in large sizes which can far exceed the current 6-inch diameter limit of high-pressure PCD (Fig. 2). They can be produced in simple low-pressure hot-filament equipment or even with high-energy, high-growth rate methods (139) (see Sec. 4.3). The grain size and structures of such sheets can also be varied within certain limits and can, in particular, be extended towards finer grains.

(iii) Production of multilayer sheets seems relatively easy

b) Cons

(i) Similar to the situation for high-pressure PCD tools, the freestanding low-pressure diamond layer bonding concept has the disadvantage of being limited to a flat geometry or at best to relatively simple 3-D shapes.

(ii) Producing the tools within the necessary narrow tolerances requires several operational steps:

- dummy surface preparation
- diamond sheet fabrication
- surface polishing
- cutting
- brazing to the tool
- final grinding and polishing operations

(iii) It is relatively inflexible as far as "grade nuances" are concerned which accommodate specific performance requirements demanded of the hard surface during tool use. Multilayer sheets are still under development (147).

5.4 Economic Considerations

Whichever method is used must fulfill not only the technical requirements noted above but also the economical ones. Apparently certain sophisticated, unique diamond products such as thin membranes (6) can be sold at very high prices per carat (139). However, most wear products will have to be available at much lower costs in order to be competitive.

Despite the large number of methods available, the choices for industrial production methods are limited if economic and technical requirements are taken into account. Critical evaluations of the raw materials and the energy yielded, as well as the cost of the investments are necessary. An exact assessment demands a thorough knowledge of the processes, therefore such information is still scarce in the literature. Recently Bigelow (139) pointed out the importance of the specific energy per carat deposited. Comparing the two representatives of a high- and a low-growth rate method, he came finally to the conclusion given in Table 8 (148). Such important methods as hot-filament, DC-plasma and flames are unfortunately not included in their considerations. It is clearly too early to give a full evaluation here. However, cost efficiency will play a key role in the selection of the best scaling-up method and will decisively influence the further developments in this technology.

		Microwave	Arc Jet
Methane flow	[sccm]	1	80
Power	[kW]	1	9.5
Deposition rate	[µm/h]	1	500
Deposition area	[cm²]	20	0.1
Carbon supplied	[g/h]	0.032	2.6
Carbon deposited	[g/h]	0.0071	0.0176
Carbon Conversion Efficiency	[%]	22	0.68
Specific yield	[kWh/g]	140	540

Table 8. Comparison of the specific energy (kWh/g deposited) for a high- and a low-growth rate method of diamond deposition (143).

6.0 MARKET POTENTIALS AND FUTURE PROSPECTS

6.1 Market Potential

Figure 29 (149) shows the potential market for CVD diamond products (150)(151). According to this estimate, from a total market of \$6.5 billion, thin-film products (in situ CVD coatings) will account for \$4.5 billion and the thick-film diamond products (freestanding films and sheet materials) are expected to amount to $2 billion. Applications for wear parts, material-removal products and abrasives will be the one of most important market sectors for CVD diamond in the near future.

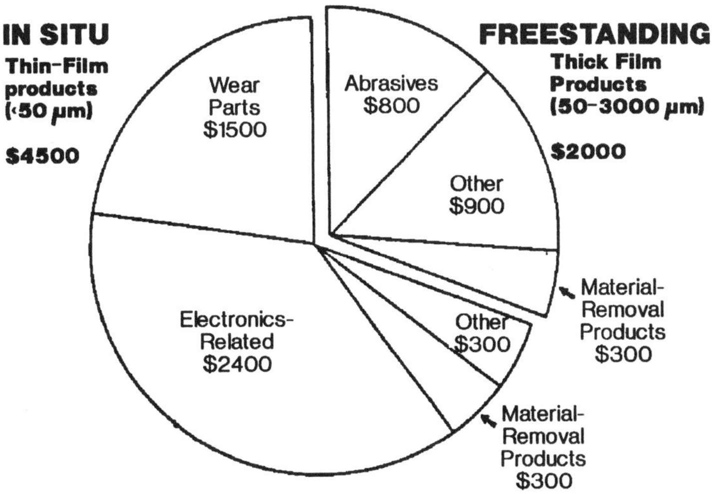

Figure 29. The potential market for CVD diamond products (in million US-dollars) (149).

Table 9 shows a typical distribution of the operations and the tool materials used in the modern automotive industry (42). Si-rich hypereutectic Al-Si alloys, used more and more for engine parts, have to be turned, milled, honed, etc. These specific applications are taken as an example and commented briefly in the following.

Figure 30 and Table 9 shows that currently normal PCD and PCBN materials represent together just a small fraction (< 0.1% !!) compared to the other tool materials (152). The high percentage of CVD- or PVD-coated cemented carbide tools (multilayers of TiC, TiN, Al_2O_3, etc.) compared with the uncoated tools demonstrates the enormous impact the in situ CVD coating has had on tooling applications since it was first applied some twenty years ago (16)(17)(153). The wide range of potential applications for superhard surfaces in the automotive industry is illustrated schematically in Fig. 31 (154).

Although the performance standard of PCD products is undoubtedly the goal to be reached, the object is not merely substitution with these products. To reach the potential indicated in Fig. 29, much broader use is needed. As discussed above, low-pressure diamond wear tools can be less expensive but should allow the realization of more complicated tool shapes (155) such as positive cutting edge angles or special geometries such as spiral shaped shaft mills with large twisting angles (40)(154).

Working Al	proportion [%]	Cutting material HSS	HM	PCD
plane milling	20	--	+	+
shaft milling	5	--	+	--
drilling	25	+	+	--
fine drilling, spindling	15	--	+	+
grating	10	--	+	+
threading (cutting, milling,forming)	20	+	+	--
turning	5	--	+	--

-- no
+ yes

HSS = high speed steel
HM = cemented carbide
PCD = polycrystalline diamond

Table 9. Todays use of cutting materials for metal cutting operations (Volkswagen VW Salzgitter, Germany) (42).

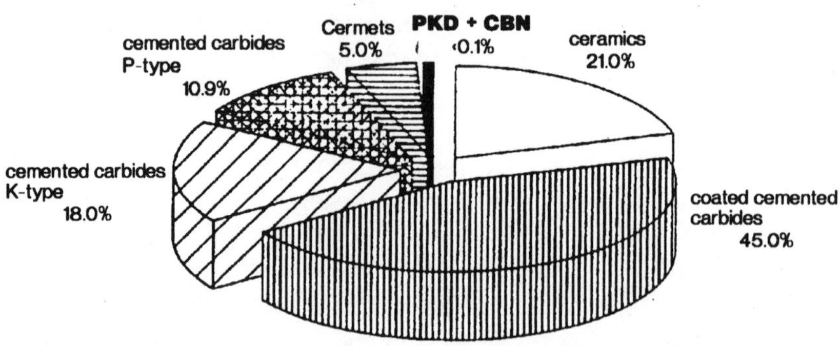

Figure 30. Typical distribution of cutting tool materials used in a modern automotive industry (152).

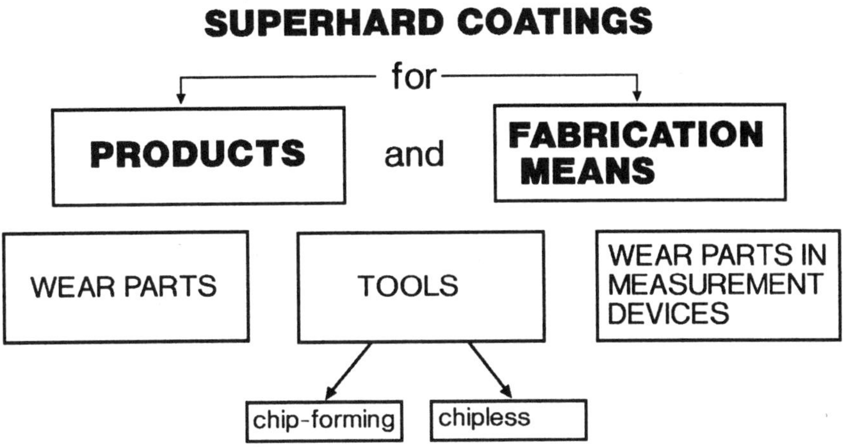

Figure 31. The broad range of potential interest of the automotive industry in superhard coatings (154).

The future market share of low-pressure diamond cutting inserts and other tools will thus be based mainly on the technical advantages of the CVD technology discussed above combined with the already well-known benefits of PCD products. These benefits include:

(a) long-lasting tools with less downtime, leading to high cutting performances characterized by high chip volumes/tool/time,

(b) low tendency to form built-up edges during machining (156), leading to a higher surface quality and more precise tolerances after machining, with no ridges, etc.

In the end, the decisive reason for a mass product producer to chose low-pressure diamond tools will be the lower global costs per piece produced.

In this case the tool costs themselves are not the most important factor. A tool with the superiority of pure diamond—having a much greater lifetime and giving excellently machined surfaces with exact tolerances—means more reliable products which require fewer efforts in quality control. This is an important contribution towards reducing the two-thirds of the wage costs which are, as shown in Fig. 32, only indirectly linked with the manufacturing of the products.

Moreover, significantly higher cutting speeds and the higher chip volume produced per tool save production time per piece manufactured.

They permit a more efficient use of highly complex, automated and expensive machines by reducing their non-productive time periods (less downtime for replacements of tools, adjustments, maintenance, etc.). This also lowers the investment costs for equipment, capital costs for building, etc. Thus an enormous amount of money is free to pay for excellent and highly reliable tools which can substantially reduce the overall production costs (154).

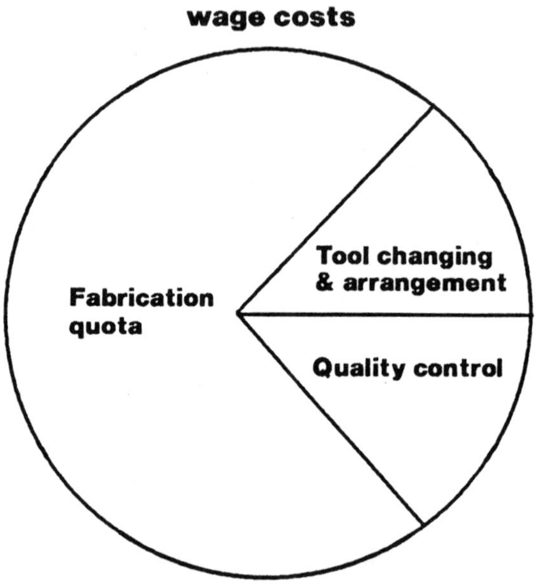

Figure 32. Wage costs and their linkage with the manufacturing process (154).

6.2 New Superhard Composite Powders and Compacted Products

Abrasive natural or high-pressure diamond powders have always held an appreciable share of the world's diamond market. Low-pressure diamond could become a strong competitor in the future (157). Recently, the feasibility of diamond *composite powders* was demonstrated (6)(132)(158)-(160). These consist of hard particles enveloped by a low-pressure diamond coating, as shown in Fig. 33 for a cubic BN particle which is enclosed by a diamond shell (161). Such powders will permit new combinations of materials previously impossible to realize. Besides their use as abrasives, composite powders mixed with binders, compacted and sintered at ultrahigh pressures, or even conventionally compacted at low pressures prior to a classical sintering operation, could result in exciting, novel PCD-type products. Since abrasive products are relatively inexpensive, a low-pressure powder fabrication in large reactors (132) with excellent yields is required to bring the production costs down to competitive levels. However, materials with outstanding properties and performances—as such composite material tools might well possess—usable for unique, sophisticated applications are not subject to stringent price limitations (6)(160).

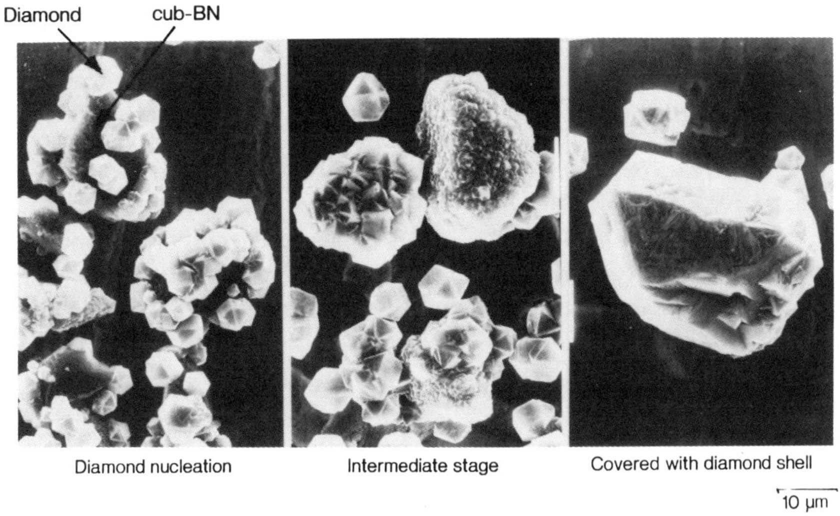

Figure 33. Different stages during production of a diamond composite powder (cub-BN core with a diamond shell) (161).

6.3 The Superhard BN Phases

Many analogies exist for the boron nitride materials. The superhard BN phases are extremely interesting for wear applications. They offer an alternative set of diamond's chemical properties in such applications. As discussed for diamond, utilizing mainly the superhardness means that polycrystallinity and lattice imperfections are acceptable. Therefore, once appropriate low-pressure synthesis methods are established, the wear applications of coated BN tools are likely to be the initial commercial products.

Up till now, however, attempts to synthesize cubic BN at low pressures have not been as successful as those to produce low-pressure diamond.

Major basic problems have apparently not yet been satisfactorily solved (162)(163)-(166). A direct analogy to the low-pressure diamond CVD synthesis appears to be an obvious possibility, however (163)-(166).

The recent overall annual market growth rates of superhard BN products, as experienced for example in the USA, were high and are projected to increase during the next couple of years (150). Superhard BN films and coatings should rapidly gain significant industrial importance and reinforce this trend. BN exhibits an excellent inertness to hot steel surfaces under conditions where diamond is violently dissolved and therefore cannot be used. As known from the ultrahigh-pressure PCBN, this fact alone promises extraordinary possibilities for cutting tools, drawing dies and other sliding friction operations important for the shaping of hot and cold steels. In view of this potential market for highly relevant commercial applications, more and more research groups are working on the low-pressure CVD synthesis of superhard BN (163)-(168).

Besides the synthesis of the pure BN phases, the combined synthesis of diamond and BN could lead to new superhard materials with interesting properties complementing those of pure diamond and pure BN. Technically they could consist either of

(a) "mechanically" mixed diamond/BN composites, i.e., multilayer coatings (70)(161) or composite powders as shown in Fig. 33, or

(b) "ternary phases" with crystal structures similar to diamond, consisting predominantly of the elements B/N/C and/or P (169).

7.0 CONCLUSIONS AND OUTLOOK

The above wear and cutting applications utilize mainly the superhard properties of diamond. Today, product development for these applications has already reached a relatively advanced stage, since single crystallinity and a perfect diamond crystal lattice are not primary requirements for their proper functioning. Their technical feasibility has been demonstrated and has already led to industrial products.

The production of technically and economically feasible products for superhard applications is, however, still only in its early stages. The rapid advancement in these "simpler" application areas will most certainly continue and even gain momentum. Some of the major problems remaining to be solved are considered below.

7.1 The Short-Range Outlook

Crucial Points for In Situ Diamond Coatings: Even better adherence and general reproducibility are of greatest importance. A number of rather successful performance tests published recently (see Sec. 3) are encouraging and prove that the diamond coating adhesion is not an unsolvable problem. However, though good adhesion could be achieved in laboratory tests, the reproducibility and quality assurance in larger production units seem now to be the major obstacles to a more widespread industrial use of diamond layers.

Reproducible substrate surface preparations, proper selection of the chemistry and constant quality of appropriate substrates together with a reduction of internal and interfacial stresses will, if combined with well-controlled synthesis parameters, improve not only nucleation, diamond growth and structures, but subsequently will also increase the reliability of the interface bonding.

Parallel to these more fundamental developments, industrial metal-cutting field tests and construction of larger units for diamond deposition are now being performed (see Sec. 3).

Crucial Points for Application of Freestanding Diamond Layer: At present, besides the economic aspects and minor bonding problems connected with different substrate materials, the optimization and structural control of the diamond deposits themselves seem the most important challenge.

The optimal diamond quality needed for a specific application is now determined by laboratory and field testing under real industrial conditions (see Sec. 3). Questions such as how to obtain the best diamond sheet structure for the specific application by choosing crystals within well-defined size limits, special crystal orientations, sheets of extreme thinness or thickness with adequate surface roughness, the application of the multilayer concept, are matters requiring further investigation.

Topics Important for Both Approaches:

a) Quality definition and assurance may at first sound like seemingly trivial matters. However, maintaining a constant high quality for a large number of specimens prepared simultaneously over a relatively large deposition area in an industrial reactor is a most important task. Not only the achievement of, but also the quality control of the uniform adhesion, surface roughness, layer thickness, sharp edges, etc., can pose serious problems.

b) Multilayer approaches will allow different "diamond surface layer grades" which most likely can be achieved by combining the diamond layers with other hard materials in thin laminated multilayer structures.

Scaling up: A number of individual, as yet unpublished larger-scale production setups are now being simultaneously developed and the scaling up of laboratory equipment to production units which should allow viable and economical production is already in progress. From the available methods, the right one must be chosen. That and the equipment will depend largely on the product. Simple production parameters such as shape (simple or complex), size (large or small) and quantity of the final products (many or few) will decisively influence the selection of the "best" method and determine the construction of the most suitable equipment.

The optimization of the two contradictory parameters, size of the usable deposition surface area and uniformity of the coating, is quite important for industrial production. It is a function as much of the specific equipment design as of the chosen synthesis methods and the geometries and sizes of the products.

The challenge now is to apply the high growth rates reported in the literature, rather than to increase them even further.

Although it is also desirable for instance to improve the carbon yield or to recycle the hydrogen, reducing the energy needed per carat of diamond deposited seems to be one of the most important parameters for the economy of processes to be applied industrially.

The CVD reactor for normal hard coatings is traditionally a batch-type unit. It seems that the initial in situ coating unit for low-pressure diamond coatings will be the same, using a deposition process with an easily controllable low growth rate and a large usable surface area for deposition.

This is valid for both in situ and freestanding diamond layers, which currently are apparently both produced this way. Despite being more complex, in the future continuous production units combined with high-rate methods might well be an interesting alternative, if the process parameters can be reproducibly controlled and the production costs can be kept low.

7.2 The Long-Range Outlook

Composite Powders. Composite powders of low-pressure diamond surrounding cubic BN or other hard material particles, as well as compacted products from such powders, are still in the early stages of development. They are, however, feasible within the technology available now, and will most likely be further improved. Their possibilities depend largely on whether their properties and applications can justify their production costs.

Superhard Boron Nitrides. Applying analogous low-pressure methods to the synthesis of superhard boron nitrides or other new superhard phases is still in the basic research stage. A number of different methods have been described recently. Though different from diamond, there is no principle obstacle known yet which would prevent the synthetic creation of superhard boron nitrides at low pressures similar to those in the diamond syntheses. This is an exciting challenge for the near future, especially in view of the considerable market potential of these materials for wear applications.

ACKNOWLEDGMENTS

We wish to express our gratitude to the Austrian "Fonds zur Förderung der Wissenschaft und Forschung" which has been sponsoring work at the Technical University in Vienna (Project P6031, P7274).

Thanks also to those in the management of Sandvik Stockholm, who believed in these ideas years ago and had sponsored parts of our research work.

We would like to thank all groups and scientists who sent documentary information, pictures and materials used in this work:

Audi AG, Germany; Dr.J.Motzet

Asahi Diamond Ind., Japan; Dr. Fuminori Okozumi

De Beers Industrial Diamond Division,

Ascot, England; Dr.P.J.Heath

Dü sseldorf, Germany; Mr. Weiland,

Gühring KG, Germany; Dr.Gottlieb Gühring and Dr. Jan Ebberink

Mercedes-Benz AG, Germany; Dr. Kirmse, Dr. P. Johannsen

Mitsubishi Materials Corporation, Japan; Dr. Yoshio Fujiwara

Sandvik Coromant, Sweden; Dr. Mats Sjöstrand and Dr. Staffan Söderberg

Sumitomo Electric Industries, Japan; Dr. Tetsuo Nakai, Dr. Tetsuo Yashiki

Toshiba Tungaloy Co., Japan; Dr. Kosuke Saijo

Last but not least, the authors want to thank all members of the research team at the Technical University in Vienna, who have devoted their time and energy to successfully solving the many challenging problems encountered in this work.

REFERENCES

1. *1st Int. Conf. Diamond Sci. and Tech. (JNDF),* Tokyo, Japan. (Oct.24-26, 1988)
2. *Proceedings of the NATO-Advanced Study Institute on Diamond and Diamond-Like Films and Coatings* NATO-ASI Series B: Physics, Vol. 266, Plenum, New York (1991)
3. *2nd Int. Conf. on the New Diamond Sci. and Tech., ICNDST-2,* Washington, DC (Sept.23-27, 1990)
4. *1st Euro. Conf. on Diamond and Diamond-like Carbon Coatings, Diamond Films 90,* Crans-Montana, Switzerland (Sept.17-19, 1990)
5. Sumiya, H., Tsuji, K. and Yazu, S., *2nd Int. Conf. on the New Diamond Sci. and Tech.,* ICNDST-2, Washington, DC (Sept.23-27, 1990)
6. Pinneo, J. M., *Proceedings of the NATO-Advanced Study Institute on Diamond and Diamond-Like Films and Coatings,* NATO-ASI Series B: Physics, Vol. 266, Plenum, New York (1991)
7. Lux, B., Haubner, R. and Wohlrab, C., *Surf. Coat. Tech.,* 38: 267-280 (1989)
8. Lux, B. and Haubner, R., *R&HM* 8(3):158-174 (1989)
9. Sjöstrand, M. E. and Thelin, A. G., *Metal Powder Report* 41(12) (1986)
10. Bonetti, R. S. and Happel, O. F, *3rd Int. Congress of Heat Treatment,* 328-337 (1983)
11. Schubert, W. D., Wruss, W., Neumann, W. and Lux, B., *J. Ref. Hard Metals,* 5(1):38-42 (1986)
12. Smith, A. B. and Whittle, N. C. *Proc. 25th Int. Machine Tool, Design and Research Conf.,* pp. 261-68, London (1985)
13. Smith, A. B. and Williams, R., *Powder Metallurgy Group Meeting,* 19.1-11, London (1986)
14. Holleck, H., *Z. f. Werkstoffk.* 17:334-41 (1986)
15. Altena, H., Colombier, C., Lebl, A., Lindström, J. and Lux, B., *4th EURO-CVD Conf.,* pp. 428-434, Eindhoven (1983)
16. Lux, B. and Schachner, H., *High Temp.-High Press.,* 10:315-323 (1978)
17. Lux, B., Funk, R., Schachner, H. and Triquet, C., CH Expose D'Invention 540990 (1971)

18. Iiyoshi, H., Doctoral Thesis, TU, Vienna (1988)
19. Bonetti, R., and Wahl, H., *4th EURO-CVD Conf.,* 510-16, Eindhoven (1983)
20. Lux, B., Haubner, R., Altena, H. and Colombier, C., *Fischmeister/ Jehn,* pp. 71-112, Verlag DGM, Oberursel, Germany (1987)
21. Colombier, C. and Lux, B., *J. Mater. Sci.*, 24:462-470 (1989)
22. Dreyer, K. et al., German Patent P2917348.6-45 (1979)
23. Lindström, J. et al., German Patent P2443160.5-14 (1974)
24. Loisel, B., Haji, L. and Guendouz, M., *Journal de Physique,* Colloque C5, suppl.au no5, Vol.50 (1989) 467-478
25. Schintelmeister, W., Wallgram, W. and Gigl, K., *High Temp.-High Pressure,* 18:211-222 (1986)
26. Glaski, F. A., *Proc. 4th Int. CVD Conf.,* pp. 521-35 (1973)
27. Karlsson, B. et al., *Phys. Scr.,* 25:775-9 (1982)
28. Lux, B., Colombier, C. and Altena, H., *Thin Solid Films,* 138:49-64 (1986)
29. Lux, B., Funk, R., and Schachner, H., Swiss Patent 1494157 (1975)
30. Reiter, N. et al., *VDI Berichte* 624:167-84 (1986)
31. Knotek, O., Bosch, W., Leyendecker, T., Münz, W. D. and Falkenstein, R., *8th ICVM,* Linz, C11.2, 340 (1985)
32. Tsukamoto, T., Sasaki, K., Shibuki, K., Momma, H. and Takatsu, S., *Metal Powder Report Conference,* Luzern, 1(21):1-24 (1983)
33. Lindström, J., Johnsson, B. F. and Ohlson, F. J., Swedish Patent 14479/71 (1971); US Patent 3,837,896 (1974); German Patent P2443160.5-14 (1974)
34. Metallwerk Plansee, Austrian Patent 312952 (1973)
35. Oakes, J., Pan, X. X., Haubner, R., and Lux, B., *1st Euro. Conf. on Diamond and Diamond-like Carbon Coatings, Diamond Films 90,* Crans-Montana, Switzerland (Sept.17-19, 1990)
36. Bergmann, E., *8th ICVM,* Linz, C12(1):370-86 (1985)
37. Bergmann, E. and Vogel, J., Balzers AG; private communication (1987)
38. Nishida, N., U.S. Patent 4,252,862 (1981)
39. Bergmann, E. and Vogel, J., *J. Vac. Sci. Tech.* A5:70 (1987)

40. Ebberink, J., Presentation at: *The Pursuit of Diamond Coatings,* Arhus, Denmark (April 9th, 1991)
41. Heath, P. J., *VDI-Berichte* 762:37-59 (1989)
42. Schweitzer, R., *VDI-Berichte* 762:293-303 (1989)
43. Prekwinkel, H., *Diamond Information Brochure M41,* De Beers Industrial Diamond Division, Germany, pp. 12-15 (1986)
44. Lach, H., *Diamond Information Brochure M41,* De Beers Industrial Diamond Division, Germany, pp. 16-18 (1986)
45. Heimbrand, E., *Diamond Information Brochure M41,* De Beers Industrial Diamond Division, Germany, pp. 19-23 (1986)
46. Moitzi, H., *Diamond Information Brochure M41,* De Beers Industrial Diamond Division, Germany, pp. 24-33 (1986)
47. Grob, U., *Diamond Information Brochure M41,* De Beers Industrial Diamond Division, Germany, pp. 34-35 (1986)
48. Walz, F., *Diamond Information Brochure M41,* De Beers Industrial Diamond Division, Germany, pp. 36-38 (1986)
49. Schimmel, P., D*iamond Information Brochure M41,* De Beers Industrial Diamond Division, Germany, pp. 46-47 (1986)
50. Schröder, W. and Gaipl, W., *Diamond Information Brochure M41,* De Beers Industrial Diamond Division, Germany, pp. 48-51 (1986)
51. Grün, L., *Diamond Information Brochure M41,* De Beers, Industrial Diamond Division, Germany, pp. 52-53 (1986)
52. Spur, G. and Wunsch, U. E., *Diamond Information Brochure M41,* De Beers Industrial Diamond Division, Germany, pp. 39-45 (1986)
53. Shafto, G. R., *Diamond Information Brochure M42,* De Beers Industrial Diamond Division, Germany, pp. 2-10 (1986)
54. Tomlinson, P. N., Pipkin, N. J., Lammer, A. and Burnand, R. P., *Diamond Information Brochure M42,* De Beers Industrial Diamond Division, Germany, pp. 11-22 (1986)
55. Champion, K., *Diamond Information Brochure M44,* De Beers Industrial Diamond Division, Germany, pp. 7-9 (1988)
56. Jennings, M., *Diamond Information Brochure M44,* De Beers Industrial Diamond Division, Germany, pp. 13-17 (1988)
57. Lammer, A., *Diamond Information Brochure M43,* De Beers Industrial Diamond Division, Germany, pp. 2-7 (1987)

58. Kübler-Tesch, G., *Diamond Information Brochure M43,* De Beers Industrial Diamond Division, Germany, pp. 8-9 (1987)
59. Gött, W., *Diamond Information Brochure M43,* De Beers Industrial Diamond Division, Germany, pp. 12-16 (1987)
60. Ebberink, J., Gottlieb Gühring KG, Germany, private communication (1991)
61. Vigneau, J., *VDI-Berichte,* 762:321-330 (1989)
62. Ogata, M., *VDI-Berichte,* 762:271-279, (1989)
63. Broska, T.J., *VDI-Berichte* 762:281-292 (1989)
64. Sato, Y., *Workshop on the Sci. and Tech. of Diamond Thin Films,* Cleveland, Ohio (May 20-24, 1990)
65. Badzian, A. R., Bachmann, P. K., Hartnett, T., Badzian, T. and Messier, R., *E-MRS Meeting,* Vol.XV, pp. 63-77 (June 1987)
66. Tomlinson, P. N.and Wedlake, R. J., *IDR* 17 (4):234-241 (1983)
67. DeVries, R. C. and Robertson, C., *J. Mat. Sci. Lett.* 4:805-807 (1985)
68. Milaschow, W. A., "Diamanten," *VEB Deutscher Verlag für Grundstoffindustrie,* Leipzig (1979)
69. Colombier, C., Lux, B., Riahi, A., Puchhammer, M. and Störi, H., *Fresenius Z. Anal. Chem.,* 329:355-360 (1987)
70. Bichler, R., Peng, J., Haubner, R. and Lux, B., *Materials Science and Engineering,* A105(6):543-547 (1988)
71. Schachner, H., Tippmann, H., Lux, B., Stjernberg, K. and Thelin, A., Swedish Patent 8403428 (1984)
72. Nakamura, T. and Nakai, T., Japan Patent Application, Tokugan-hei 3-5368 (1991)
73. Tanabe, K. and Fujimori, N., Japan Patent Application, Tokugan-hei 2-81777 (1990)
74. Asahi Diamond Industrial Co.; private communication (1989)
75. Nakai, T. and Yagashi, T., Sumitomo Electric Industries; private communication (1991)
76. Yashiki, T., Nakamura, T., Fujimori, N. and Nakai, T., *Int. Conf. on Metallurgical Coatings and Thin Films (ICMCTF),* San Diego, California (April 22-26, 1991)
77. Anthony, T., private communication (1989)

78. Bichler, R., Doctoral Thesis, TU-Vienna (1988)

79. Otake, N. O., Tokura, H., Yoshikawa, M. and Yang, C. F. *1st Int. Conf. Diamond Sci. and Techn.*, P2-25, pp.140-141, Tokyo, Japan (Oct.24-26,1988)

80. Yoshikawa, M. and Tokura, H., *New Diamond,* pp. 86-88, Japan New Diamond Forum (1990)

81. Okuzumi, F., *Tool Technik,* pp. 3-7 (1988)

82. "TAB Bonding Tool," Sumitomo Electric Industries, LTD. EV-32R (commercial documentation) (1990)

83. Yagi, M., Saijo, K., Katsumura, Y. and Suzuki, H., *Keikinzoku,* 39(6):417-422 (1989)

84. Söderberg, S., Gerendas, A., Sjöstrand, M., *Vakuum,* 41(4-6):1317-1321 (1990)

85. Söderberg, S, private communication (1991)

86. Reineck, I., Söderberg, S., Westergren, K. and Shahani, H., *2nd Int. Conf.on the New Diamond Sci. and Tech., ICNDST-2,* Washington DC (Sept.23-27, 1990)

87. McCune, R. C., Hoffman, D. W., Whalen, T. J., and McHugh, C. O., *Mat. Res. Soc. Symp. Proc.,* 130:261-266 (1989)

88. Katsumura, Y., Yagi, M., Suzuki, H. *Keikinzoku,* 39(9):634-638 (1989)

89. Kikuchi, N., Komatsu, T., Yoshimura, H., *Materials Science and Engineering,* A105/6:525-534 (1988)

90. Murakawa, M., Takeuchi, S., Hirose, Y., Komaki, K., Yanagisawa, Y., *2nd Int. Conf. on the New Diamond Sci. and Tech., ICNDST-2,* Washington DC (Sept.23-27, 1990)

91. Saijo, K., (Toshiba Tungaloy) private communication (1991)

92. Saijo, K., Yagi, M., Shibuki, K. and Takatsu, S., *J. Materials and Manufacturing Process* (in press)

93. Fujiwara, Y., (Mitsubishi) private communication (1991)

94. Murakawa, M. and Takeuchi, S., *Proceedings of the NATO-Advanced Study Institute on Diamond and Diamond-Like Films and Coatings,* NATO-ASI Series B: Physics, Vol. 266, Plenum, New York (1991)

95. Nakai, T., Sumitomo Electric Industries; private communication (1990)

96. König, W. and Boehmke, H., RWTH Aachen, private communication

97. Kikuchi, N., *Gorham Advanced Materials Institute,* Marco Island, FL (Oct. 15-17, 1989)

98. Lux, B. and Haubner, R. *Proceedings of the NATO-Advanced Study Institute on Diamond and Diamond-Like Films and Coatings,* NATO-ASI Series B: Physics, Vol. 266, Plenum, New York (1991)

99. Dismukes, J. P., *Workshop on the Sci. and Tech. of Diamond Thin Films,* Cleveland, Ohio (May 20-24, 1990)

100. Spitzin, B. V., Bouilov, L. L. and Derjaguin, B. V. *J. Cryst. Growth,* 52:219-26 (1981)

101. Matsumoto, S., Sato, Y., Tsutsimi, M. and Setaka, N., *J. Mat. Sci.* 17:3106-12 (1982)

102. Anthony, T., *Proceedings of the NATO-Advanced Study Institute on Diamond and Diamond-Like Films and Coatings,* NATO-ASI Series B: Physics, Vol. 266, Plenum, New York (1991)

103. Williams, B. E., Glass, J. T. and Davis, R. F., *SPIE,* 877:56-63, Micro-Optoelectronica Materials (1988)

104. Joffreau, P. O., Haubner, R. and Lux, B., *J. Ref. Met.-Hard Metals,* 7(4):186-194 (1988)

105. Lindlbauer, A., Doctoral Thesis, TU-Vienna, (1992)

106. Lindlbauer, A., Diploma Thesis, TU-Vienna (1989)

107. Badzian, A. R., Badzian, T., Pilione, L., *3rd Int. Conf. on Surface Modification Technologies,* pp. 51-61, Neuchatel, Switzerland (Aug. 28-Sept.1, 1989)

108. Hirabayashi, K., Ikoma, K., Taniguchi, Y. and Kurihara, N. I. *1st Int. Conf. Diamond Sci. and Techn.,* P 2-03, Tokyo, Japan (Oct. 24-26,1988)

109. Yugo, S., Kimura, T. and Kanai, H., 1st *1st Int. Conf. Diamond Sci. and Tech.,* P 2-03, pp. 130-131, Tokyo, Japan (Oct. 24-26,1988)

110. Yoshikawa, M., Otake, N. O., *Bull. Japan Soc. of Prec. Eng.,* 22(3): 171-176 (1988)

111. Wild, C., Herres, N., Wagner, J. and Koidl, P., *1st Int. ECS Symp. on Diamond and Diamond Like Films,* Los Angeles (May 7-12, 1989)

112. Yarborough, W. A. and Roy, R., *MRS Spring Meeting,* Reno, Nevada (April 4-9, 1988)

113. Murakawa, M. and Takeuchi, S., *1st Euro. Conf. on Diamond and Diamond-like Carbon Coatings, Diamond Films 90,* Crans-Montana, Switzerland (Sept. 17-19, 1990)

114. Meier, O. and Schlaak, U., Härtereitechnik-Bremen, unpublished measurements (1988)

115. Bichler, R., Haubner, R. and Lux, B., *High Temperature-High Pressure,* 22:99-118 (1990)

116. Söderberg, S., Westergren, K. and Sjöstrand, M., *1st Euro. Conf. on Diamond and Diamond-like Carbon Coatings, Diamond Films 90,* Crans-Montana, Switzerland (Sept.17-19, 1990)

117. Okoli, S., Haubner, R. and Lux, B., *J. de Physique,* 50(5)159-168 (1989)

118. Kikuchi, N., Kamatsu, T. and Yamaschita, H., *Proc. 9th. Symp. on ISIAT 85,* p. 233, Tokyo, Japan (1985)

119. Akatsuka, F., Hirose, Y. and Komaki, K., *Jpn. J. Appl. Phys.* 27(9) L1600-1602 (1988)

120. Kawarada, M., Kurihara, K., Sasaki, K., Teshima, A. and Koshino, N., 1st *1st Int. Conf. Diamond Sci. and Techn.,* P 2-03, pp. 42-43, Tokyo, Japan (Oct. 24-26,1988)

121. Hirose, Y., 1st *1st Int. Conf. Diamond Sci. and Techn.,* P 2-03, pp 38-39, Tokyo, Japan (Oct. 24-26, 1988)

122. Carrington, W. A., Hanssen, L. M., Snail, K. A., Oakes, D. and Butler, J. E., *Material Transactions A,* 20A:1282-4 (1989)

123. Hanssen, L. M., Carrington, W. A., Butler, J. E. and Snail, K. A., *Materials Letters,* 7(7,8):289-292 (Dec.1988)

124. Sato, Y. and Kamo, M. *Surf. Coat. Tech.,* 39/40:183-198 (1989)

125. Liou, Y., Inspektor, A., Weimer, R., Knight, D. and Messier, R., *J. Mat. Res.* 5(11):2305-2312 (1990)

126. Nyaiesh, A. R., Kirby, R. E., King, F. K. and Garwin, E. L., *J. Vac. Sci. Tech.* A3(3):610 (1985)

127. Suzuki, K., Sawabe, A., Yasuda, H. and Inuzuka, T., *Appl. Phys. Lett.* 50(12) (1987)

128. Matsumato, S., Hino, M. and Kobayashi, T., *Appl. Phys. Lett.* 51(10):737-739 (1987)

129. Li, C., Lau, Y. C. and Girshick, S. L., *179th Meeting of the Electrochemical Society,* Washington, DC (May 5-10, 1991)

130. Mitsuda, Y., Yoshida, T. and Akashi, K., *1st Int. Conf. Diamond Sci. and Techn.,* pp. 44-45, Tokyo, Japan (Oct. 24-26,1988)

131. Kawarada, M., Kurihara, K., Sasaki, K., Teshima, A. and Koshino, N., *1st Int. Conf. Diamond Sci. and Tech.,* pp. 42-43, Tokyo, Japan (Oct. 24-26,1988)

132. Litos, R., Doctoral Thesis, TU-Vienna (1992)

133. Anthony, T., *MRS Fall Meeting,* Boston, N3.8 (Nov. 30 - Dec.3, 1987)

134. Okoli, S., Doctoral Thesis, TU-Vienna (1990)

135. Satrapa, D., Haubner, R. and Lux, B., *1st Euro. Conf. on Diamond and Diamond-like Carbon Coatings, Diamond Films 90,* Crans-Montana, Switzerland (Sept. 17-19, 1990)

136. Sawabe, A. and Inuzuka, T., *Thin Solid Films,* 137:89-99 (1986)

137. Suzuki, K., Sawabe, A. and Inuzuka, T., *Jpn. J. Appl. Phys.,* 29:153-157 (1990)

138. Ohtake, N., Mashimo, Y., Yoshikawa, M., *2nd Int. Conf. on the New Diamond Sci. and Tech., ICNDST-2,* Washington DC (Sept. 23-27, 1990)

139. Bigelow, L. K. and Henderson, N. M., *2nd Int. Conf. on the New Diamond Sci. and Tech.,* ICNDST-2, Washington DC (Sept. 23-27, 1990)

140. Pan, X. X., Doctoral Thesis, TU-Vienna (1989)

141. Patterson, D. E., Bai, B. J., Chu, C. J., Hauge, R. H., Margrave, J. L., *2nd Int. Conf. on the New Diamond Sci. and Tech.,* ICNDST-2, Washington DC (Sept. 23-27, 1990)

142. Patterson, D., Chu, J., Bai, B., Xiao, Z., Komplin, N., Margrave, J. L. and Hauge, R. H., *179th Meeting of the Electrochemical Society,* Washington, DC (May 5-10, 1991)

143. Rudder, R. A., Thomas, R. E., Markunas, R. J., *179th Meeting of the Electrochemical Society,* Washington, DC (May 5-10, 1991)

144. Haubner, R. and Lux, B., *Surf. Mod. Tech. III,* The Minerals, Metals & Materials Society, pp. 835-839 (1990)

145. Matsubara, H. and Kihara, J., *1st Int. Conf. Diamond Sci. and Tech.,* pp. 114-115, Tokyo, Japan (Oct. 24-26,1988)

146. Pacher, O., Doctoral Thesis TU-Graz (1975)

147. Kikuchi, N. 59-166670 (A) Jpn. Patent Appl. 5840286 (1984); 59-166671 (A) Jpn. Patent Appl. 5840287 (1984); 59-166672 (A) Jpn. Patent Appl. 5841706 (1984)

148. Bigelow, L. K., private communication (1991)

149. Genasystem, Worthington, Ohio; *Materials and Processing Report,* 5(4/5), Elsevier Science Publishing Co., Inc (July/Aug 1990)

150. "Superhard Market Projections in USA," *Leading Edge Reports,* Cleveland, Ohio (Oct 1990)

151. "Diamond Depositions," *Science and Technology* 1(3) (Dec 1990)

152. Johannsen, P., *VDI-Berichte* 762:259-271 (1989)

153. Kornmann, M., Schachner, H., Funk, R. and Lux, B., *J. Cryst. Growth* 28:259-262 (1975)

154. Motzet, J., AUDI AG; private communications (1990)

155. Okuzumi, F., Matsuda, J. and O'Oka, K., *1st Int. Conf. Diamond Sci. and Tech.,* P 2-02, pp. 94-95, Tokyo, Japan (Oct. 24-26,1988)

156. Schedler, W., *Hartmetall für den Praktiker,* VDI-Verlag GmbH, Düsseldorf (1988)

157. Recker, K., *Z. Dt. Gemmol. Ges,* 1/2:5-34 (1984)

158. Matsumoto, S., Sato, Y. and Kamo, M., *Jpn. Kokai Tokkyo,* Koho JP 59-137211 (1983)

159. Komaki, K., Yanagisawa, M., Yamamoto, I., Arashida, K. and Fujimaki, T., *1st Int. Conf. Diamond Sci. and Tech.,* P 2-03, pp. 112-113, Tokyo, Japan (Oct.24-26,1988)

160. Lux, B., European Patent, EP 0226898 A2 (1986)

161. Haubner, R., *J. Ref. Met.-Hard Mat.,* 9(2):70-77 (1990)

162. Rödhamer, P., *12th Int. Plansee-Seminar,* C1, Reutte, Austria, (May 8-12, 1988)

163. Yoshida, T. and Mieno, M., *Int. Conf. on Metallurgical Coatings and Thin Films (ICMCTF),* San Diego, CA (April 22-26, 1991)

164. Ikeda, T., Satou, T., Satoh, H., *Int. Conf. on Metallurgical Coatings and Thin Films (ICMCTF),* San Diego, CA (April 22-26, 1991)

165. Doll, G. L., *Int. Conf. on Metallurgical Coatings and Thin Films (ICMCTF),* San Diego, CA (April 22-26, 1991)

166. Matsui, N., Enomoto, S., Ishiguro, T. and Ichinose, A., *Int. Conf. on Metallurgical Coatings and Thin Films (ICMCTF)*, San Diego, CA (April 22-26, 1991)

167. Murakawa, M., Watanabe, S. and Miyake, S., *PSE '90*, Garmisch-Partenkirchen (Sept. 10-14, 1990)

168. Saitoh, H. and Hirose, T., Matsui, H., Hirotsu, Y., Ichinose, Y., *Surf. Coat. Tech.*, 39/40:265-273 (1989)

169. Badzian, A. R., *Advances in X-Ray Analysis*, 31:113-127, Plenum Publishing Corporation (1988)

6

Characterization of Diamond Films

Wei Zhu, Hua-Shuang Kong and Jeffrey T. Glass

1.0 INTRODUCTION

Carbon occurs in many crystalline and disordered solid forms. Among these are a wide range of pure and mixed phases which are markedly diverse with respect to electrical, optical, chemical and mechanical properties. The chemical bonding of various carbon allotropes varies in hybridization from the sp^1 chain structure of carbyne to the sp^2 layered structure of graphite to the sp^3 covalently bonded cubic structure of diamond. Numerous disordered forms of carbon also exist which are composed of a mixture of hybridizations. This wide diversity of the carbon hierarchy, especially the stability of graphite (sp^2) relative to diamond (sp^3) at atmospheric pressure and room temperature, makes growth and characterization of diamond complicated. To distinguish between the many structurally and compositionally different types of carbon films which could be deposited from the vapor phase is not an easy task, especially to convince oneself that a second phase material is absent. Early skepticisms about low pressure vapor phase deposition of diamond stemmed largely from the well known metastability of diamond with respect to graphite and the inadequate characterization of the deposits. Therefore, a complete characterization of diamond and related materials deposited in a pressure and temperature regime where diamond is metastable relative to graphite presents a non-trivial and yet very important and interesting task. Consequently, sophisticated characterization techniques are needed.

As with any material, there are numerous techniques which can be useful in the characterization of diamond films. This makes a review of these characterization techniques very difficult, if not impossible, in any single article. In fact, even if this review is limited to those techniques already utilized to examine natural diamond and chemical vapor deposited (CVD) diamond, the length would be prohibitive. Therefore, only the major categories of CVD diamond analysis which have proven useful in the recent past will be covered, that is, chemical characterizations which identify and distinguish different carbon phases in terms of composition and chemical bonding, and physical characterizations which investigate the morphological and structural properties of the films such as defects and interfaces. These techniques include vibrational spectroscopies, electron spectroscopies, electron microscopies, and various diffraction techniques. A brief catalog of these techniques is given in Table 1 along with the types of information they can yield. Many of the various measurements have been related to those critical and most interesting areas such as identification of the diamond phase vs. graphite and amorphous carbon, nucleation and interfacial phenomena, surface morphology and internal defects. The techniques for the characterization of growth processes and specific physical property measurements are not covered in this review. Interested readers should refer to other chapters in this book and references therein.

The function of this chapter is to review these various methods used in the characterization and analysis of CVD diamond. Characterization of natural diamond will also be discussed for comparison purposes. The treatment is intended to be as broad and practical as possible, with little emphasis on detailed descriptions of instrumentations, physical principles involved and associated mathematical theories. Instead, emphasis is placed on abundant structural and chemical information on CVD diamond which are gathered from these various techniques. However, sufficient background information is provided for an understanding of the basic principles of the techniques. Many of the phenomena characterized are complicated and not yet fully understood. For this reason, different types of experiments are given even though they may yield approximately the same information. In this way, some generalizations can be made regarding both the selection of specific characterization techniques and the understanding of detailed properties of CVD diamond.

2.0 VIBRATIONAL SPECTROSCOPIES

The vibrations in the bulk structure or on crystal surfaces can be studied by Raman spectroscopy or infrared (IR) spectroscopy. In both

Table 1. Techniques for the Characterization of CVD Diamond Films

Technique		Basic physical process	Type of information
Vibrational spectroscopy	Raman spectroscopy	inelastic light scattering with lattice vibration which changes the polarizability	carbon bonding (sp^3/sp^2), structural perfection, stress state, crystallite (domain) size
	infrared spectroscopy	inelastic light interaction with lattice vibration which changes the existing dipole moment	chemical bonding (C-C, C-H, C-Si), absorption coefficient, refractive index, impurity
Electron spectroscopy	x-ray photoelectron spectroscopy (XPS)	excitation and emission of electrons by incident x-rays	surface species and chemical bonding (C, C-O, Si-C), surface electronic state
	Auger electron spectroscopy (AES)	creation of an inner hole by incident electron beams which is subsequently filled by an outer electron accompanied by the release of an Auger electron	surface composition and species, carbon bonding (sp^3/sp^2) (fine structure)
	electron energy loss spectroscopy (EELS)	electron energy losses by core level ionization, interband transitions or bulk and surface plasmon excitation	carbon bonding (sp^3/sp^2), surface electronic structure
Electron microscopy	transmission electron microscopy (TEM)	high energy (50-200 keV) electron beams passing through a thin sample and forming an image, the contrast of which is determined by crystallographic orientation or atomic mass	internal structural defects (twins, stacking faults, dislocations, secondary phases, interfaces, etc.)
	scanning electron microscopy (SEM)	scanning and focusing an electron beam (5-30 keV) over a surface and sensing the secondary electrons emitted to form an image	surface morphology

Diffraction methods	x-ray diffraction (XRD)	x-ray elastic scattering with atoms (Bragg diffraction)	phase (structure) identification, lattice constant measurement, crystal orientation or texture, residual stress
	electron selected area diffraction (SAD)	high energy (50-200keV) electrons elastic scattering with atoms (Bragg diffraction)	phase (structure) identification, lattice constant measurement, crystal orientation or texture, esp. when the sample is thin or both the size and amount are small
	low energy electron diffraction (LEED)	low energy (10-200eV) electrons elastically backscattered on the surface (Bragg diffraction, surface specific)	surface structure (reconstruction)
	reflection high energy electron diffraction (RHEED)	high energy (30-100keV) electrons elastically backscattered on the surface (Bragg diffraction, grazing incidence and emergence angles, surface specific)	surface structure
Others	secondary ion mass spectrometry (SIMS)	ionized surface atoms ejected by impact of high energy (1-20keV) ions and subjected to mass analysis	elemental identification and distribution (including H and He), chemical concentration, depth profile
	luminescence spectroscopy	excitation of electrons by photon or electron radiation resulting in emission of photons	electronic state, point defects
	Rutherford backscattering spectrometry (RBS)	collision of He^{+} ions (1-3 MeV) with "target" atoms leading to energy transfer	quantitative atomic composition (including H), depth profile
	spectroscopic ellipsometry (SE)	ellipsometry angles or pseudomorphic dielectric functions are modelled by linear regression analysis (LRA) and multilayer optical techniques to deduce photon energy independent structural parameters	microstructure (volume fraction of sp^2/sp^3 bonded carbon), layer thickness, densities, refractive index

techniques, the inelastic process exploited results from the same physical entity, the vibrating dipole, with complementary selection rules applied for Raman and IR activities. Both techniques have found tremendous uses in the characterization of CVD diamond films due to their ability to distinguish different chemical bonding types. Particularly in this respect, Raman spectroscopy has become the most popular and powerful technique in the analysis of the quality of CVD diamond films.

2.1 Raman Spectroscopy

Raman spectroscopy is one of many light-scattering phenomena. All these phenomena originate from the principle that the intensity of a beam of light decreases measurably when it passes through a nonabsorbing medium. The Raman effect belongs to the class of molecular-scattering phenomena. It is based on inelastic light-scattering arising from the interactions of photons with lattice vibrations or phonons. A laser tuned to a single line is used to excite the Raman scattering. The phonons are coupled to the photons through the polarization induced in the crystal by the electric field of the intense light beam. They are related by the polarizability as:

$$P = \alpha E$$

where P is the induced electric moment, E is the electric field and α is the polarizability. A lattice vibration is Raman active when the vibration changes the polarizability (1)(2). Some photons will be emitted from the induced oscillating dipoles which are either of the same frequency as the incident beam (elastic Rayleigh scattering) or have been frequency shifted by an amount equal to the vibrational frequency of the lattice (inelastic Stokes or anti-Stokes scattering):

$$P = E_o\alpha_o\cos(2\pi\nu_o t) + \tfrac{1}{2}E_o\Sigma\alpha_n\{\cos2\pi(\nu_o - \nu_n)t + \cos2\pi(\nu_o + \nu_n)t\}$$

where E_o is the electric field of incident light, ν_o is the frequency of the incident light, ν_n is the phonon frequency, α_n represents the polarizability as a function of the vibration mode, and t is the time. The shifted frequency is the Raman scattering which was named for Sir C. V. Raman, an Indian Nobel laureate, who discovered the effect in 1928 (3). The Stokes scattering (the normal Raman effect) is produced when energy is extracted from the light beam to the crystal and sets it into vibration, whereas the anti-Stokes scattering arises from the annihilation of the existing thermally-excited vibration. The vibrational modes are dependent on the atomic bonding in the solid. Therefore, Raman spectra can be used to interpret the structures in solids.

Raman scattering has been very useful for studying the chemistry and physics of carbon, particularly of CVD diamond films (4)(5). It is very sensitive to the nature of the carbon bonding. Therefore, it is able to distinguish various types of carbon: diamond, graphite, amorphous carbon and hydrogenated carbon. It has far exceeded other techniques in popularity for this purpose. The intense first order Raman peak for diamond is located at 1332 cm^{-1}, and a second order feature centered at 2458 cm^{-1} is often continuous and similar to the vibrational density of states (6). For highly oriented pyrolytic graphite, the first and second order peaks are located at 1580 cm^{-1} and 3240 cm^{-1}, respectively (7). An additional peak at 1355 cm^{-1} is observed for microcrystalline graphite which is related to the small domain size causing the momentum selection rules of the Raman scattering process to be relaxed (8). For amorphous carbon, there are two broad band features at approximately 1550 cm^{-1} and 1355 cm^{-1} (5)(9)(10). If the amorphous carbon is hydrogenated, the broad band at 1355 cm^{-1} becomes a shoulder of the 1550 cm^{-1} band (9). Since the Raman scattering efficiency for graphite is much greater than for diamond (11), small amounts of graphitic carbon co-deposited in CVD diamond films can readily be detected. Thus, if one is interested in a fast, sensitive and qualitative test for the quality of a film in terms of the bonding types, Raman spectroscopy is an excellent choice. Furthermore, information about domain (crystal) size, stress state and structural perfection can be obtained, because they cause broadening and shifts of the Raman peaks. However, the incorporation of low concentrations of sp^2 or π-type bonding may destroy any evidence of diamond in a Raman spectrum due to the large differences in scattering cross-sections. Therefore, if a small amount of non-diamond component is added to the film, a complex situation may arise in the interpretation of the Raman spectra. For example, in the authors' laboratory, certain films displaying mostly an sp^2 peak in the Raman spectra with only minor evidence of a diamond peak have been shown to yield EELS (electron energy loss spectroscopy) spectra essentially identical to diamond. Unfortunately, the exact "small amount" of the graphitic component in the film which is sufficient to distort the Raman spectra is not currently known.

Typical examples of Raman spectra for diamond films deposited under different conditions are shown in Fig. 1. For films deposited under optimal

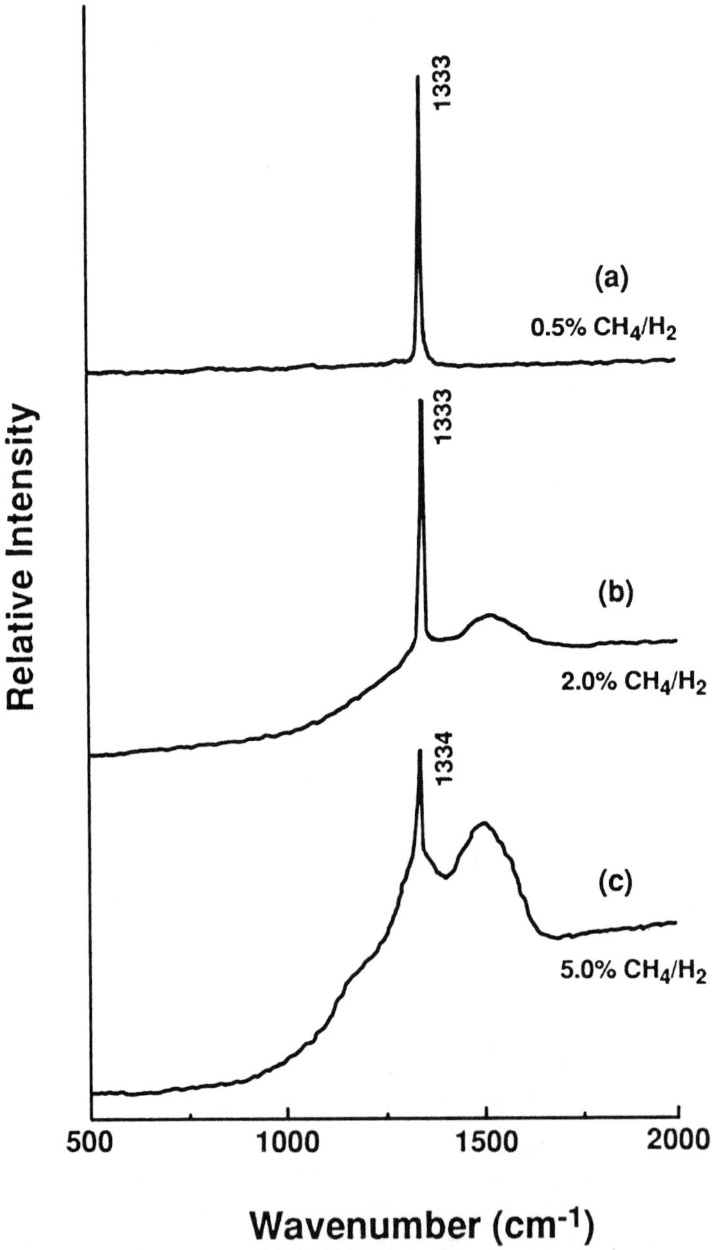

Figure 1. Raman spectra of CVD diamond films deposited at (a) 0.5% CH_4/H_2; (b) 2.0% CH_4/H_2; and (c) 5.0% CH_4/H_2. Other major deposition parameters include substrate temperatures of 800 - 950°C, pressures of 50 - 90 torr, and flow rates of 100 sccm (103).

conditions, the spectrum is dominated by a sharp feature at 1332 cm^{-1} which indicates the presence of crystalline diamond (Fig. 1a). A significant broad feature centered at ~1500 cm^{-1} is observed in addition to the 1332 cm^{-1} diamond signature for films deposited at non-optimized conditions (Fig. 1b). The intensity of this broad feature in CVD diamond can be fifteen times as strong as the peak intensity of pure graphite or microcrystalline graphite (12). This feature has been attributed to the presence of disordered or amorphous sp^2 carbon, although this assignment has not been universally accepted. As films become less like true diamond and more similar to "diamond-like" or amorphous carbon films (for example, as methane concentration increases in the feed gas during the deposition), this peak increases, and the 1332 cm^{-1} peak decreases or even disappears as shown in Fig. 1c. Furthermore, a new feature at 1140 cm^{-1} has also been observed (4). This new feature is still quite controversial and not well understood. It increases as the size of diamond crystallites decreases and has been observed most strongly for the diamond deposited at the beginning of a growth experiment (i.e., during the nucleation and initial growth phase). Nemanich et al. (4) indicated that the 1140 cm^{-1} mode is at a frequency similar to that proposed for disordered or microcrystalline sp^3 bonded carbon. Therefore, it is attributed to the small crystal sizes and/or disordered structure of diamond similar to the assignment of the 1355 cm^{-1} peak to microcrystalline graphite. An important aspect which should be noted from the comparison of the different spectral signatures is that all features associated with the sp^3 diamond bonding occur at frequencies lower than 1332 cm^{-1}, whereas the peaks due to sp^2 or graphitic bonding occur at higher frequencies than 1332 cm^{-1}. This is because the bond between sp^2 carbon is stronger than the sp^3 carbon.

It would be extremely useful to quantify the relative amount of sp^3 to sp^2 bonding in a carbon film via Raman spectroscopy. It would seem that if the relative scattering strengths are known, then it would be straightforward to determine the relative amounts of diamond and graphite in a "composite" CVD diamond film, assuming that the relative intensities of the two are directly related to the relative volume fractions of diamond and graphite present. Unfortunately, there are actually two important parameters related to the bonding structure which can affect the absolute Raman intensity: the scattering cross-section and the absorption coefficient. The absolute Raman cross-sections for diamond to graphite have been reported to be about 1:50 per unit volume (11) or 1:100 in terms of scattering per nearest-neighbor bond (1:75 per carbon atom) (12). On the other hand, graphite is highly absorbing, and diamond is effectively transparent in the visible range.

If samples containing graphite and diamond regions are illuminated with visible light, they will not be uniformly sampled. The transparent regions will be totally illuminated while only the surface of the absorbing regions will be measured. The effect of different absorptions becomes important when the domain size of the absorbing regions (graphite) is equal to or larger than the absorption length in graphite (~20 nm). Therefore, both concentrations and crystallite sizes of graphite have significant effects on the resulting Raman spectra. For CVD diamond, the size effect can be ignored for films in which there is no microcrystalline graphitic component. Films produced at optimal conditions seem to fulfill this requirement. In contrast, films produced at non-optimized conditions (for example, the methane concentration in hydrogen greater than 2.0%) contain features attributable to microcrystalline graphite, and quantitative analysis of these films may require a correction due to the domain size. Other considerations related to the quantitative analysis of sp^3/sp^2 bonding include (12)(13): accounting for the anomalously strong Raman intensity of the sp^2 region at 1500 cm^{-1} (a model of the sp^2 bonding configurations in which the π bonds in the graphitic regions are localized or attached to the sp^3 sites has been proposed by Shroder et al. to account for these features at ~1550 cm^{-1}) (12); considering the effect of crystallographic orientations on the intensity of the diamond Raman line (especially in the case of microfocus Raman spectroscopy); separating the photoluminescence background and accurately determining the integrated area of the spectral features (this requires careful spectroscopy and computer deconvolution); and determining the relations between the peak intensity and the peak width. Therefore, to obtain a quantitative analysis of the relative percentage of sp^3 to sp^2 bonded carbon, not only the exact amount and nature of the sp^2 regions are important, but also their distribution within the crystals and the domain sizes of the sp^2 regions. It is believed that quantification of the Raman results will be a high priority. Continuing efforts toward quantification of sp^3/sp^2 in CVD diamond films via Raman spectroscopy are currently underway.

2.2 Infrared Spectroscopy

IR spectroscopy is a complementary technique to Raman spectroscopy. The selection rule for IR active vibration modes is that the vibration produces a finite change in the existing dipole moment (2), as opposed to the change of polarizability for Raman active vibration modes. That is, only transitions involving vibrational states coupled by a dipole operator can be observed by IR spectroscopy. Not all vibrational modes lead to a change of electric dipole moment, and consequently they will not be IR active. The measurement of absorbed radiation is the basis of IR spectroscopy. An IR spectrum can be measured either by dispersive or interferometric methods (2)(14). The dispersive method involves dispersing radiation from an incoherent source via a prism or grating and is, therefore, an inefficient use of source intensity which requires monochromator entrance and exit slits. In contrast, FTIR (Fourier Transform IR) samples all the frequencies of radiation from the source using an interferometric method resulting in improved signal-to-noise ratios. IR is useful for detecting various carbon bonds and impurities in diamond films, since it is very sensitive to the nature of chemical bonding, type and distribution of defects, crystallite sizes, morphologies and grain boundaries.

Diamond is known to be transparent through most of the infrared as there are no first order (single phonon) absorptions permitted (below 1330 cm^{-1} or above 7.5 μm) (15). However, there are intrinsic second order (two phonon) absorptions (1600 - 2650 cm^{-1} or 3.8 - 6.0 μm) both predicted theoretically and observed experimentally (16). In addition, all diamond shows additional or extrinsic IR absorptions as a consequence of the defects and impurities inherent which destroy the translational symmetry of the crystal. Indeed, these are the characteristics that lead to the use of IR properties as a basis for classifying both natural and synthetic diamonds into the various type designations. For instance, the absorptions in the range of 1000 - 1600 cm^{-1} (or 6.0 - 10.0 μm) due to the nitrogen impurity are used to distinguish the type I and type II diamonds (16)-(18).

As IR spectroscopy has been the method of choice for typing natural and synthetic diamond, it is no less applicable to thin film diamond. IR can provide information about the carbon-hydrogen bonding (stretching mode 2700 - 3300 cm^{-1}; bending and other deformation mode 800 - 1650 cm^{-1}) in the diamond films as well as the carbon-carbon bonding (19). In general, C-H stretches with sp^1 hybridized carbon occur at 3100 - 3250 cm^{-1}. The sp^2 and sp^3 C-H stretching modes are more complex, and generally, bands occurring above 2960 cm^{-1} can be assigned to sp^2 C-H bonds and those below to sp^3 C-H bonds (20). Ideally, by summing up the areas of the absorption bands in these regions, the proportion of sp^2 and sp^3 bonded carbon as well as hydrogen content can be deduced. However, there is currently limited quantitative data on infrared properties of CVD diamond, and an accurate assignment of each IR spectral feature from CVD diamond still presents a challenge. The reasons are that the C-H stretching and bending modes are usually weak absorptions in IR spectra and are, therefore, complicated by the presence of interference bands due to the limited film thickness (a film must be made thick enough to make reliable absorption measurements) and scattering from surface roughness and grain boundaries (diamond films grown with a variety of techniques exhibit typically a surface roughness of 1 - 3 μm which induces severe optical scatter). These problems make interpretation and quantitative absorption measurements of the IR spectra difficult. Nevertheless, numerous reports on the measurements of IR spectra from rough CVD diamond films have been published (21)-(24). These IR spectral features have provided valuable information on the type and amount of impurities present and relative absorption coefficients of CVD diamond films.

Yarbrough et al. (21) measured both transmission and reflection spectra of a 22.5 μm thick diamond film over a middle IR range from 400 - 4000 cm^{-1} (or 2.5 - 25 μm). They reduced the raw data by using the method of Cody et al. (25) to obtain the absorption coefficient (α) and refractive index (n). They found two distinct absorption bands, one broad and weak at ~1300 cm^{-1} (or 7.8 μm) due to C-H bending vibrations plus possible defect induced one-phonon absorption, and the other, much stronger, due to C-H stretching bands at about 2800 cm^{-1} (or 3.5 μm). Both of these bands were attributed to hydrogen bonded to sp^3 hybridized carbon. No evidence of IR absorptions at 3000 - 3200 cm^{-1} due to hydrogen bonded to sp^2 or sp^1 carbon sites was found. The intrinsic two phonon absorption of diamond was only weakly observed due to scattering from the rough growth surface and the relatively small sample thickness employed. The absorption coefficient was found to be extremely high (in the range of 212 cm^{-1} to 20 cm^{-1}) and varied with wavelength. The highest effective energy loss occurred at the high energy end of the spectrum. The effective refractive index also varied from a low of 2.36 to a high of 2.73. This relatively high "absorptivity" was an indication of the energy losses arising from all sources including scattering from the rough as-grown surface which was suspected to be a primary source of energy loss in the case of unpolished CVD films. It should be noted that non-diamond carbon also contributes to energy losses, particularly if the scale of the microstructure is of the same order as the wavelength of interest.

Similar IR analysis has been performed on a thick free-standing diamond film (400 μm thickness) by Wild et al. (22). In this case, the intensity of the scattered light as a function of wavelength (which is a smooth and featureless curve, e.g., a cubic spline) was subtracted from the measured transmission spectrum leaving only the relatively sharp absorption bands. They observed the intrinsic two-phonon absorption bands of diamond (1600 - 2650 cm^{-1}), together with extrinsic absorptions, e.g., carbon-hydrogen stretching bands from 2800 - 2970 cm^{-1} and absorptions in the diamond single-phonon region (< 1330 cm^{-1}) due to C-H bending plus possible contributions of defect-induced one-phonon absorption. The absorption coefficient in the C-H stretch region was about 15 cm^{-1} and in the single phonon region was about 4 cm^{-1}. They estimated the hydrogen content to be less than 1.0% based on the absorption intensities.

Bi et al. (23) have been able to extract absorption coefficients from diamond films as thin as 10 μm by mathematically correcting the spectra for the contribution of surface roughness induced scatter. The two major peaks in the C-H stretch region at 2923 cm^{-1} (or 3.4 μm) and 2854 cm^{-1} (or 3.5 μm) were characterized as C-H_2 (methylene groups) asymmetric and symmetric stretching modes based on reference molecular spectra data. They indicated that the absorptions at or within ± 5 wavenumbers of these values were typical for CVD diamond deposited by all reported deposition methods.

Hartnett and Miller (24) used a microbeam FTIR instrument to acquire spectra and used its data manipulation capabilities to subtract the effects of surface scatter. They examined the role of hydrogen incorporation in CVD diamond in relation to its responsibility for absorption in the 7 - 12 μm region. They found that the incorporation of small amounts of hydrogen during the growth process resulted in absorption in the "C-H stretch" region of ~2900 cm^{-1} as well as absorption in the extrinsic defect related single-phonon region from 750 to 1430 cm^{-1} (Fig. 2). The maximum absorption in the single-phonon region occurred at 7.5 μm (or 1332 cm^{-1}), and the tail of the absorption fell rapidly towards longer wavelengths. Approximate absorption coefficients at the maximum 7.5 μm range were from 4.2 to 33 cm^{-1} for the sample studied. Hartnett and Miller (24) postulated that the hydrogen in CVD diamond was associated with a planar structural defect in the diamond lattice such as a twin or stacking fault. The amount of hydrogen necessary to cause significant absorption in both the C-H stretch region and the single phonon region was less than 1.0%. Furthermore, the magnitude of absorption in the C-H stretch region changed very little when the film was annealed to temperatures up to 1500°C, indicating that the hydrogen was strongly bonded in the diamond lattice. They also correlated the IR absorptions in the C-H stretch region (2700 - 3100 cm^{-1}) with the Raman signatures of these films and found that with the increase of absorption in the C-H stretching region, the IR absorption in the single phonon region (750 - 1430 cm^{-1}) increased, the Raman linewidth at 1332 cm^{-1} increased, the Raman intensity at 1536 cm^{-1} increased, and the luminescence background increased. The luminescence background suggested that hydrogen in diamond may be electrically active. Such a conclusion is also supported by electrical measurements on CVD diamond films (26). However, more studies on the temperature dependence of the absorptions and photoconduction measurements are needed to further determine the nature of the hydrogen impurities and the defects with which they are associated.

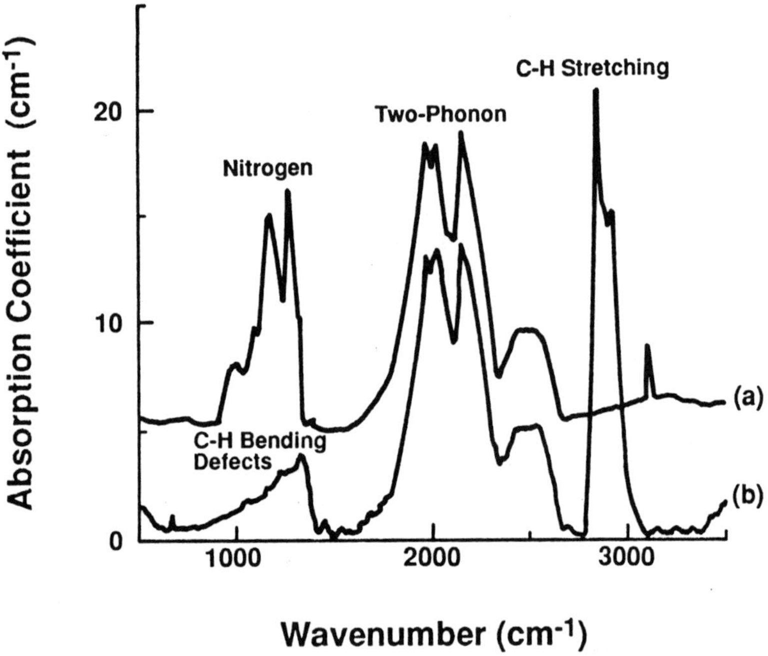

Figure 2. IR absorption spectra obtained from (a) natural polycrystalline diamond (ballas); and (b) a CVD diamond thin film (24).

It is clear that the rough surface and polycrystalline nature of diamond films present a formidable problem for obtaining accurate IR data. Various methods have been used to reduce or eliminate this problem and to improve the IR measurements (27). The techniques for reducing optical scatter involve modifying surface morphologies during the growth process by controlling nucleation density, re-nucleation frequency and/or the orientation of crystal faces at the film surfaces. However, techniques for post-deposition polishing of the surface of CVD diamond are more realistic at the present time, and they include abrasive polishing (28), chemical polishing (29) and ion/sputter polishing (30). Wang et al. (31) used a heated cast iron scaife (350°C) to polish the rough growth surface. The polishing by this method was believed to be chemical in nature and to proceed via a reaction between carbon and iron to form iron carbide. The resulting iron carbide oxidized in air at the high temperature and was easily removed by mechanical abrasion. Several methods to accelerate the polishing rate have also been tested. The polished diamond surface was a very smooth, mirror-like surface with a roughness of about 10 nm. For these polished films, nearly 70% transmittance was obtained over the range of 600 - 4000 cm^{-1} which approached the theoretical value of transmittance for diamond (assuming n = 2.41). The transmittance for non-polished films was much lower and varied strongly with the wavenumber. Besides the C-H stretching absorption bands near 2900 cm^{-1}, an absorption band at ~800 cm^{-1} was observed which corresponded to the stretching vibration of Si-C presumably formed at the interface between the diamond film and the silicon substrate. The absorption coefficients were found to vary from as high as 150 cm^{-1} to as low as 7 cm^{-1} depending on the film quality and the wavenumber. The refractive index was found to be between 2.41 and 2.49. These values were still much larger than the corresponding values of the natural type II diamond, but were closer than the values for the non-polished film reported by Yarbrough et al. (21). The sources of energy losses in the polished films were the presence of a substantial amount of non-diamond carbon and various structural imperfections such as grain boundaries, stacking faults, etc.

IR spectroscopy has also been successfully used by Meilunas et al. (32) to examine the early stage of the diamond nucleation and growth on silicon to detect the presence of a possible carbide layer. They observed the time evolution of the SiC layer from the IR measurement of the Si-C stretching located at 800 cm^{-1}. The SiC signal increased up to a point and then saturated. Interestingly, the SiC absorption saturated before the diamond crystals fully covered the substrate, possibly due to diffusion limitations on the SiC growth. The thickness of the SiC layer was estimated at 10 nm, much less than that of the Mo carbide layer (~1.5 µm) formed during diamond growth on Mo (32). The growth rate of the SiC layer was also much slower than in the Mo carbide case.

3.0 ELECTRON SPECTROSCOPIES

If a surface is irradiated with a probe beam of photons, electrons or ions, one generally finds that photons, electrons and ions are produced in various combinations. A given method consists of using a particular type of probe beams and detecting a particular type of produced species. If the phenomenon is studied as a function of the energy of the produced species or the energy of the probe beam, the method becomes a spectroscopic technique. Quite a few combinations are possible, but only electron spectroscopies are considered here due to their widespread use in the analysis of diamond. Such spectroscopies involve the detection of electrons emitted or scattered from the surface in the energy range of 5 - 2000 eV. Although the methods which will be discussed differ in various aspects, they all employ energy analysis of emitted electrons to gain information about the electronic energy levels in the surface regions and typically allow analysis of the composition and chemical bonding at the surface. Of these spectroscopic techniques two stand out because of their exceptional popularity and widespread use in surface analysis: x-ray photoelectron spectroscopy (XPS) and Auger electron spectroscopy (AES). Both techniques are primarily concerned with determining the atomic species present on the surface and the local chemical environment. Electron energy loss spectroscopy (EELS), which has also been very useful in studying the chemical bonding and electronic structure, is also discussed. As will be shown, these spectroscopies which can monitor changes on the growing diamond surfaces are essential to improve our understanding of the chemical state and structure in the diamond nucleation and growth processes.

3.1 X-ray Photoelectron Spectroscopy

XPS is, in principle, a very simple process. A mono-energetic x-radiation (typically Al or Mg K_α emissions at 1486.6 eV or 1253.6 eV, respectively) of energy $h\nu$ (h: Planck constant and ν: frequency) penetrates the surface and is absorbed by an electron (1s, 2s, 2p, etc.) with a binding energy E_b below the vacuum level, which then emerges from the surface with a kinetic energy ($h\nu$ - E_b). Therefore, the energy distribution of photoemitted electrons is simply the energy distribution of electronic states in the surface shifted up in energy by an amount $h\nu$, although this simple picture is complicated in practice (33). Because core level energies of electrons are essentially characteristic of the atomic species, the observation of binding energy peaks in an XPS spectrum is an indication of the presence of a particular elemental species in the surface region. The additional information on exact peak positions is capable of indicating the chemical state of some of these elements. Since the method is often used for chemical analysis, it is sometimes termed ESCA for Electron Spectroscopy for Chemical Analysis.

Diamond surfaces are particularly susceptible to irreversible graphitization, and even common surface analytical tools such as AES can induce surface reactions on the time scale of a single experiment. Thus, one major advantage of XPS is the non-damaging nature of the incident x-rays. It has been widely employed to study the atomic and electronic structures of diamond surfaces. Pate et al. (34)(35) obtained C1s spectra as a function of hydrogen adsorption and desorption on natural diamond surfaces. They found that if a polished (111) surface was annealed at temperatures lower than 900°C, a (1x1) LEED (low energy electron diffraction) pattern existed, and no occupied surface states were observed by XPS. After annealing at temperatures higher than 900°C, the LEED pattern reconstructed to (2x2/2x1), and occupied surface states were observed near the top of the valence band. It was found that the polished surface, devoid of surface states, was covered with chemisorbed hydrogen, whereas hydrogen was absent from the surface annealed at temperatures above 900°C. A surface-induced binding energy shift of 0.95 ± 0.1 eV to lower binding energy for carbon atoms on the (111) (2x2/2x1) surface vs. bulk atoms was observed (the C1s bulk bind energy is 284.7 eV below the Fermi energy, E_F) (36). While the magnitude of the energy shift associated with the surface core-level peak was related to the nature of the bonding in the (111) (2x2/2x1) reconstructed surface, the relative bulk-to-surface intensity ratio was determined by the number of reconstructed special surface atoms and the mean free path of the photoelectrons in the material. By comparing the C1s photoemission intensity from a clean diamond (111) surface to those obtained from a fluorine-covered surface, Morar et al. (37) determined that only a single (111) layer of atoms participated in the diamond (2x2/2x1) surface reconstruction.

As has been shown by McFeely et al. (36), the binding energy of core electrons in diamond vs. graphite or amorphous carbon shifts less than 1 eV. That is, the value of the C1s binding energy for pure carbon phases (C-C bonding) is only slightly affected by the hybridization of the atoms. Unfortunately, many conventional XPS systems which do not utilize crystal monochromators for the incident x-rays do not have sufficient resolution to distinguish such a small shift. Thus, for many conventional systems, in the absence of static charging (the charging effect can be eliminated by illumination of the sample with a xenon discharge lamp), diamond, graphite and amorphous carbon have essentially the same C1s binding energy for practical purposes. Although the valence band XPS spectra can reveal noticeable and significant differences between diamond and graphite or amorphous carbon (36), the photoelectron intensities are so low that accurate measurements can be difficult. However, the C1s binding energies can easily be used in many cases to distinguish pure carbon phases such as diamond, graphite, amorphous or glassy carbon from carbon compounds such as carbonates and carbides. Belton and Schmieg (38) used this method to characterize carbon deposited via hot filament CVD. By comparison of the C1s binding energies to a highly oriented pyrolytic graphite standard, they were able to distinguish pure carbon phases, such as diamond and graphite (C1s = 284.7 eV), from compounds, such as SiC (C1s = 283.4 eV) and oxygenated carbon phases (C1s > 285.5 eV). It should be noted that the C1s peak from hydrocarbons (C-H bonding) is actually very close to that of a pure carbon phase (C-C bonding) (33)(37). Therefore, the C1s peak for C-H bonding usually cannot be resolved experimentally from that for C-C bonding. Since the binding energy shifts are small in a sample containing a mixture of carbon phases, multiple peaks from different phases are almost always poorly resolved or unresolved. Under such conditions, they used curve fitting to interpret the data, and as many as seven parameters (binding energy, full width at half maximum (FWHM), peak height, Gaussian/Lorentzian ratio, and three parameters to define the tail) were used for each peak to arrive at a best fit. Furthermore, by comparing the FWHM of the unknown C1s to that for graphite, it can be determined whether more than one type of carbon is present. However, the final identifications of the pure carbon phases contained in the film must be accomplished with another technique such as EELS as discussed later.

XPS has also been used to examine the growing diamond surface at selected intervals during the nucleation and growth processes in both hot filament CVD system (39) and microwave plasma CVD system (40) without exposing the sample to air. The experiments were capable of offering a time-dependent monitoring of the surface during diamond growth, and thus allowing characterization of the surface on which diamond was actually growing. As shown in Fig. 3, before growth, XPS showed that the Si substrate was covered by a thin layer of SiO_2 (Si2p = 102.7 eV) and some oxygenated carbon contaminants (the C1s peak for C-O bonding is located at higher binding energy, > 285.5 eV); no diamond particles were detected on the surface pretreated with diamond paste. However, after 0.25 hour of growth, a distinct SiC phase (Si2p = 100.3 eV and C1s = 283.4 eV) was formed at the expense of some of the native oxide and surface contaminants. Between 0.25 hour and 1 hour, the SiC layer grew, and diamond nucleation began. Approximately 15% diamond (C-C) and 40% carbide (C-Si) bonding was observed after 0.5 hour growth (determined by measuring the peak areas and applying known sensitivity factors) (41). After 1 hour, the XPS signal (C1s = 284.7 eV) showed an increasing amount of diamond component, and by 4 hours of growth, only diamond bonding was observed (the identification of diamond was confirmed by Raman spectroscopy). Therefore, for the silicon substrate, three distinct phases of diamond film growth were identified: removal of initial substrate surface contaminants, formation of a SiC layer, and nucleation and growth of diamond. However, the XPS results cannot determine the actual site upon which the diamond is nucleating, but only those phases present on the surface. Therefore, whether diamond nucleates on Si or SiC is still an open question. A technique which has spatial capabilities, such as scanning tunneling microscopy (STM), is necessary to determine the answer to this question. Recent results suggest that, regardless of Si or SiC as the substrate surface, diamond actually nucleates on some types of carbon-hydrogen (C-H) precursors on the surface (41).

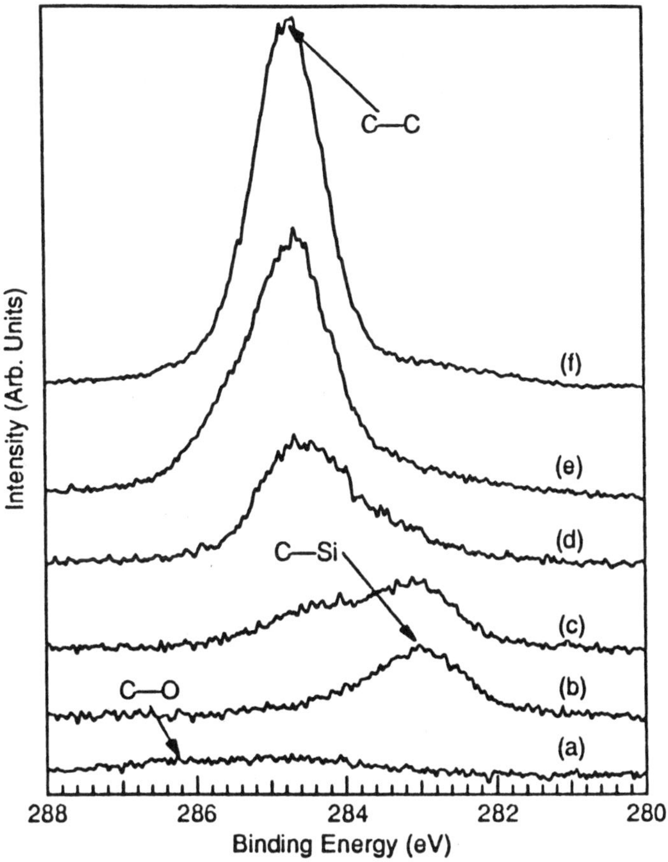

Figure 3a. XPS spectra of the C1s region for different growth times of a diamond film grown on Si by microwave plasma CVD: (a) Before growth; (b) 0.25 h; (c) 0.5 h; (d) 1 h; (e) 2 h; and (f) 4 h (41).

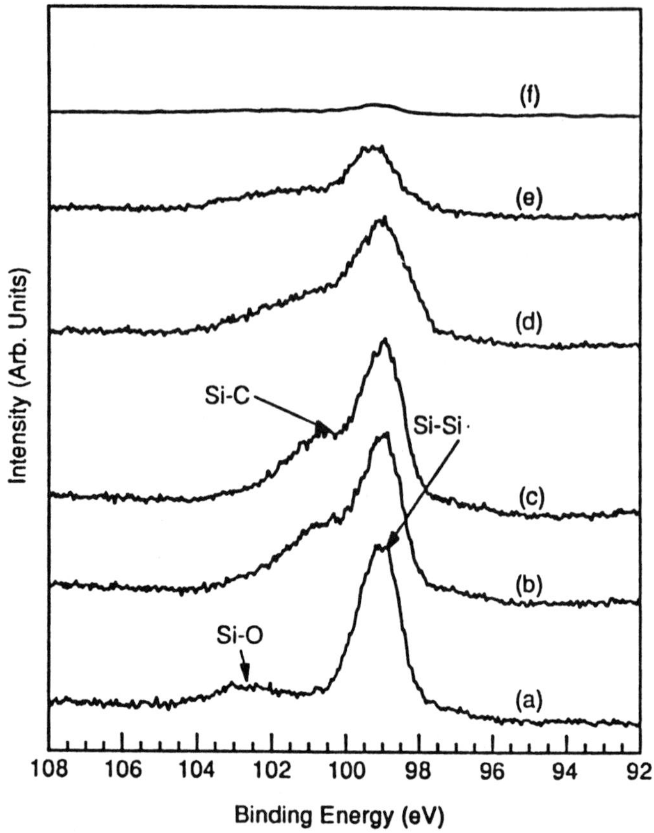

Figure 3b. XPS spectra of the Si2p region for different growth times of a diamond film grown on Si by microwave plasma CVD: (a) Before growth; (b) 0.25 h; (c) 0.5 h; (d) 1 h; (e) 2 h; and (f) 4 h (41).

For diamond growth on Ta substrates, similar results were obtained by in vacuo XPS analysis (40). It was found that a layer of Ta carbide was formed very rapidly at the beginning of the growth (<0.25 hour). Diamond then nucleated and grew on this Ta carbide layer. In another study of the hot filament CVD of diamond films on ordered Ni (100) substrate, Belton and Schmieg (42) used XPS together with EELS, x-ray excited AES (XAES), LEED and SEM to examine the sample at selected intervals during the growth process. On Ni (100), they found that several distinct changes in the substrate preceded the formation of diamond particles. The sample was prepared prior to growth by a series of sputtering and annealing cycles to obtain a well-ordered surface. The first stage of growth involved the deposition of a graphite layer which was in good registry with the substrate. Following this layer was a thicker layer of disordered glassy carbon upon which the diamond particles were nucleated. The deposited diamond particles were not in registry with the substrate because of the intermediate disordered layer. Deposition of diamond on nickel in the author's laboratory has also resulted in the formation of a graphite layer which impedes the nucleation of diamond particles directly on the nickel surface. However, a recent report indicated that diamond can be grown heteroepitaxially on Ni substrates (43). This report, if confirmed, represents a breakthrough in the CVD diamond technology. XPS can certainly play an important role in elucidating the growth mechanisms of such a heteroepitaxy.

XPS has many favorable aspects as a surface sensitive spectroscopic technique. It is widely available and relatively easy to perform. It is surface sensitive while generally nondestructive. The disadvantages are that it usually averages over a large region (~20 mm^2) making nucleation difficult to study; it is not very sensitive to minority species (sensitivity about 0.5% of the surface atoms); signals from the valence band that might distinguish local carbon bonding (sp^2 vs. sp^3) are weak; and carbon core levels are insensitive to subtle changes in local bonding configurations. This last disadvantage has limited the usefulness of XPS for the study of carbon films. However, information contained in the characteristic loss electrons associated with the C1s core level is often valuable (Sec. 3.3). This combination of XPS and EELS processes will become an effective method for identifying the number and kinds of carbon phases present on the surface during diamond nucleation and growth.

3.2 Auger Electron Spectroscopy

AES utilizes a relatively high energy electron beam (2 - 3 keV) to ionize a core electron (for example, K electron) of a surface atom, creating a hole. An outer electron (for example, an L_1 electron) may drop into this hole while simultaneously emitting energy (E_K - E_{L1}). This energy may appear as a characteristic x-ray photon emission or as kinetic energy given to another outer electron (for example, an L_{111} electron) which is emitted from the atom with energy (E_K - E_{L1} - E_{L111}). These competing processes are dominated by the photon emission only when the initial core hole is deeper than about 10 keV which is indeed the physical process used in a conventional x-ray generator. The alternative radiationless emission of the energy as the electron kinetic energy is the Auger effect, named after its discoverer Pierre

Auger (44). In either case, the emitted particles are characteristic of some combinations of atomic energy levels of the sample and so form the basis of a core level spectroscopy. While the emerging photons have long mean-free paths in solids and give rise to a bulk analytical tool (x-ray microprobe analysis), the detection of Auger electrons provides a surface sensitive probe of chemical composition (33)(45). Although Auger electron emission, being a three level process, is intrinsically more complex than the photo-electron emissions discussed in the preceding section, its strength lies in the fact that it can be generated by incident electron beams and that the production, focusing and deflection or scanning of electron beams is a well-developed technology. Most AES studies simply use the Auger spectrum as a fingerprint of the chemical composition and are not concerned with a detailed analysis of the basic processes.

It is typical for AES to detect 0.1% of almost any impurity species (other than hydrogen and helium) (46). Because it is easy to produce high intensity electron beams relative to x-ray sources, AES is typically faster and more sensitive than XPS because of the higher incident fluxes and comparable excitation cross-sections. For the same reason, AES is much more likely to induce sample surface damage. The incident electron beam can generate empty states in the band gap and loss of structure in the valence band via surface damage or the desorption of surface species. Another great strength of AES, in comparison with XPS, is the high spatial resolution associated with the use of high current electron beams of small size. Surface compositional imaging using Auger electron emissions is also possible. This chemical imaging in AES, particularly combined with ion sputtering to produce depth profile images, is of great value in the characterization of semiconductors and semiconductor devices.

AES has proven to be very useful in distinguishing the diamond phase from graphite and amorphous carbon. Since various measurements for the C1s binding energy of pure carbon phases fall within 1 eV of the value 284.7 eV, the difference in the binding energy of core electrons in diamond compared with graphite or amorphous carbon is too small to observe in most conventional electron spectrometer systems. In addition, charging of the substrate during the spectroscopic measurement causes energy shifts and is difficult to assess quantitatively, thus adding further ambiguity to binding energy measurements. However, Lurie and Wilson (47) have shown that the AES fine structure of the C KLL peak is sensitive to the three types of pure carbon phases (i.e., diamond, graphite, amorphous carbon) and is not affected significantly by charging. The change in the chemical environment of the carbon atom in diamond, graphite and amorphous carbon is reflected in a variation of the line shape of the carbon Auger signals. This is analogous to those fine structure patterns which have been used to identify the presence of metal carbides vs. elemental metal phases (48). In the case of diamond vs. graphite, this fine structure has been explained by a band structure model correlating the experimental peaks with the calculated density of states in the respective band structures. A detailed calculation of the density of states of diamond by Painter et al. (49) gave three occupied maxima in the valence band (V_1, V_2, V_3) and two empty maxima (C_1, C_2) in the conduction band. Although a full and unambiguous interpretation of the Auger transition is not yet available, it is possible to make a comparison between the Auger fine structure and predominant features in the valence band density of states. For example, there are a total of four peaks for diamond in the fine structure of the C KLL peak, while graphite has three, and amorphous carbon has only two. The main Auger peak A_0 (268 eV) for diamond is assigned as KV_1V_1, the peak A_1 (256 eV) as KV_2V_2 and other peaks as resulting from surface and bulk plasmon losses from A_0(47). In addition, a high energy satellite peak has also been observed in Auger spectra for diamond which possibly arises from an Auger transition from a doubly ionized carbon atom.

Therefore, in an AES system with moderate resolutions (~1 eV for the data shown here), the shape of the fine structure on the lower energy side of the C KLL peak can be used to distinguish diamond from graphite and amorphous carbon. Differentiated C KLL Auger spectra which illustrate the C KLL fine structure for a typical CVD film are shown in Fig. 4a. Inset in this figure are the spectra for natural diamond, graphite, and amorphous carbon from Lurie and Wilson (47). It can be seen that diamond, which contains only sp^3 bonds, yielded a very different fine structure from graphite and amorphous carbon which both contain sp^2 bonding. Direct comparisons of these spectra with the spectrum obtained from the film revealed that the film yielded a fine structure very similar to natural diamond. That is, the intensity of the fine structure peak closest to the primary C KLL peak was greater than the peak located at a slightly lower energy. The situation was reversed for graphite as seen by Lurie and Wilson's results in the inset. Amorphous carbon yielded a "smeared" fine structure relative to diamond and graphite as also shown in the inset. Therefore, it appeared that the surface of this CVD film was composed of diamond (sp^3) type bonding.

An interesting change in the chemical bonding near the film surface was observed after argon sputtering. This is significant because argon sputtering is commonly used for substrate cleaning prior to the vapor deposition of many materials. Figure 4b is an energy scan taken about the

a

b

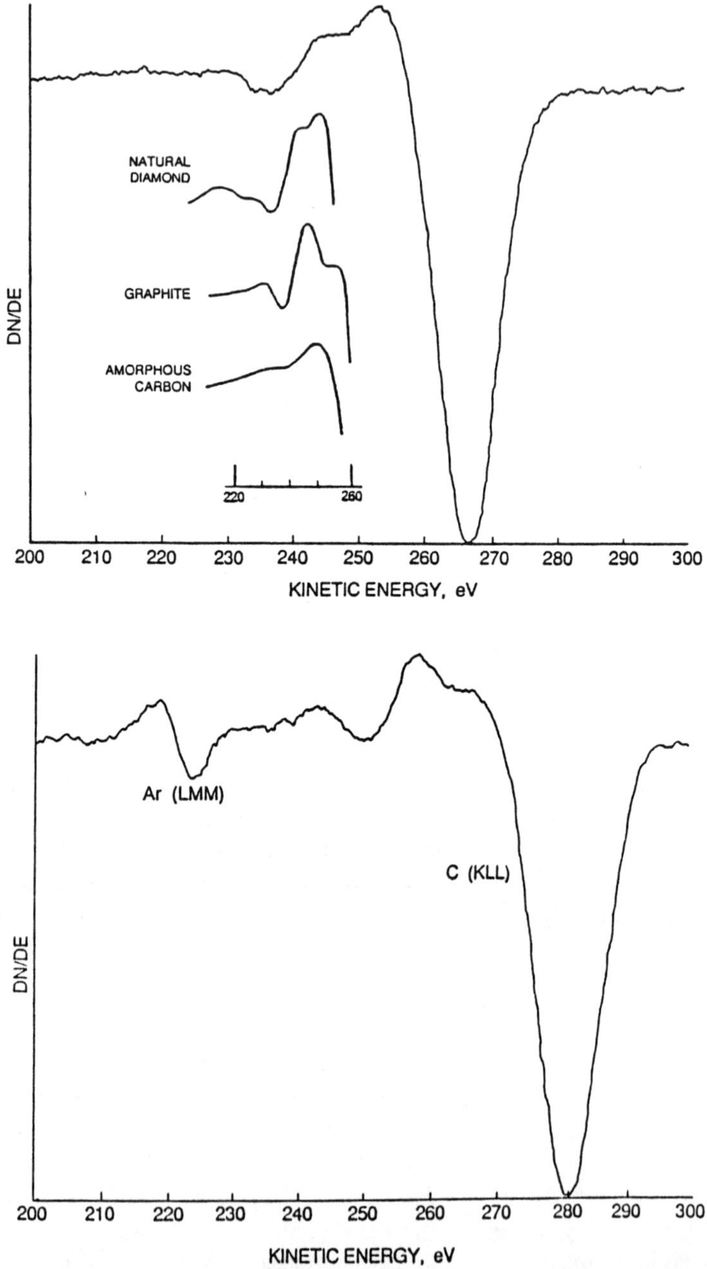

Figure 4. AES spectra near the carbon KLL peak obtained on (a) a diamond film with inset spectra from Lurie and Wilson (47); and (b) a diamond film after argon ion beam sputtering (53).

C KLL peak after sputtering for 30 seconds with a rastered (9x9 mm^2) Ar^+ beam of relatively low accelerating voltage and beam current (V = 1.5 kV, I = 2.5 A). Comparisons of this spectrum to that shown for graphite in the inset revealed that the argon sputtering had caused the shape of the AES fine structure to become more graphitic. That is, the lower energy peak intensity was much greater than the intensity of the peak closest to the major carbon peak, which was reversed from what was observed for diamond. This was an indication that the argon sputtering had increased the sp^2 bonding component at the surface of the film. This result has recently been confirmed by real time spectroscopic ellipsometry measurements (50).

The AES fine structure of the C KLL peak has also been utilized to monitor the phase changes in vacuo as a function of time in the initial stage of diamond growth. Differentiated C KLL Auger spectra obtained after different growth times on silicon are shown in Fig. 5. Prior to growth, the carbon Auger lineshape resembled that of hydrocarbon contamination. After 0.25 hour, the fine structure at lower energies of the major C KLL peak was indicative of the presence of silicon carbide. This is in agreement with both XPS results (discussed in the above section) and TEM results (Sec. 4.1). By 1 hour, definite changes in the fine structure associated with the C KLL had occurred, and the spectrum resembled a mixture of diamond and silicon carbide. The spectrum obtained at 4 hours of growth had the fine structure expected for a nearly continuous diamond film.

Because of the relative ease of producing sufficiently energetic (1.5 - 5 keV) electron beams of high intensity (1 - 100 μA), AES is often performed with incident electron beams. Nevertheless, Auger electrons can, of course, be produced by x-ray radiation, and their study in conjunction with direct photoelectron emissions can sometimes give special complementary information. For example, Belton and Schmieg (42) used x-ray (Al K_α 1486.6 eV) excited C KLL AES transition to study the diamond nucleation and growth on Ni (100). Since the Auger transition was excited with x-rays, effects due to electron-beam damage (e.g., electron stimulated adsorption and desorption of surface species and charging) were not a problem (x-ray damage to diamond and graphite has not been observed). Although the resolution was poor, the resulting spectra did reveal different fingerprints of deposits at different growth intervals.

3.3 Electron Energy Loss Spectroscopy

An electron of energy ~5 to 2000 eV passing through a solid can lose energy in a number of ways. The main processes involved are plasmon

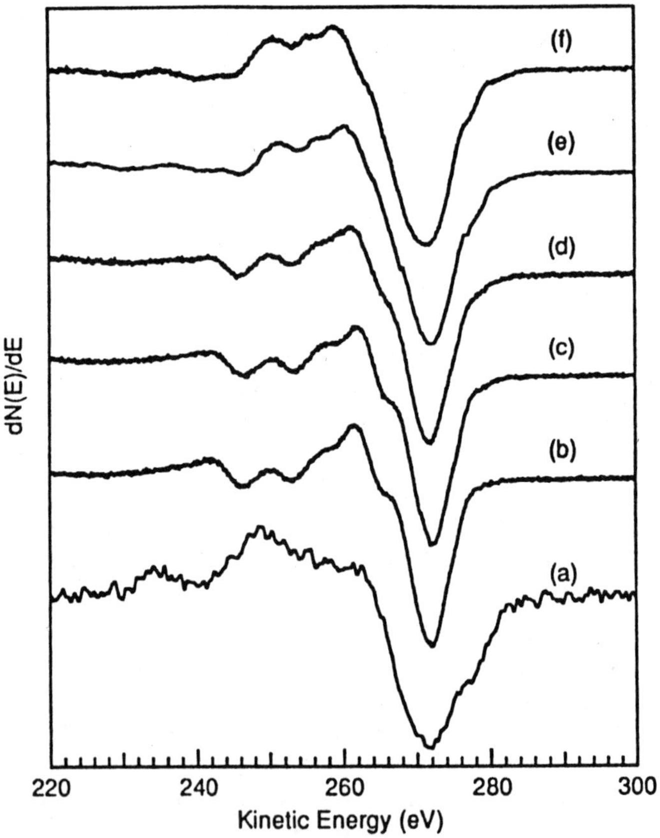

Figure 5. AES fine structure of the carbon KLL peak for a diamond film at different growth times: (a) Before growth; (b) 0.25 h; (c) 0.5 h; (d) 1 h; (e) 2 h; and (f) 4 h (41).

scattering, single-particle electron excitations involving valence electrons and ionization of atoms via core level excitations (51). These phenomena form the physical basis of EELS which can be performed in several different ways, such as: *(i)* the technique utilized by Egerton and Whelan (52) and Williams et al. (53) using high energy electrons (80 - 200 keV) in the transmission mode in a transmission electron microscope, *(ii)* the characteristic loss features detected in the back-scattering direction from 132 - 144° with respect to the normal incident electron beam (47), and *(iii)* the loss spectra obtained via an x-ray photoelectron spectrometer from x-ray excitation of the C1s level (36)(38)(42). In the analysis of carbon phases, EELS has the ability to distinguish changes in the carbon bonding configuration (54). A series of loss peaks are observed which can be attributed to various interband transitions between conduction and valence band maxima as well as plasmon losses. For example, the plasmon losses at lower energies near the zero loss peak can be used to identify different phases of carbon, although the differences are usually not as obvious as those shown by McFeely et al. (36) (They showed peaks P_1, P_2 and P_3 for diamond which had energy differences from the C1s peak of 11.3, 25.4 and 34.1 eV, respectively; and graphite was seen to yield a significantly different energy loss spectrum with the P_1, P_2 and P_3 positions at 6.3, 28.1 and 33.3 eV, respectively. The P_1 peak at 6.3 eV was unique to graphite and at 5.6 eV to microcrystalline graphite or glassy carbon. Diamond had a very low loss intensity at ~6 eV, and its P_1 peak at 11.3 eV was usually poorly resolved). Other useful EELS features which can be used for this purpose are the fine structures created during the ejection of an electron from the core K level of a carbon atom (i.e., K-shell ionization loss spectra). The fine structure is due to interband transitions and plasmon losses which appear at higher energies than the threshold energy for K-level ionization (284.7 eV). The position and shape of this region of the spectrum is strongly dependent on the type of bonding present. Therefore, it can be used as a fingerprint for distinguishing between diamond, graphite, and amorphous carbon and carbides.

For the natural diamond (111) surface, a fair amount of information is available about occupied surface states from XPS data, but little is known about unoccupied surface states. The existence of unoccupied surface states has been demonstrated using EELS to observe transitions from occupied to unoccupied surface states and transitions from the C1s core level to unoccupied surface states. Both core level EELS and EELS from valence band excitation have been used by Pepper (55)(56) to study unoccupied surface states on the diamond (111) surface annealed at temperatures higher than 900°C. A feature found at 2.1 eV loss energy was attributed to an excitation from occupied surface states with energies within the valence band into unoccupied surface states of energy within the bulk band gap. A surface band gap was estimated to be about 1 eV. It was found that the feature could be removed from the spectrum by exposing the annealed surface to excited hydrogen. This is consistent with the hypothesis that a reconstructed diamond surface contains π bonded chains, but that this surface will be returned to the bulk structure when terminated with hydrogen (35)(57).

For CVD diamond films, Williams et al. (53) used EELS spectra near the carbon K level ionization threshold to examine the bulk bonding character in the film. The spectrum shown in Fig. 6 was obtained using 100 kV electrons in a Philips 400 transmission electron microscope. Compared with the reference spectra from natural diamond obtained by Egerton and Whelan (52) inset in this figure, there were two important features in terms of identifying the diamond bonding. Firstly, the prominent loss peak near the carbon K-edge in the diamond film (measured at one-third of the peak height of the absorption edge) was located at 291 eV which was characteristic of natural diamond (the first prominent loss peaks near the K-edge energy for natural diamond, graphite, and amorphous carbon observed by EELS in the transmission mode are approximately 291, 286, and 284 eV, respectively) (52). Secondly, the shape of the K-shell ionization region for the spectrum obtained on the diamond film was characteristic of that of diamond. That is, relative to graphite, no peak was observed below 291 eV, and an additional peak was observed at a higher energy. Therefore, it was concluded that the CVD film contained diamond sp^3 type bonding.

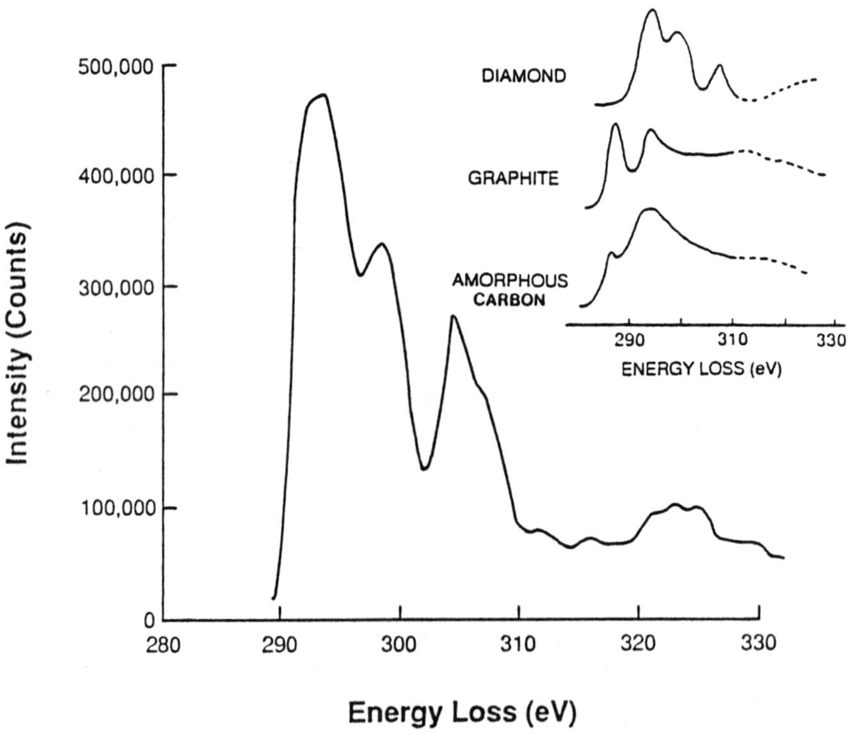

Figure 6. EELS spectra near the K-shell ionization energy for a diamond film with reference spectra obtained by Egerton and Whelan (52) from amorphous carbon, graphite and diamond (53).

As previously mentioned, EELS spectra can also be obtained from losses of the x-ray excited C1s peak for CVD diamond films. The characteristic energy losses observed are sensitive to the density of states in the valence band, Therefore, XPS-EELS has the ability to provide the same bonding information as provided by electron excited EELS in the TEM mode. For samples with mixed deposits of diamond and graphite, or carbide and diamond, curve fits of the EELS data can qualitatively determine which phases are present. A detailed XPS-EELS study was recently reported by Belton and Schmieg (38). In their research, they were able to distinguish films of diamond, graphite, and amorphous carbon all grown in a hot filament assisted CVD system. Mixtures of pure carbon phases in CVD films were identified by qualitatively fitting the EELS spectrum of the unknown with combinations of standard spectra. For films on Si, fits to the EELS data clearly demonstrated that the deposit was a mixture of SiC and diamond. For the growth on Ni, the EELS data were fit with combinations of graphite and diamond, showing that these two phases were present in the films. For the growth on Pt, hydrocarbon was detected in addition to diamond. However, it should be emphasized that although EELS is a very powerful tool for the analysis of diamond growth, there are several constraints on the analysis of EELS data. It has to rely on comparison to standards, and the analysis is only qualitative in most cases. It is also insensitive to minor components. In addition, the experimental sensitivity of EELS for non-diamond components has not been established.

To summarize, electron spectroscopic characterization of CVD diamond has shed considerable light on the chemical and physical processes which control the nucleation and growth of diamond films. XPS, AES and EELS are by far the most widely used surface analytical techniques. A major virtue of these techniques is that they can be performed in vacuo without exposing the sample to air after growth, thereby avoiding atmospheric contamination. Such experiments allow the growth process to be characterized sequentially and provide accurate measurements of the surface chemistry in a time dependent manner. Many of the differences in the application of XPS, AES or EELS relate more to instrumentation and traditional attitudes than to intrinsic differences. The best analysis of the carbon films is obtained when XPS, AES and EELS are used in concert to identify the carbon phases in a film. For example, the C1s binding energy (284.7 eV), when taken alone, is insufficient to distinguish between graphite and diamond; however, when the C KLL Auger transition and energy loss spectra are obtained in conjunction with the C1s core level data, the identification of diamond is straightforward. On the other hand, the spatial resolution capability of AES coupled with the availability of intense electron beams makes this technique superior for fast, routine analysis, particularly in the case of inhomogeneous surfaces. The more advanced state of knowledge and cataloging of chemical effects in XPS favors this technique for the study of the chemical state of surface species. While AES provides information on the filled valence band density of states, EELS can probe empty conduction band density of states. The ultimate usefulness of any of these surface analytical techniques will depend on how quantitative the technique can be made.

4.0 ELECTRON MICROSCOPIES

The goal of microscopy is to obtain enlarged pictures of objects with the best possible resolution. Electron microscopies are techniques that employ electrons to resolve physical features in sufficient detail, in contrast to optical microscopy using visible light. The techniques discussed below are two of the most widely used electron microscopic techniques: transmission electron microscopy (TEM) and scanning electron microscopy (SEM). Both techniques are capable of imaging microstructural features at magnifications of 1,000 to 450,000 X with very high resolutions. While TEM is used mainly for revealing internal structures, SEM is ideal for observation of the surface structural features.

4.1 Transmission Electron Microscopy

Basic Principles and Sample Preparation. TEM is a powerful technique for thin film structure characterization. In contrast with an optical microscope, a transmission electron microscope employs an accelerated electron beam associated with a much shorter wavelength which enables the operator to obtain fundamentally better resolutions and substantially higher useful magnifications. The electron beams can be deflected and focused by electrostatic or electromagnetic lenses in complete analogy to the focusing of light waves by glass lenses. Most transmission electron microscopes have similar designs. They are commonly composed of an electron source, a series of focusing and enlarging electromagnetic lenses and photographic systems. All these components are contained in a high vacuum (10^{-4} - 10^{-5}torr). TEM has very high magnifications (normally in the range of 8,000 - 300,000X) associated with a high resolution (0.2 - 0.3 nm using 100 kV acceleration voltage), allowing very detailed study of localized regions of the sample. Selected area diffraction (SAD) allows detailed correlation between the image and the electron diffraction pattern. Specifically, TEM is capable of detecting and analyzing various internal structural defects, such as lattice imperfections of stacking faults, twins and dislocations, very small precipitates or second phase inclusions, grain boundaries and interfaces, etc., because they produce variations in lattice orientation and are, therefore, distinguishable by their unique image contrast. In addition, the microscope can be operated in a dark field imaging mode (where the image background is dark, and only regions with a particular orientation appear bright) which is useful for determining the distribution of regions of a particular orientation in great detail. This allows differentiation of certain crystal types or orientations from the remainder of the crystals, even when they are not so clearly distinguishable in a bright field image. For detailed instrumentation and quantitative analysis of contrast theories and their applications in analyzing structural defects, books of Heidenreich (58), Grundy and Jones (59), Hirsch et al. (60) and Thomas and Goringe (61) should be consulted.

The basic requirement for a TEM sample is that it is thin enough to allow the electron beam to penetrate through to form images. Usually samples approximately 50 - 300 nm thick, depending upon their atomic weights, are more or less transparent to electrons. The preparation of such thin samples is the most time-consuming part of using the TEM technique for studying film structures. There are many special apparatuses and techniques developed for the preparation of TEM thin foils (62)-(64). Electrolytic polishing, such as the window method, the Bollman technique and the jet-polishing method are often used for preparing metallic materials. Other methods, such as mechanical procedures including forging, cleaving, cutting with a microtome, grinding and ion milling, and chemical methods using suitable acids to chemically thin materials, are of great importance for nonmetallic materials. Replica and powder techniques also have been used.

For the TEM study of natural diamond crystals, diamond samples were thinned by oxidation with carbon dioxide (CO_2) at about 1350°C (65). However, this method suffered from disadvantages that a surface carbon layer had to be removed by boiling in an acid mixture, and there was also the possibility that defects present in the diamond were being thermally annealed at the high thinning temperature. Therefore, most diamonds were thinned by passing a slow stream of dry oxygen (O_2) at atmospheric pressure over diamonds at 750°C. No surface carbon layer was formed, and the danger of changing the defect distribution was reduced at this lower thinning temperature.

For CVD diamond films, TEM samples are usually prepared by mechanical polishing and ion milling. The experimental procedures for preparing a diamond TEM plan-view sample are illustrated in Fig. 7a. A 3 mm disc sample is ultrasonically cut from an as-grown film. This sample is then mounted on a specimen lapping holder with precise scales and mechanically polished to a thickness of 100 µm from the substrate side using 600 grit sandpaper. It is further thinned to 20 µm in the center of the sample with a dimpling machine to facilitate subsequent ion thinning. Then it is put into an ion milling machine for final thinning. In the ion miller, argon ion beams are accelerated with a voltage of 5 - 10 kV and a current of 3 - 5 mA, bombarding the sample at a small angle (~10°). The samples are thus ion milled until perforation occurs. The area around the perforated hole is normally thin enough for TEM observations. During the final stage of ion milling when perforation is about to occur, a reduced voltage and current is usually applied to the argon ion beam to avoid possible damage to the sample surface by the ion bombardment (e.g., graphitization or amorphization).

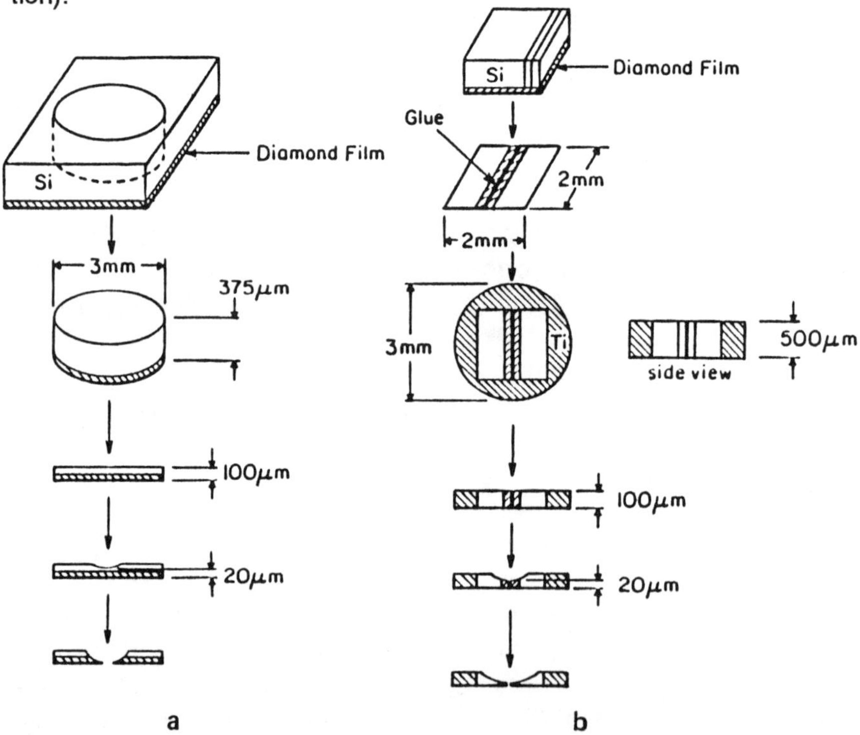

Figure 7. Schematic of the sample preparation procedures for (a) plan-view TEM; and (b) cross-sectional TEM observations (66).

A method for preparation of TEM samples, without any thinning procedure required after film deposition, has also been reported (66): a 75 µm thick silicon wafer was dimpled and ion milled prior to diamond deposition until a tiny hole appeared; the perforated substrate wafer was then put into the deposition system for 2 hours to obtain a film with a thickness of 1 - 2 µm; the diamond film formed near the edge of the hole was thin enough for electron transmission. This method was developed to test for possible artifacts introduced by argon ion bombardment in the final ion thinning stage of sample preparation. Care must be taken when making holes in the substrates to assure that the holes are very small and have a low angle wedge around them.

Analysis of Defect Structures by Plan-View TEM. TEM has shown that structural defects exist extensively in CVD diamond. These defects include planar defects (microtwins and stacking faults), line defects (dislocations) and second phase inclusions (graphite) and have been characterized by various research groups. Previous investigations have shown that planar defects including {111} stacking faults and {111} twins are quite common in natural and HPHT synthesized diamonds (67-72). Dislocations and platelet defects associated with the presence of aggregates of foreign atoms (e.g., nitrogen) also have been examined in great detail (73)-(77). For CVD diamond films, Matsumoto and Matsui used TEM in early 1983 (78) to investigate the defect structures of CVD diamond. They reported extensive microtwins formed in diamond crystals. Later, extensive TEM examinations of diamond defect structures were reported (53)(79)-(83). It was found that the majority of diamond crystals had very high defect densities comprised of {111} twins, {111} stacking faults and dislocations as shown in a series of TEM micrographs in Fig. 8. Most of the defects originate at the center of diamond particles and radiate outward. The density of these defects increases and their sizes become smaller as the methane concentration increases during the deposition of the films. These defects can be distributed either uniformly in the grains, only at the grain boundaries, or in bands within the grains.

The nature of multiply twinned particles (MTPs) of diamond also has been studied via TEM (84). In Fig. 9a, a bright field micrograph of a fivefold MTP shows the five growth sectors comprised of tetrahedra bounded by {111} planes sharing a common [110] axis and twinned relative to each other. The electron beam is parallel to a [110] direction. Figure 9b is a weak beam dark field micrograph imaged by using a (111) diffraction spot. Twinned regions within the sectors which are in the diffracting condition appear bright (as streaks). Twins on intersecting {111} planes parallel to the

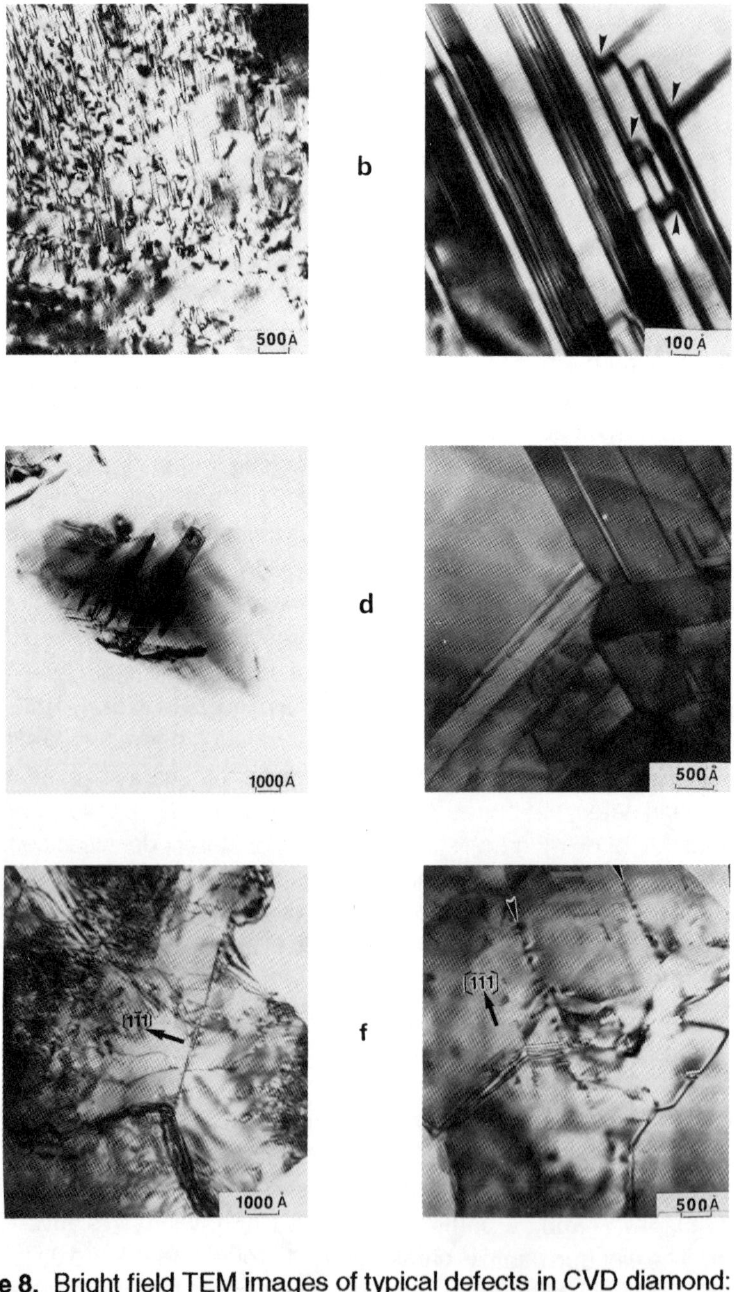

Figure 8. Bright field TEM images of typical defects in CVD diamond: (a) and (b) {111} stacking faults; (c) and (d) {111} twins; and (e) and (f) dislocations. Arrows in image (b) indicate interactions between stacking faults, and arrows in image (f) indicate aligned end-on dislocations (80).

major twin boundaries separating the tetrahedra are most visible. A small secondary sector is also shown twinned relative to its neighbors. Perhaps the most interesting observation from these micrographs is the absence of defects near four of the five major twin boundaries which separate the sectors. In general, many defects of the types shown in Fig. 8 are found within each sector. In the regions near at least four of the five major twin boundaries, the defect density is significantly reduced. This phenomenon has also been found by Tamou et al. (85). This defect reduction is significant because if its origin can be determined, this information is useful in producing high quality diamond films. Detailed examinations of twins and stacking faults as well as their intersections have been conducted by Williams et al. (84) with high resolution TEM (see next section). In addition, the atomic structures and energetic levels of dislocations, twins and grain boundaries also have been investigated and modelled by Narayan (86).

a

b

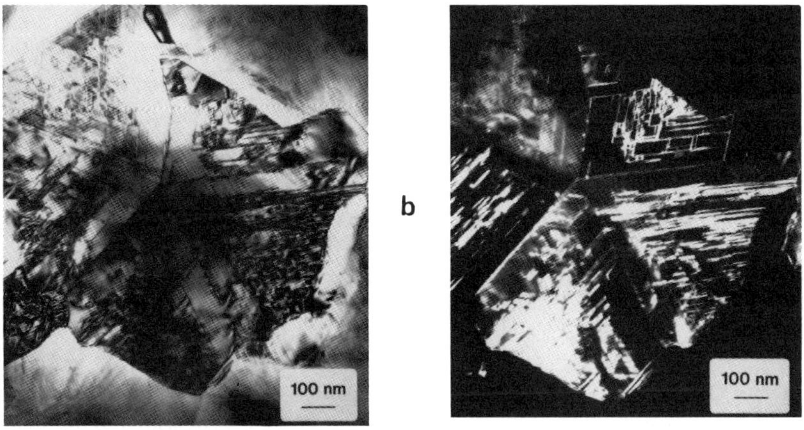

Figure 9. (a) A bright field TEM micrograph of a multiply twinned particle (MTP); and (b) a dark field TEM micrograph imaged by using a diamond {111} diffraction spot (84).

Ma et al. (87) recently correlated the biasing condition in a hot filament CVD system with the defect densities in diamond films. It was found that the defect density was substantially reduced under zero and reverse bias (substrate negative relative to the filament) as compared to forward bias (substrate positive relative to the filament). Furthermore, the diamond/ silicon interface of the reverse and zero biased films consisted of a single thin interfacial layer, whereas multiple interfacial layers existed at the interface of films deposited under forward (positive) bias. They concluded that the biasing conditions, which influenced the role of charged species in the growth of diamond, affected the defect density and the nature of the diamond/silicon interface. The forward biasing was not favorable for growing high quality, low defect density diamond films, although the growth rate was believed to be higher. More recently, Ma et al. (88) also reported a general reduction of defect density in diamond films grown on Mo substrates by an oxyacetylene combustion flame as the substrate temperature was increased (from 500 to 870°C). They attributed this to the enhanced surface mobility of carbon atoms at the high substrate temperature.

One very interesting aspect of CVD diamond films which has been illucidated by TEM is the close relationship between crystalline defects and surface morphologies (66)(81)(82). Figure 10 shows three diamond films grown by microwave plasma CVD with typical surface morphologies of {111} square faces, {100} triangular faces and a cauliflower appearance composed of small crystal aggregates. The {111}-faceted film showed extensive twinning and stacking fault features throughout its volume which were mostly visible on the {111} planes (Fig. 10a). The crystal shown here was multiply twinned with a pseudo five-fold symmetry. Since there is only a relatively slight energy difference between the normal and twinned diamond structures, diamond is predisposed toward twinning. The diamond growth is believed to proceed via frequent twinning on {111} planes which provides easy addition of new atoms to existing crystals (89)(90). If these twinning bands extend and intersect with the surface, striking twin features will be recorded in the final morphology which has been frequently observed by SEM. However, unlike the internal structure of natural diamond crystals of cubic habit which possess fibrous or cuboid features (91), the crystals in the {100}-oriented film were planar-defect free (Fig. 10b). Also, only nanometer scale roughness could be discerned on the {100} square faces. These features were different from the microscopic roughness which were usually observed on either {111} or {100} facets by SEM (e.g., steps, secondary nucleation, etc.) and were, therefore, interpreted as the sites where atoms could attach and result in growth in the normal direction. Along the boundaries of the {100}-oriented crystals where the {111} planes lie, planar defect features again could be detected extensively. For the film with cauliflower morphology deposited at a high methane concentration, no obvious structural features other than small defective crystals were present (Fig. 10c).

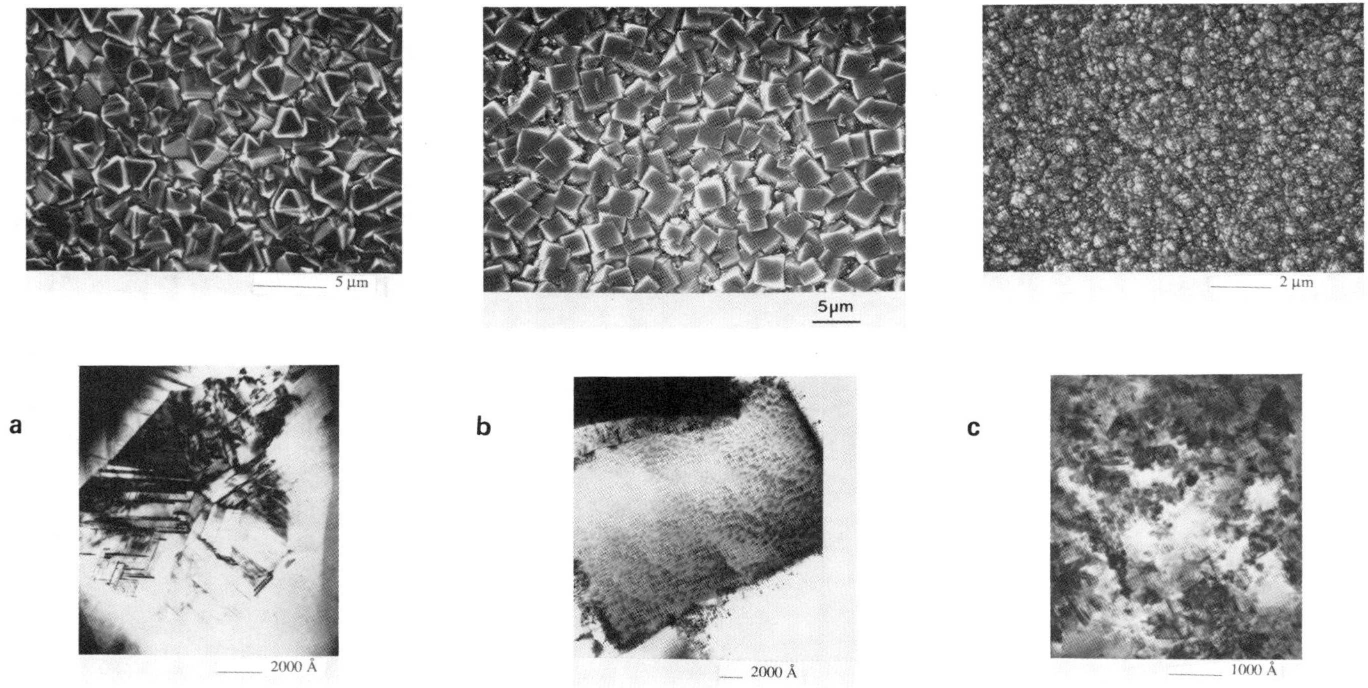

Figure 10. TEM and SEM micrographs of internal structures and corresponding surface morphologies of (a) a diamond crystal with the {111} surface morphology which is highly defective with {111} twins and stacking faults; (b) a diamond crystal with the {100} surface morphology which is planar-defect free; and (c) a film with a cauliflower morphology containing numerous small defective crystals (66).

The above relationship between the external surface morphologies and internal structural defects is of great importance in designing experimental conditions to develop structurally perfect diamond films. Obviously the {100} planes are the desired growth surfaces for defect-free structures. This has been verified experimentally in the homoepitaxial diamond films (92)-(94). Perfect diamond crystals could be epitaxially grown on the {100} diamond surfaces, whereas highly defective structures were developed on the {111}-oriented diamond substrates. However, the growth rate was much faster on the {111} surfaces than on the {100} surfaces. It can be speculated that this is because the defects on the {111} surfaces promote nucleation and growth. Recent oxidation studies also found that the {111} surfaces of diamond are more defective than the {100} surfaces (95). As shown in Fig. 11, the oxidation preferentially occurred around defect sites on the {111} faces, whereas the {100} faces were more resistant to oxidation.

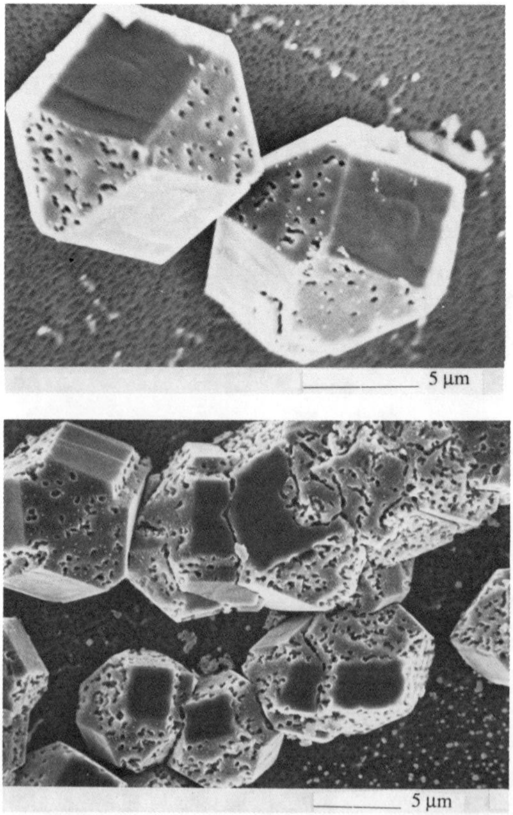

Figure 11. Preferential oxidation on the {111} faces of diamond, indicating that the {111} faces are more defective than the {100} faces (95).

Zhu et al. (96) have also characterized the graphitic inclusions incorporated in diamond films grown at higher methane concentrations (>2.0%) by dark field imaging techniques. They found that graphite existed in a turbostratic disordered structure with dimensions ranging from 50 Å to 100 Å in the ***a*** direction on the basal planes and 20 Å to 50 Å in the ***c*** direction as shown in Figs 12b and 12c. In the film deposited at a methane concentration of 0.5% (Fig. 12a), diamond crystals contained planar defects of twins and stacking faults on the {111} planes. No graphite phase or other non-diamond components could be detected. However, in the film deposited at a methane concentration of 2.0% (Fig. 12b), graphite crystallites formed on or close to the {111} planes of diamond and were in a highly oriented state with the hexagonal graphite axis aligned with or close to the trigonal diamond axis. Related to the textured electron diffraction pattern, graphite crystallites exhibited a domain distribution in the diamond matrix. Each domain corresponded to a group of graphite crystallites lying on or close to a particular {111} plane of diamond. In the film deposited at a methane concentration of 5.0% (Fig. 12c), graphite crystallites were randomly distributed in the diamond matrix with no preferential orientation of the hexagonal basal planes with the diamond {111} planes. The wide range of orientations of graphite crystallites was believed to be associated with the disturbance produced by the rapid formation of new graphite crystallites and the small sizes of increasingly disoriented underlying diamond crystals. This study also indicated that while low substrate temperatures introduced more amorphous phases in the films, high substrate temperatures favored the graphite formation with graphite tending to nucleate at boundaries and to grow uninhibited to larger sizes, up to tens of nanometers. Although both planar defects (twins and stacking faults) and graphite formed preferentially on the {111} planes of diamond, no evidence for any selective graphite nucleation on such planar defects or any spatial relation between them could be found. This same study also determined that the film quality improved as the film grew. The regions close to the substrate contained more amorphous carbon, graphite and defects than the material closer to the film surface.

In general, crystalline defects like stacking faults, microtwins and dislocations, as well as second phase, non-diamond components such as graphitic and amorphous carbon, are common features in CVD diamond films. They can degrade film performance and, therefore, are critical to many applications (especially to electronics and opto-electronics). Indeed, the thermal and electrical properties of CVD diamond films have been found to be inferior to those of natural diamonds (97)(98). This is due largely to the

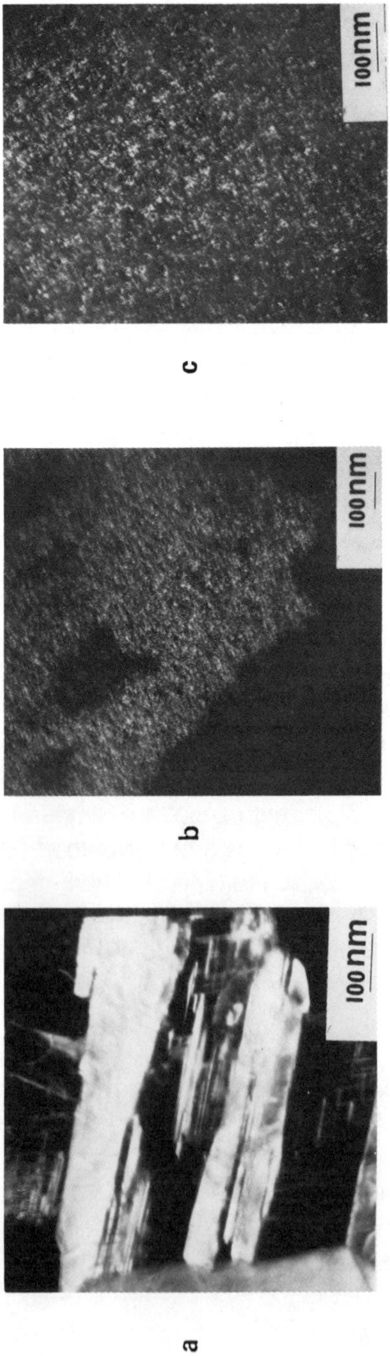

Figure 12. TEM dark field images of diamond films deposited at different methane concentrations: (a) 0.5%, imaged by a diamond {111} diffraction spot; (b) 2.0%, imaged by a graphite {00,2} diffraction arc; and (c) 5.0%, imaged by a graphite {00,2} diffraction ring (95).

extensive existence of the various structural defects discussed in this section. Whether these structural imperfections, including extensive planar defects, grain boundaries, surface roughness and impurities, can be effectively controlled or eliminated remains a primary goal of many investigations. TEM will remain a key characterization tool for the evaluation of structural quality of diamond films.

High resolution TEM. High resolution TEM (HRTEM) imaging is based on amplitude phase contrast in which both transmitted and diffracted beams are used to form the image by interfering with each other (99). This is quite different from the conventional images discussed in the preceding section in which the intensities in the images have been the square of the amplitude of the single beam allowed through the apertures. The contrast of HRTEM images is, therefore, a composite of interference effects and amplitude changes produced by defects. By suitable choice of imaging conditions, it is possible to image directly the crystallographic lattice itself which is particularly useful for ascertaining crystallographic details of interfaces and the nature of the defect structures. This technique has been profitably used to analyze interfaces and defect structures of CVD diamond at hitherto unprecedented spatial resolutions. It is of great value in analyzing lattice defects such as twins and stacking faults in CVD diamond, because the defect density is so high that conventional bright field TEM can not yield detailed images without significant overlap of the defects.

In general, the experimental conditions necessary to yield enough resolution for lattice imaging are very stringent. To optimize image contrast, it is essential to choose the optimal conditions for the specimen (thickness, orientations) and the instrument (defocus, illumination, objective aperture). The interpretation of HRTEM lattice images is also non-trivial and depends on a number of experimental parameters. Williams et al. (84) investigated various defect structures and their intersections as well as MTPs of CVD diamond by HRTEM on an atomic scale. Examples of HRTEM micrographs of {111} twins and {111} stacking faults are shown in Fig. 13. Since the diamond grain was in a [011] orientation, {111} cross-fringes of 2.06 Å spacing were visible. In Fig. 13a, it can be seen that within 100 Å, three twin planes and a stacking fault were present. This was representative of the high defect density observed in certain regions of the CVD diamond; however, their distribution was quite inhomogeneous, and accurate quantification of the twin and stacking fault densities was not possible. Figure 13b shows the intersections of twins in diamond. A twin labeled A appeared to have climbed across several planes to avoid crossing another twin labeled B. Although there were various distortions and apparent secondary twins and stacking faults which complicated the situation, the primary observation was that twin B terminated at the position labeled C instead of crossing the other twin boundary. It is interesting to note that there was no evidence for the presence of a hexagonal component formed in these twin intersections.

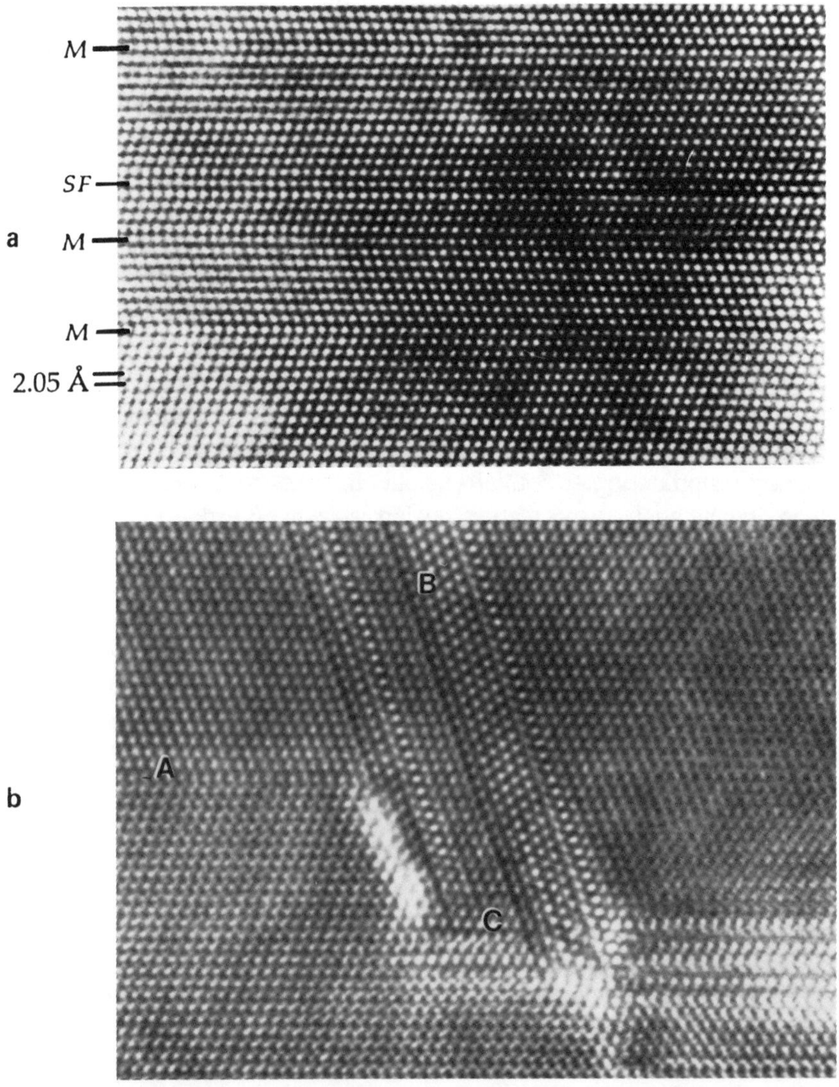

Figure 13. HRTEM images of (a) a region in a diamond crystal where a single {111} stacking fault (indicated as SF) and several {111} twin boundaries (indicated by M) were observed; and (b) a region in a diamond crystal where twins on different {111} planes are intersecting (84).

For MTPs, the common forms found in CVD diamond include a single twin separating particles into two halves which are the mirror image of each other, pseudo-pentagonal particles (known as decahedral MTPs) containing five tetrahedra sharing a common [110] axis and each twinned relative to its neighbor, and an icosahedral particle bounded by 20 {111} planes composed of 20 tetrahedra sharing a common apex. An HRTEM micrograph of the core of a fivefold MTP and its schematic diagram are shown in Fig. 14a. A very irregular core was observed due to interactions with secondary twins and apparent "wandering" of the primary twins. One of the five major twin boundaries was, in fact, not a single twin boundary at all. Rather, the boundary between regions 3 and 4 was an extended boundary consisting of multiple parallel twins. It was speculated that at the true nucleus of the MTP a perfect core existed but became distorted as the particle grew. In this way, the overall fivefold symmetry was maintained by the earlier growth. Because the five {111} sectors which make up the MTP fill only 352.5° of space around the common [110] axis, 7.5° of misfit must be accommodated. This misfit has been shown (Fig. 14b) to be accommodated in two of the five major twin boundaries present rather than by elastic deformation as was the case in reports by Narayan et al. (100). These two twin boundaries contained series of dislocations as imaged in Fig. 14c. Approximate locations of dislocations were indicated by the arrows. The frequency of these dislocations was related to the amount of misfit which was accommodated at the twin. Based on a repeat distance of eight spacings between these dislocations, a misfit angle of approximately 7.18° was calculated by simple geometric considerations. The presence of dislocations along the twin boundary may be viewed as the superposition of a low angle grain boundary and a twin boundary which was more descriptively named by Williams et al. (84) as a "tilted twin boundary" (TTB).

Cross-sectional TEM. Cross-sectional TEM (XTEM) has proven itself to be very useful for the study of the evolution of thin film structures. It allows detailed study of the interface between the film and substrate. However, the preparation of cross-sectional samples containing the desired interface can be extremely difficult. In the case of diamond films, Zhu et al. (66) prepared the samples in the following manner (Fig. 7b): two diamond specimens (0.5 x 2 x 0.5 mm^3) were sliced from a sample with a wire saw; they were bonded together face-to-face; this bonded piece was then put on a 3 mm titanium disc support which had a slot for its insertion; it was then thinned in the same way as that described earlier for the plan-view TEM sample preparation, that is, mechanical polishing, dimpling and ion beam thinning but from both sides. Because of large differences in sputtering rates

a

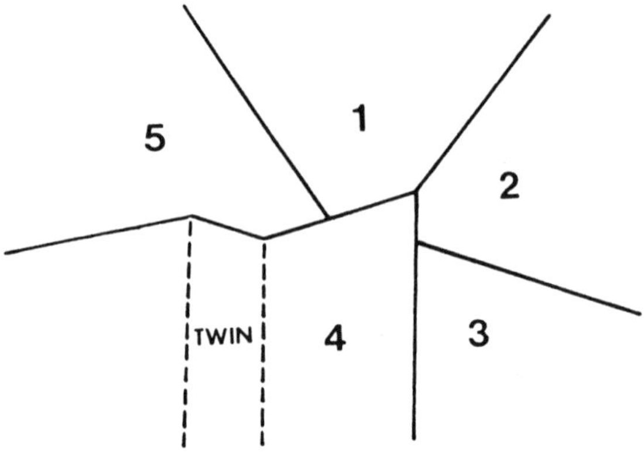

b

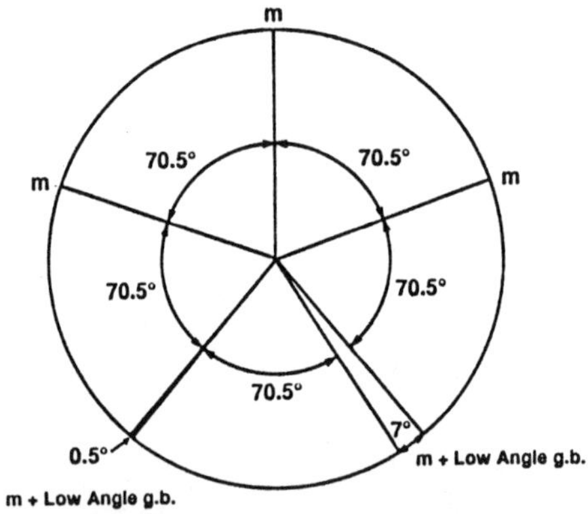

c

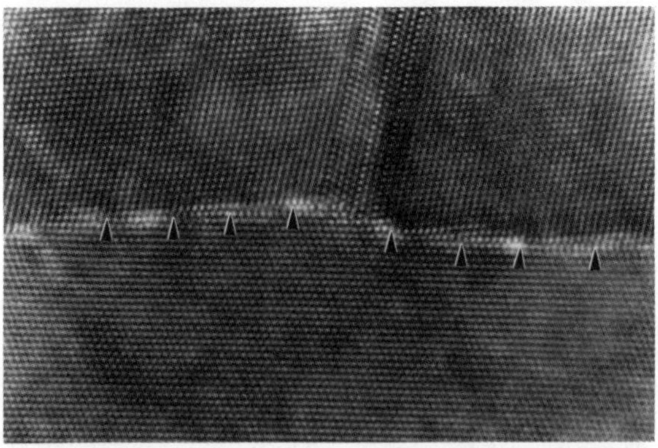

Figure 14. (a) A HRTEM image and its schematic diagram of the core of a five fold MTP; (b) schematic of low angle grain boundaries in the MTP with angles between different {111} planes; and (c) a HRTEM micrograph of a titled twin boundary (84).

among the diamond film, silicon substrate and bonding glue, special precautions must be taken before and during the ion milling process for the cross-sectional TEM sample to achieve homogeneous sputtering. For example, fine carbon powder can be added to the bonding glue to lower its sputtering rate, and a very small sputtering angle can be used.

Figure 15 shows XTEM micrographs of diamond films grown at different methane concentrations obtained by Williams et al. (53). The dark areas which were associated with defects and/or strain in the films appeared to fan out from a small region at the substrate interface, implying a single nucleation site for the grain. This supports the general observations that the CVD diamond undergoes three-dimensional rather than two-dimensional growth. It was also found that the film grown at 0.3% CH_4 (Fig. 15a) appeared to contain a lower defect density than that grown at 2.0% CH_4 (Fig. 15b) as seen by dense, dark regions in the TEM. This was probably due to the lower growth rate at the lower methane concentration which allowed a less defective and/or less strained film to form. This observation might also be related to a decreased sp^2 bonding component which was observed by Raman spectroscopy.

Closer examinations of the diamond/silicon interfaces of the films grown at 0.3% and 2.0% CH_4 revealed significant differences in the nature of the interfaces. In Fig. 15b, the apparent interfacial layer appeared to be caused by ion milling damage or a thin amorphous carbon layer. No interfacial carbide layer was found for the film deposited at the high methane concentration (2.0%), presumably because the carbide layer was too thin to be imaged. However, the film grown at 0.3% CH_4 yielded an obvious interfacial layer approximately 50 Å thick as shown in Fig. 15c. This was easily distinguished from an ion milling damage layer by the defect contrast observed as dark lines in the layer and the depth which was apparent parallel to the electron beam in this region, indicating it is not simply a surface artifact. The electron diffraction pattern from this interfacial region indicated that this interlayer was β-SiC as discussed in Sec. 5.2.

4.2 Scanning Electron Microscopy

SEM is a technique available to examine thin film surfaces with sub-micron size features. It operates by scanning a focused electron beam over a surface and sensing the secondary electrons emitted from the surface (101). The beam can be focused to an extraordinarily small diameter of only 10 - 20 nm. This beam size essentially determines the resolution. The brightness, i.e., the intensity of the secondary electron emission, depends

a

b

c

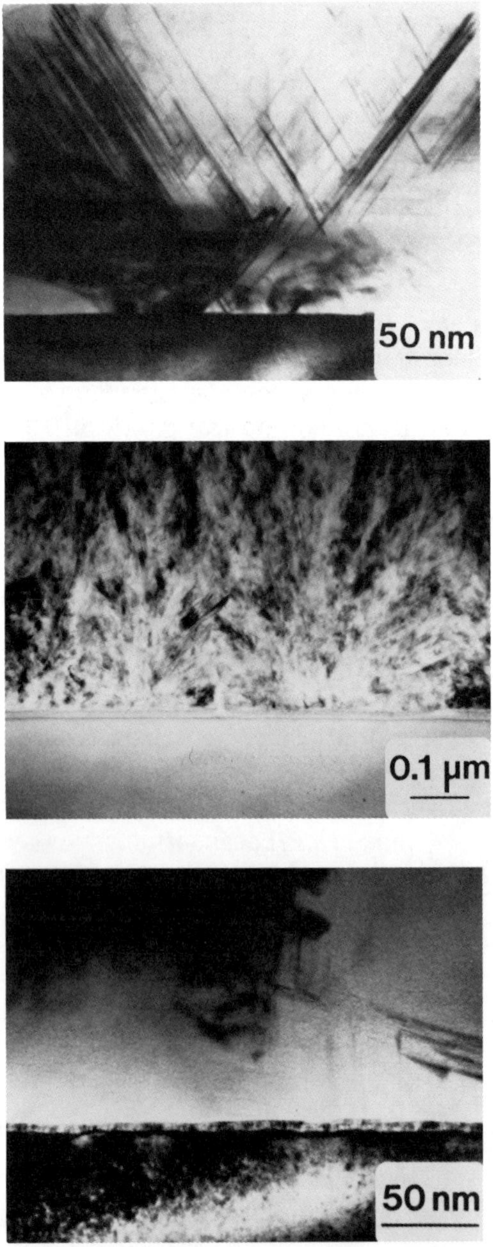

Figure 15. XTEM micrographs of (a) a diamond film grown at 0.3% CH_4/H_2, showing defects in the diamond film fanning out from the diamond-Si interface; (b) a diamond film grown at 2.0% CH_4/H_2, showing an increased density of defects in the diamond film; and (c) a diamond film grown at 0.3% CH_4/H_2, showing a ~50Å layer of β-SiC at the interface (53).

on the local surface geometric relief because of the asymmetric location of the detector. The sample surface topography is, therefore, magnified into an image on the CRT (cathode-ray tube). SEM has many advantages over optical microscopes in its high magnifications (normally in the range of 50 - 40,000X), high resolution (2.5 - 10 nm), its extraordinary depth of field (500 times greater), and the three dimensional appearance of pictures. In addition, the EDS (energy dispersive spectrometry) mode allows elemental identification and distribution to be performed. Regarding a comparison with TEM, the sample for TEM must be so thin that it is transparent to electrons, while for SEM, solid bulk samples can be used. A conducting sample can be investigated immediately in its original condition without any preparation. Non-conducting samples generally need to be pre-coated with a thin layer of conducting materials (gold or carbon). For example, free-standing diamond films and diamond films grown on SiO_2 need to be coated with gold prior to the SEM observation, while diamond films grown on Si can be directly observed by SEM without any conducting coatings being deposited on the surface (102). However, it is always useful to utilize low beam voltage and current to avoid charging of the films.

SEM has proven to be, by far, the most widely utilized technique to characterize the surface morphology of vapor phase deposited diamond. It is also one of the most useful because it has now been correlated with other techniques such as Raman spectroscopy. It was probably those beautiful SEM micrographs showing crystals with well-defined crystallographic facets, more than any other factor, that captured the attention of researchers throughout the world early in the development of this technology.

To date, various preparation and characterization efforts for CVD diamond have yielded various types of films, including homoepitaxial single crystalline, large grained polycrystalline, and sub-microcrystalline films. It has been recognized that CVD diamond films have a wide range of surface morphologies which are highly dependent upon the experimental conditions. Generally, a diamond film grown on a non-diamond substrate (the most common of which is silicon) will have a rough, highly faceted, large-grained (>1 µm) polycrystalline morphology, as shown in Figs. 16a and 16b. These surfaces generally correlate with a relatively high quality Raman spectrum (i.e., a high intensity 1332 cm^{-1} peak and a low intensity 1500 cm^{-1} peak). However, if growth parameters (especially the substrate temperature and the methane concentration in hydrogen) are not optimized, the morphology becomes sub-microcrystalline and then cauliflower-like as shown in Fig. 16c. This correlates with a deterioration of the diamond peak and an increase in the graphitic peak in the Raman spectrum (103).

a

b

c

Figure 16. SEM micrographs of diamond films with typical surface morphologies of (a) the {111} triangular faces; (b) the {100} square faces; and (c) cauliflower like small crystalline aggregates (107).

In general, the growth morphology and its evolution reflect the growth controlling variables such as temperature, supersaturation, impurities, defects, etc. If one understands all of these deposition-morphology relations, it is theoretically possible to design diamond morphologies for specific applications, or understand the growth history of a film by examining its morphological evolution. Unfortunately, such a detailed understanding is not yet available. However, there are several experimental reports on surface morphologies of diamond films which have related the morphological development to various deposition parameters (104)-(107).

Spitysn et al. (104) reported that, for films deposited by the chemical transport reaction (CTR) method, high methane concentrations in hydrogen and low substrate temperatures favored the formation of {111} faces, while low methane concentrations in hydrogen and high substrate temperatures produced mainly {100} faces. Growth steps were observed on both {111} and {100} faces. Kobashi et al. (105) indicated that, with an increase of methane concentration in a microwave plasma CVD process, the film morphology changed from a predominance of {111} faces to {100} faces prevailing. Further increases in methane concentration led to a film with fine crystalline aggregates with no microscopically visible crystallographic facets. Growth steps were only seen on {100} planes, not on {111} planes. Furthermore, they utilized SEM at selected growth intervals to construct the following scheme for the evolution of surface morphology during growth. Diamond particles initially increased their sizes almost uniformly until the substrate was eventually covered, then a secondary growth took place, followed by surface restructuring processes such as fusion and absorption among secondary crystallites. As a result, well-defined diamond faces were formed progressively on the film surface, and higher order growth followed by the restructuring processes occurred periodically as the CVD synthesis proceeded. The changes in the morphology were viewed as a consequence of elementary processes of surface restructuring.

Haubner and Lux (106) investigated the influence of inhomogeneous microwave plasmas on the morphology of diamond grown on SiAlON substrates. They observed variations due to gradients in the intensity of the plasmas and linked them with deposition parameters such as gas pressure, flow rate, methane concentration and substrate temperature. High substrate temperatures and high plasma intensities favored {111} faces, while low substrate temperatures and low plasma intensities favored {100} faces. They suggested that the formation of different diamond crystal habits was controlled by the carbon supersaturation in the gas phase, the carbon mobility on the surface, and the surface energy of the growing crystal facets.

Zhu et al. (107) performed a systematic study on the morphology of diamond crystals and films grown by microwave plasma CVD. They found that the substrate temperature, methane concentration and gas flow rate effectively changed the relative development of {111} and {100} faces, resulting in different surface morphologies. The film morphology was relatively insensitive to the gas pressure (in the range of 50 - 150 torr) in the optimum substrate temperature range (900 - 1000°C). With an increase of substrate temperature or gas flow rate in their respective appropriate ranges (substrate temperature from 850 to 1150°C; gas flow rate from 0 to 400 sccm), the surface of diamond films changed from {100} faces dominating to {111} faces prevailing. With an increase of methane concentration in hydrogen from 0.5% to 2.0%, the morphology of diamond films changed from triangular {111} faces to square {100} faces. Further increasing methane concentration to 5.0% resulted in sub-micron clustered films which contained a substantial amount of graphite. This is consistent with the results obtained by Kobashi et al. (105). Based on these experimental results, they developed a "morphology field map" on the axes of substrate temperature and methane concentration (Fig. 17) which is very useful for processing experimental data and directing the development of the diamond CVD process. For example, the {111} faces can be found most often in the films deposited at high substrate temperatures and low methane concentrations, while films with the {100} faces can be produced most probably in the ranges of medium substrate temperatures and medium methane concentrations. The morphology zones also were correlated with physical processes such as adatom mobility and gas phase supersaturation rather than only structural forms.

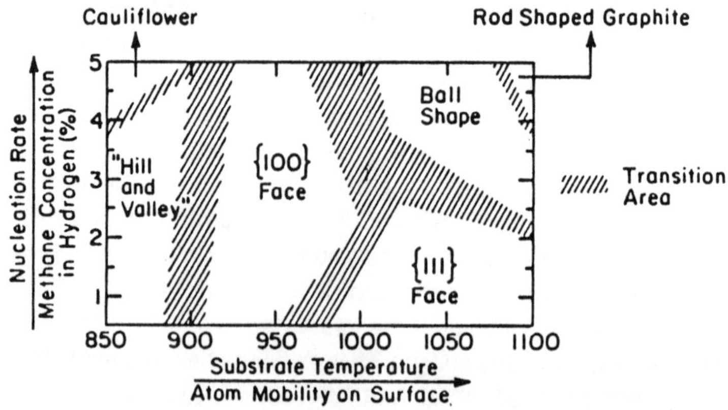

Figure 17. A field map of the surface morphology of diamond films as a function of methane concentration in hydrogen and substrate temperature (107).

It is interesting to mention some particle morphologies which have been observed prior to their intergrowth to form a continuous film as first reported by Matsumoto and Matsui (78). The embryo of these particles was postulated to be a variety of different gas species which resembled the different morphologies as schematically presented in Fig. 18. The simplest morphology (Fig. 18a) is a single crystal which generally takes the form of a cubo-octahedron. The particle is bounded by {100} and {111} planes. Twinned cubo-octahedra have also been observed where a twin separates the cubo-octahedra into two parts (Fig. 18b). Multiply twinned particles (MTPs) yielding a pentagonal profile (five-fold twinning) (Fig. 18c) also exist in diamond as previously observed only in FCC (face-centered cubic) metal particles (108)(109). Such a MTP can have relatively perfect five-fold symmetry or may contain secondary twinning creating a gap in one of the sides of the MTP, presumably due to the misfit of 7.5° which must be accommodated in five-fold particles. Finally, twenty-fold twinning creates an icosahedral MTP made up of twenty single crystal units (Fig. 18d), twinned relative to their neighbors and bounded by 20 triangular {111} planes. Zhu et al. (107) have also shown some typical forms of well-shaped diamond crystals grown by microwave plasma CVD. They are single crystal and twinned cubo-octahedra (Fig. 19a), icosahedra (Fig. 19b) and decahedral Wulff-polyhedra (Fig. 19c). It is seen that the crystals often present chamfered edges along certain directions (such as <110>), the origin of which is not understood at present. They indicated that the frequent appearance of the intermediate cubo-octahedral shapes, instead of the extreme octahedral and cubic shapes, implies that the growth rates in <111> and <100> directions were so close under those given conditions that the disappearance of a single plane due to a higher growth rate could not occur. However, changes in the deposition conditions can effectively change the relative rates and the final morphology, and as a result, in addition to those cubo-octahedral shapes, crystals with more complex shapes can develop and were observed. Intergrowth, twinning and renucleation frequently occur during the growth process, which complicates the face development and results in irregular morphologies.

In addition to the examination of growth surfaces, SEM can be utilized to evaluate the "backside" of films which have been removed from their substrates, thereby yielding information about the interface. For example, Pickrell et al. (102) used SEM to examine the interface sides of free-standing films which were previously grown on both Si and SiO_2 substrates. On Si, they found that diamond grew from nucleation sites upward into columns

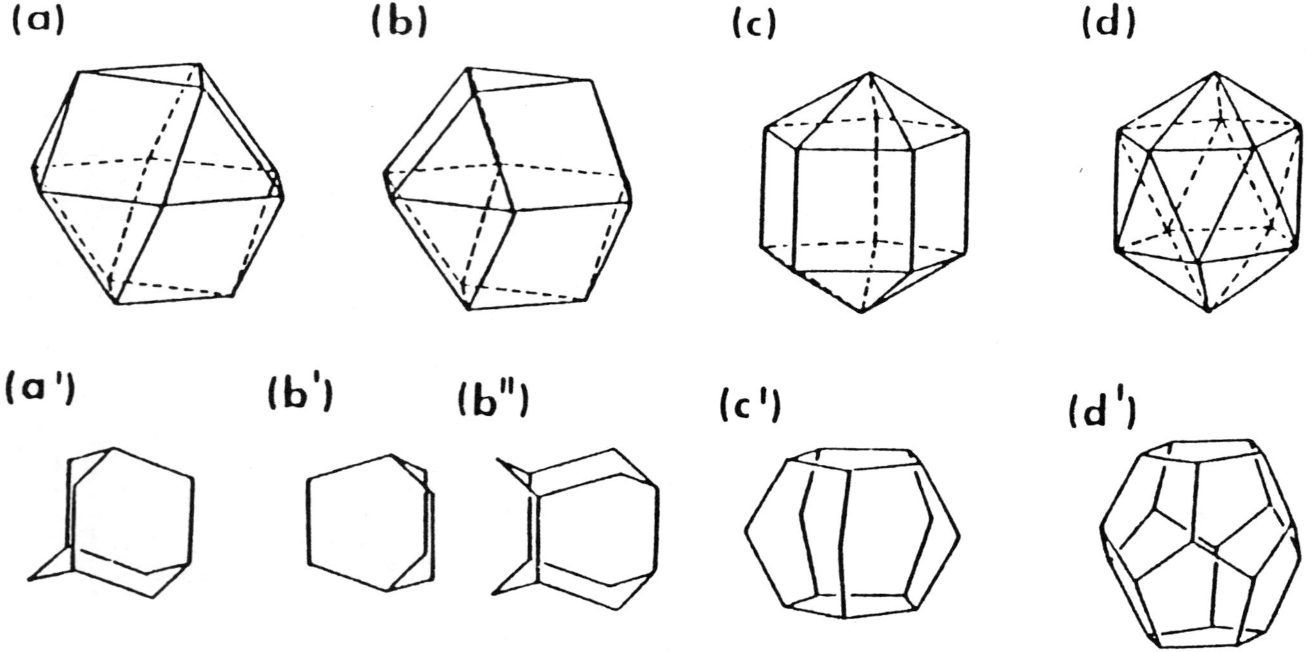

Figure 18. Schematic of particle morphologies and their possible relationship with gas species: (a) cubo-octahedron (a single crystal); (b) twinned cubo-octahedron; (c) decahedral-Wulff-polyhedron; (d) icosahedron; (a') adamantane; (b') bicyclo-octane; (b") tetracyclo dodecane; (c') hexacyclo-pentadecane; and (d') dodecahedrane (78).

a

b

c

Figure 19. SEM micrographs of typical crystal shapes observed in CVD diamond: (a) a single crystal cubo-octahedron and a twinned cubo-octahedron; (b) an icosahedron; and (c) a decahedral-Wulff-polyhedron (107).

and parallel to the substrate surface to form large contact areas (Fig. 20a). These large flat contact areas served to anchor the film, which was partly responsible for the film's relatively good adhesion. Voids were present between the column bases which were visible from the interface sides of the films. On SiO_2, a particulate type of intermediate layer formed first (Fig. 20b), and adhesion was found to be rather poor. SiC was detected in this layer, but it appeared in discrete areas as opposed to a rather continuous layer as was believed to form on Si. In addition, the particulate layer contained graphitic carbon and possibly even Si-O-C phases.

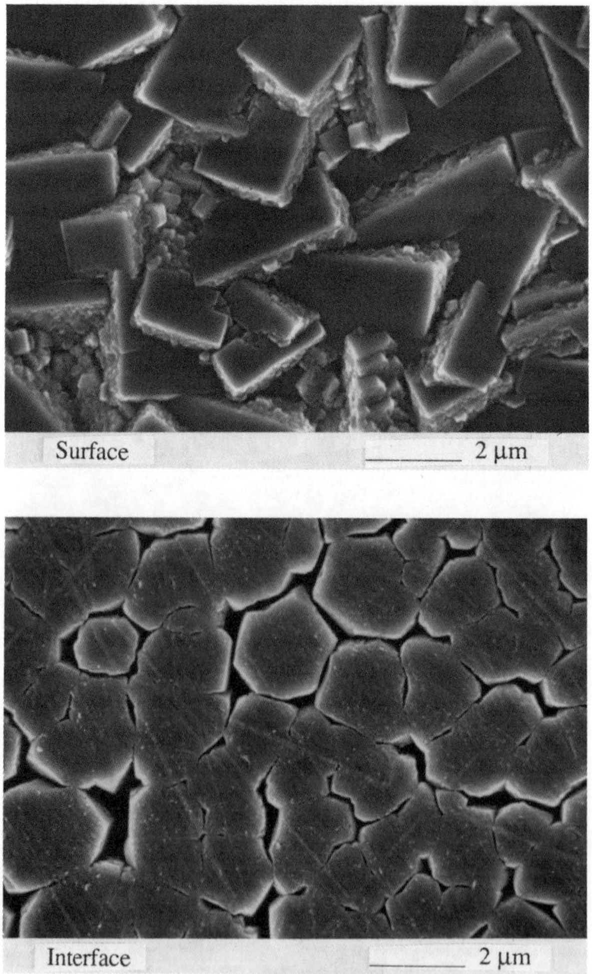

Figure 20a. SEM micrographs of the interfaces and their corresponding surfaces for diamond films grown on a Si substrate (102).

In summary, the diamond surface morphology is extremely sensitive to the exact deposition conditions. This is not surprising since the surface free energy of a face is a function not only of the atomic structure of the surface, but also of all the interactions between the surface and the gas phase environment. Therefore, the morphological development is not a constant characteristic of the substance of the crystal but will depend strongly on the nature of the growth process. A detailed mechanistic understanding of the deposition-morphology relationships which will enable the design of surface morphologies for specific applications is yet to be developed.

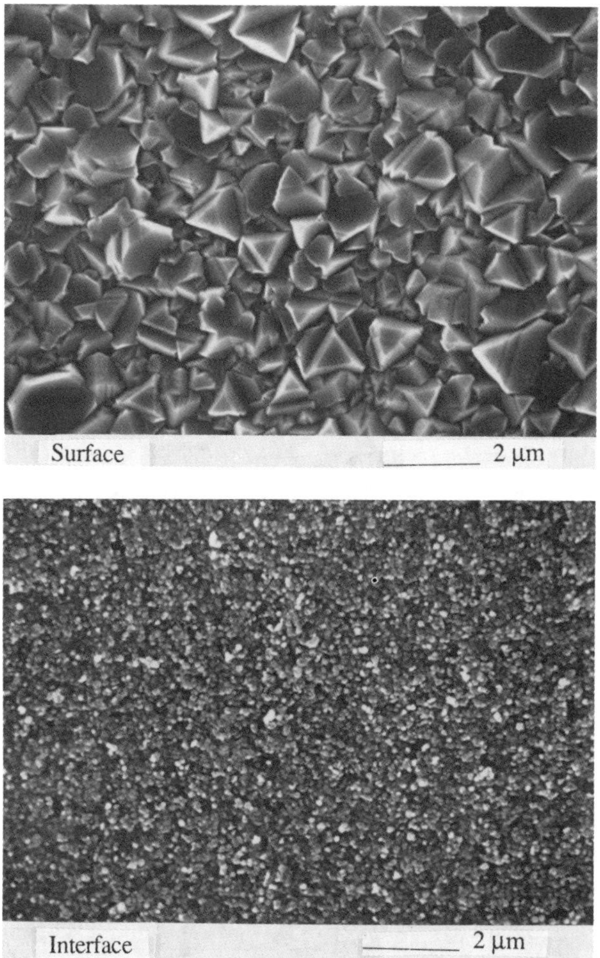

Figure 20b. SEM micrographs of the interfaces and their corresponding surfaces for diamond films grown on a SiO_2 substrate (102).

5.0 DIFFRACTION TECHNIQUES

Diffraction techniques can provide direct information on the periodicity or translational symmetry of a crystal structure. By studying the intensities of diffracted beams and relating these to quantitative scattering theories, information on the actual locations of atoms within the unit cell can be obtained. However, the process may prove to be complicated in some cases. While x-ray diffraction (XRD) and electron selected area diffraction (SAD) techniques are basically used to study the bulk structures, low energy electron diffraction (LEED) and reflection high energy electron diffraction (RHEED) provide surface specific structural information. All these techniques have found uses in the characterization of CVD diamond.

5.1 X-ray Diffraction

The principles of XRD are based on Bragg's diffraction law (110):

$$2d\sin\theta = n\lambda$$

where d is the inter-planar spacing, θ is the diffraction angle, n is an integer and λ is the incident wavelength. An x-ray diffractometer usually consists of an x-ray source, a goniometer, a counter tube and counting electronics. Results are obtained in the form of a chart of peak intensity vs. 2θ. This angle can then be converted to the d spacing between crystallographic planes using Bragg's law. The XRD pattern usually represents a good average throughout the film thickness because of the relatively deep penetration of x-rays. The intensity of the diffracted beam at a given θ is related to a number of structural and geometrical factors. The powder and rotating-crystal diffraction methods are of most general utility. The interpretation of XRD is usually simpler than electron diffraction because the kinematic approximation (i.e., strong single scattering but weak multiple scattering) is generally applicable (111). High accuracy in determining orientation is generally possible, and the sample preparation and handling is quite easy. Therefore, it is widely used as a basic tool for identification of crystal phases and determination of lattice constants and crystal orientation unless the sample is very thin or both the size and the amount of the sample are very small.

For CVD diamond films, XRD experiments have been used for both lattice parameter measurement and texture determination. X-ray diffraction patterns can be used to differentiate between the different phases of crystalline carbon. To obtain a pattern characteristic of polycrystalline diamond requires that the diamond film be sufficiently well crystallized (i.e, have crystallites of the order of ~10 nm or larger). Due to the high symmetry of the diamond cubic crystal system, structure factor calculations give very few allowable diffracting planes (allowed diffraction planes {hkl} are: {111}, {220}, {311}, {400}, {331}, etc., with h, k, l, all odd, or with both h, k, l all even and h+k+l = 4n where n is integral) (111). Furthermore, in the case of randomly oriented diamond only one of these, the {111} diffraction line, is relatively strong. The x-ray diffraction lines may be shifted, greatly broadened or show asymmetry resulting from strain, defects, or small crystallite size. For example, Badzian et al. (94) used rotating-crystal XRD to analyze the growth sectors of homoepitaxial diamond films. For the <111> sector, the diffraction showed strong diffuse scattering and extra-reflections due to high densities of planar defects. For the <100> sector, epitaxial films could be grown continuously as nearly perfect single crystals, and thus did not exhibit such effects. Since many of the diamond films are highly oriented, the relative intensities of these diffraction lines can be significantly distorted, and the diffraction pattern of an as-deposited film may not correlate with a randomly oriented reference pattern. For this same reason, information on the growth texture in the films can be deduced from the diffraction patterns. Figure 21 shows typical XRD patterns from both {111} and {100} surface dominated films which exhibit large differences in the relative diffraction intensities of crystallographic planes.

Kaae et al. (82) obtained XRD patterns of five different films with different surface morphologies. Because only crystallographic planes parallel to the substrate surface contribute to the diffraction pattern, comparisons of the diffraction intensities of low index crystallographic planes in a film with the diffraction intensities of the same planes in an isotropic powder sample allow a qualitative determination of the preferred orientation in the film parallel to the deposition plane. As indicated in the relative XRD peak intensities compiled in Table 2, two of the films (labeled A and E) showed considerably larger {111} diffraction intensities in relation to the diffraction intensities of other low index planes than the {111} diffraction peak of an isotropic sample. In film B, the {400} peak was much larger in relation to the diffraction intensities of other low index planes than the {400} diffraction peak of an isotropic sample. Therefore, films A and E possessed preferred orientation of the {111} planes parallel to the deposition plane, and in film B there was a preferred orientation of the {100} planes parallel to the deposition plane. These results were shown to be consistent with the SEM observations of the surface morphologies of these samples.

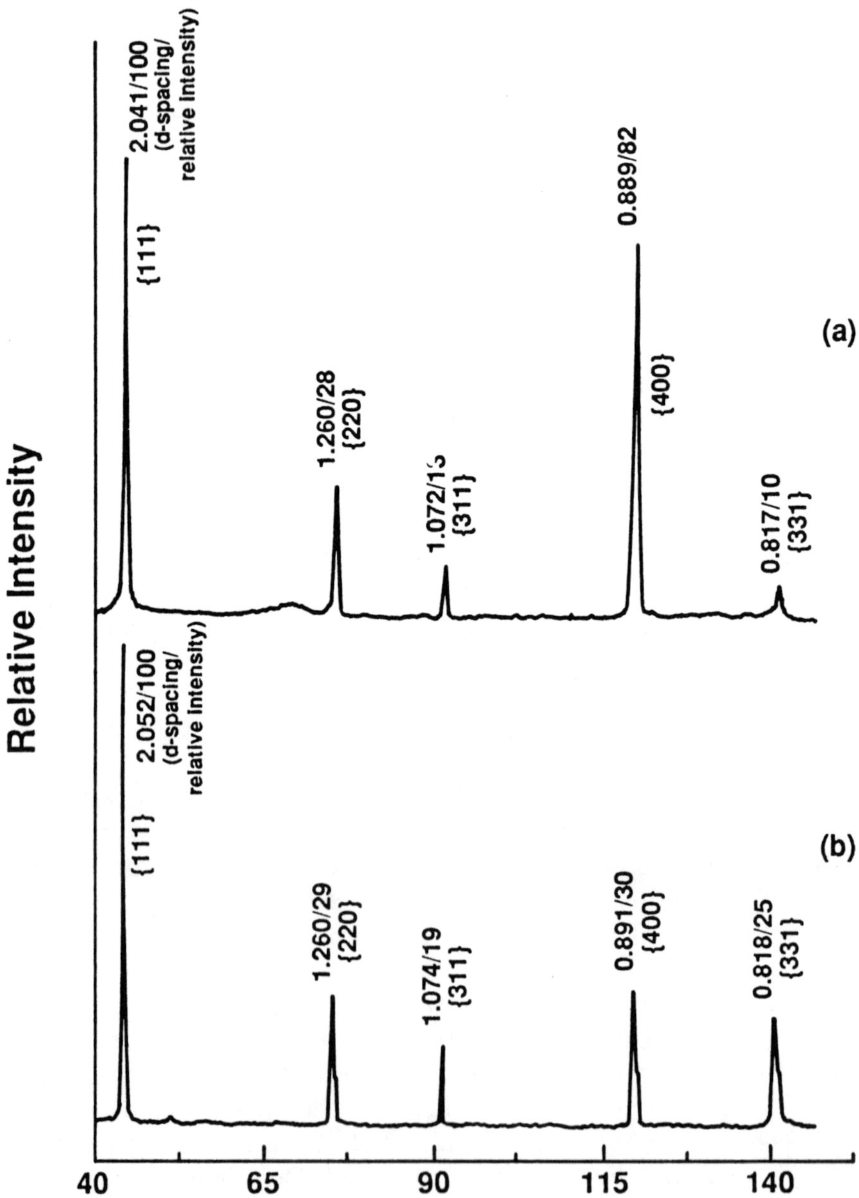

Figure 21. Typical XRD patterns for diamond films with (a) {100} facets; and (b) {111} facets on the surface.

Table 2. Relative X-ray Diffraction Peak Intensities from Crystalline Planes in Diamond Films (82).

Film	Relative x-ray diffraction peak intensity			
	{111}	{220}	{311}	{400}
Powder (ASTM)	100	25	16	8
A	100	5	3	1
B	100	8	8	100
C	100	28	10	1
E	100	7	7	5

Kobashi et al. (105) carried out x-ray diffraction to measure the lattice spacings of CVD diamond and found that the lattice constants were the same as those of natural diamond within the experimental error. They also investigated the changes of x-ray diffraction intensities with methane concentrations. At low methane concentrations, the intensities of diffraction lines were similar to the standard pattern of diamond powders. As the methane concentration increased, the intensity of the {111} line decreased, whereas the intensity of the {220} line remained the same. The {400} diffraction line became detectable near 1.0% CH_4 in accordance with the unidirectionally ordered orientations of {100} faces as observed in SEM. These results implied that the diamond component in the film decreased with the increase of methane concentrations, and the orientations of the diamond grains were random at lower methane concentrations, whereas {100} crystallographic planes tended to be more parallel to the substrate surface at higher methane concentrations.

The growth texture of CVD diamond films has further been investigated by several researchers by x-ray texture analysis (112)-(114). Most of them found a <110> growth texture as indicated by the increased intensity of the {220} diffraction in an XRD pattern. In other words, the {110} planes tended to align perpendicular to the growth direction or parallel to the substrate surface, although the {110} faces have not been identified on the final surfaces of CVD diamond films. Wild et al. (112) investigated the texture formation of a 400 μm thick diamond film with a dominant {111} surface morphology. They found that the film exhibited a pronounced <110> fiber texture. They used van der Drift's "evolutionary selection" model (115) and computer simulations to explain such a texture development in diamond films. The model predicts that random nucleation followed by competitive growth, during which nuclei with the fastest growing crystallographic direction perpendicular to the substrate envelope and overgrow the less favorably oriented ones, leads to a texture with the fastest growing direction normal to the substrate. They found that the crystals formed at the beginning of the growth were randomly oriented, and that a preferential orientation of {110} planes developed with increasing film thickness due to the competition of differently oriented crystals, which implies that <110> was the direction of fastest growth. Twinning during the nucleation stage of the film appeared to destabilize the <100> growth and result in the preferential growth along the <110> direction.

In addition to the <110> texture, <111> texture (116)(117) and <100> texture (90)(105)(114) have also been reported in CVD diamond films. Again, van der Drift's "evolutionary selection" model was applied to explain these textures in relation to the growth rates in the <100>, <110> and <111> directions under different deposition conditions. Clausing et al. (90) recently have presented direct evidence for this model and showed that the growth texture can be controlled by controlling the direction of the most rapid crystal growth.

XRD has also been used by Meilunas et al. (32) to study the time evolution of diamond nucleation and growth on molybdenum substrates in microwave plasmas by monitoring the intensities of Mo, Mo carbide (Mo_2C) and diamond diffraction signals. They found that within a minute from the start of the deposition, Mo carbide was formed. With the passage of time, the XRD signal from the Mo substrate decreased as the Mo carbide growth proceeded. Although nanocrystallites of diamond were observed after 1 min of growth using SEM, XRD from the diamond (111) plane was not detected until about 20 min after plasma ignition. At this point, the Mo carbide signal stopped increasing. The thickness of the Mo carbide layer was estimated to be about 1.5 µm.

A final note is that x-ray topography, which can reveal the internal structure of diamond to a resolution of about 1 µm, has been widely used in the studies of defects in natural and HTHP synthetic diamonds (118). Linear and planar defects such as dislocations and growth sector boundaries are readily revealed. However, it is not suitable for characterizing CVD diamond films because sufficiently large (~1 cm x 1 cm), relatively perfect, single crystal samples are necessary which are not available from CVD processes to date.

5.2 Electron Selected Area Diffraction

It is possible to operate the transmission electron microscope in a mode such that an electron diffraction pattern is produced on the final image screen. The interpretation of electron diffraction patterns is based on the Bragg diffraction in crystal planes, similar to XRD (119). The intensity and distribution of diffraction spots provide information such as lattice type, lattice constants and crystallographic orientations. In addition, in electron selected area diffraction (SAD) it is possible to insert a selector aperture such that only small regions of the sample containing interesting inclusions or structural features contribute to the diffraction. Therefore, the possibility of identifying such small areas or second phases and analyzing structural differences with high spatial resolutions is one of the strengths of the transmission electron microscope. Furthermore, the possibility of immediately switching from the diffraction pattern to the image of the same sample area is invaluable. However, unlike XRD, very thin samples must be prepared for electron SAD to allow electron transmission diffraction due to the weak penetrating power of electron beams as compared to x-rays.

Electron diffraction techniques have popularly been used to determine the structure of the diamond phase. An SAD pattern from a polycrystalline diamond film is shown in Fig. 22a which consists of diffraction rings of diamond {111}, {220}, etc. The lattice spacings of diamond as determined from the diffraction pattern are listed in Fig. 22b. The reported ASTM values for natural diamond are also shown in this figure for comparison and indicate an excellent agreement.

Figure 23 shows the diffraction pattern of a typical diamond crystal with {111} planar defects of twins and stacking faults with the incident electron beam along the diamond [110]. Because twin structures are closely related to stacking faults as can be seen in a faulted stacking sequence where monolayer twins are formed (Fig. 24), twins and stacking faults usually have similar origins. In addition, twinning on the {111} planes of diamond causes little structural disturbance and corresponds to a minimum twin energy. In the diffraction pattern shown in Fig. 23a, additional spots located at one third of the distance between the main diffraction spots along <111> directions represent the presence of {111} twins. Streaked lines joining the main spots along <111> directions indicate that either stacking faults of random repetition of {111} planes are present or the twins are very thin. For a diamond crystal which is twinned on all four {111} planes, each twin component is obtained from the matrix reciprocal lattice by a rotation of 180°

a

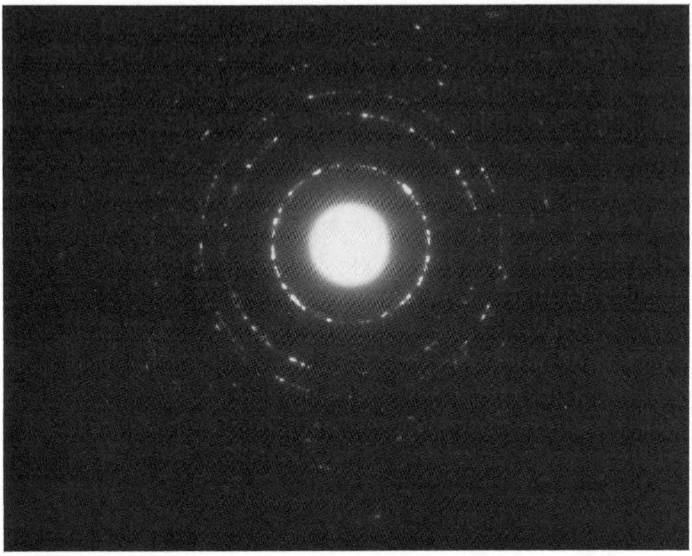

b

Observed {hkl}	Reported d (Å)	ASTM 6-675 d (Å)
111	2.060	2.060
220	1.265	1.261
311	1.073	1.0754
400	0.896	0.8916
331`	0.814	0.8182

Figure 22. (a) An electron SAD pattern from a polycrystalline diamond film; and (b) interplanar spacings calculated from the SAD pattern in (a) with reported ASTM spacings for natural diamond (53).

a

b

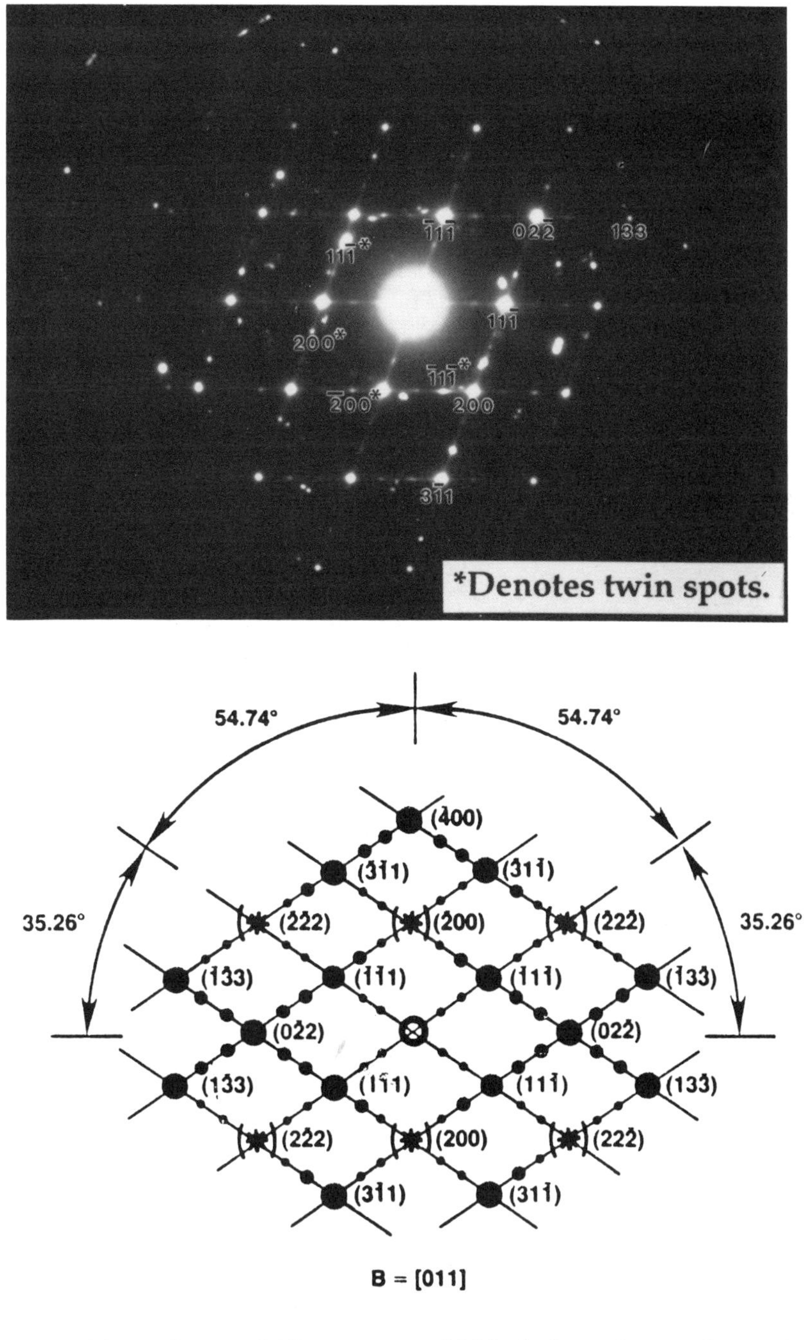

Figure 23. (a) An electron SAD pattern; and (b) its indexes from a diamond crystal containing {111} planar defects of twins and stacking faults (80).

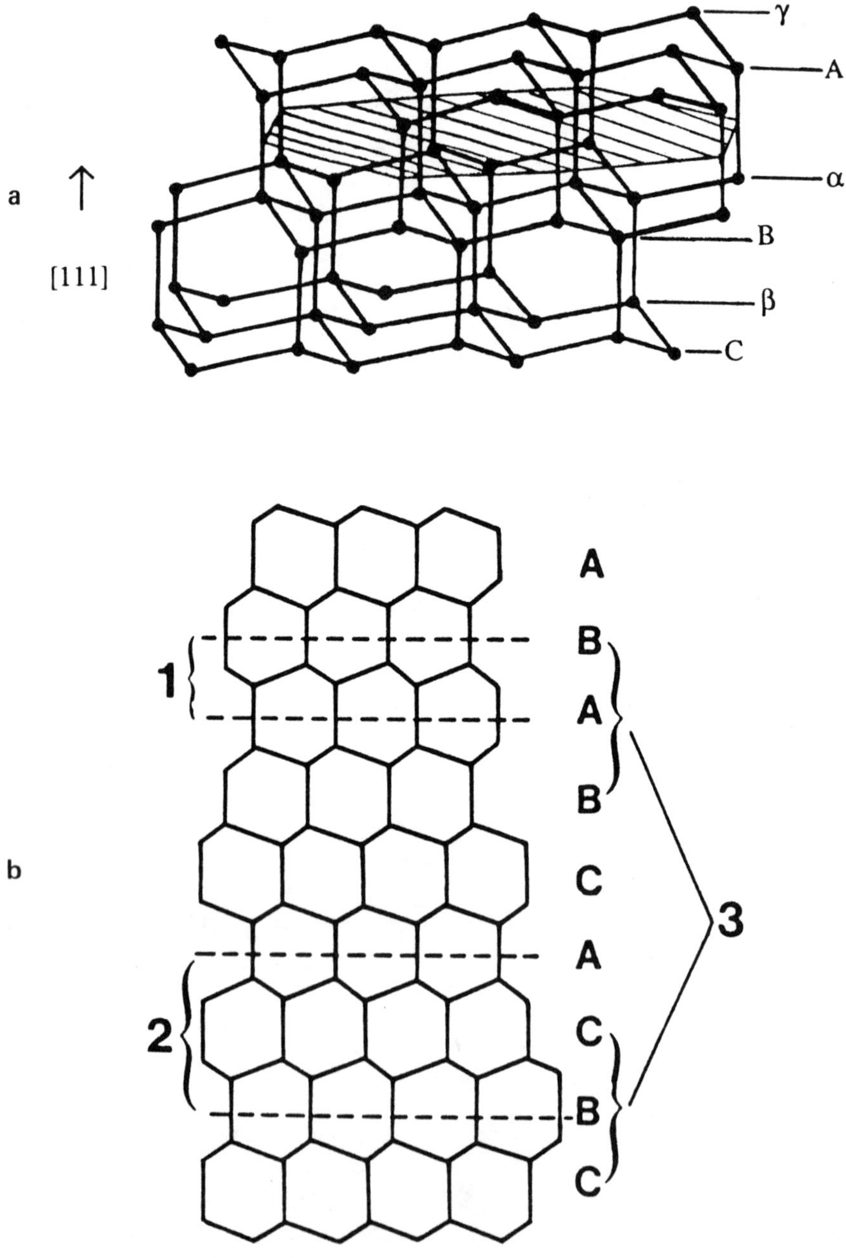

Figure 24. Schematic of (a) the {111} planes in the diamond structure; and (b) the {111} stacking sequences projected on the ($1\bar{1}0$) plane, where (1) is an intrinsic stacking fault, (2) is an extrinsic stacking fault, and (3) are monolayer twins (66).

about a <111> twin axis (60). This causes a point (hkl) of the matrix to be transformed to the point:

$$[p(ph+2qk+2rl)-h(q^2+r^2)]/(p^2+q^2+r^2)$$
$$[q(2ph+qk+2rl)-k(p^2+r^2)]/(p^2+q^2+r^2)$$
$$[r(2ph+2qk+rl)-l(p^2+q^2)]/(p^2+q^2+r^2)$$

where (pqr) is the {111} twin plane. For example, if (pqr)=(111), we have

$$(111) \rightarrow (111);$$
$$(\bar{1}11) \rightarrow 1/3(\bar{5}1\bar{1}) = (200) - 1/3(111)$$
$$(\bar{1}\bar{1}1) \rightarrow 1/3(11\bar{5}) = (00\bar{2}) + 1/3(111)$$
$$(220) \rightarrow 1/3(228) = (113) - 1/3(111)$$
$$(311) \rightarrow 1/3(177) = (022) + 1/3(111)$$

Thus, the twin spots either coincide with matrix points, or are determined by a translation of ±1/3<111>. The <111> lines connecting the matrix reciprocal lattice points are, therefore, divided into threes. Any twin point can be written as (u'v'w') = [u±(1/3)p, v±(1/3)q, w±(1/3)r], where (uvw) is a matrix point. But not all of these one-third points on the <111> lines of the reciprocal lattice are allowed to be occupied by twin spots. The selection rule which determines whether a point [u±(1/3)p, v±(1/3)q, w±(1/3)r] is occupied by a twin spot is

$$pu + qv + rw = \pm(3N + 1)$$

where N is an integer.

Double diffraction can easily occur at microtwin boundaries when a diffracted beam from the matrix passes into a twin and suffers a second diffraction, or vice versa. The direction of the double diffracted beam is determined by adding together the reciprocal lattice vectors corresponding to the two component diffractions. It can be deduced that extra spots resulting from the double diffraction are also displaced from matrix points by vectors of ± 1/3<111> (120). Thus, forbidden spots such as {200} and {222} which are not allowed from the structure factor calculations are present due to the double diffraction. A complete indexation of the diffraction pattern is thus possible as shown in Fig. 23b. These indexes can be confirmed further by imaging in the dark field mode which shows the matrix (111) orientation as bright areas and the twins as dark areas when imaging the (111) matrix spot and shows those twins as bright areas and the matrix as dark areas when imaging the (111) twinning spot (80).

Similar to XRD, electron diffraction can also supply information regarding the growth orientation or texture of the films. An SAD pattern from an interfacial region between a diamond film deposited by microwave plasma CVD at 0.3% CH_4 and the silicon substrate is shown in Fig. 25. The pattern indexes indicated that there was an interfacial layer of β-SiC contributing to the diffraction in addition to diamond and Si. The alignment of the Si (111) and SiC (111) diffraction spots suggested that in the region of this SAD, the β-SiC buffer layer was a single crystal, epitaxial with the Si substrate (i.e, β-SiC (111)//Si (111), β-SiC [110]//Si [110]). This is not surprising since epitaxial β-SiC layers have successfully been grown on Si (111) substrates by thermal CVD (121). However, the quality of the epitaxial layer of β-SiC on Si (111) formed in the microwave plasma CVD process was relatively poor. This is apparent in Fig. 25a by the smearing out of the SiC spots (indicating residual strain) and in Fig. 15c by the alternating dark and light areas in the layer observed in the corresponding bright field image, which are believed to be caused by various defects and strain associated with the poor lattice and thermal coefficient of expansion matches between SiC and Si and the fact that the growth conditions were not optimized for the formation of SiC. There are several other interesting observations which can be made from this SAD pattern. First, although the diamond film is known to be polycrystalline, this diffraction pattern indicates that one of the grains sampled here (grain #1) was actually in a twinned epitaxial relationship with the Si substrate, that is, diamond (111) ll Si (111) and diamond $[0\bar{1}1]$ ll Si $[01\bar{1}]$. Thus, two of the diamond {111} spots were seen to be aligned with Si {111} spots, while the other two {111} spots for this grain were reflected across the [111] direction from where one would expect for "conventional" epitaxy. Therefore, the Si substrate and diamond grain were related by a mirror plane located at the diamond/Si interface in this case. High resolution TEM and cross-section TEM has verified this "twinned epitaxy" and the presence of the SiC buffer layer (122). Diamond grain #2 in this micrograph was rotated ~34° about the [110] direction relative to the [110] direction in the Si substrate. This gave rise to a grain boundary where these two grains met.

As discussed previously in Sec. 4.1, graphite can be detected by electron SAD in films deposited at methane concentrations above 2.0% in hydrogen (96). A diffraction pattern obtained from a diamond film grown at 2.0% CH_4 showed graphite {00.2} diffraction rings coexisting with diamond diffraction spots as shown in Fig. 26a. Graphite {00.4} diffraction was very weak owing to the scattering nature and high structural distortion of graphite. The graphite diffraction rings showed a nonuniform intensity distribution

a

b

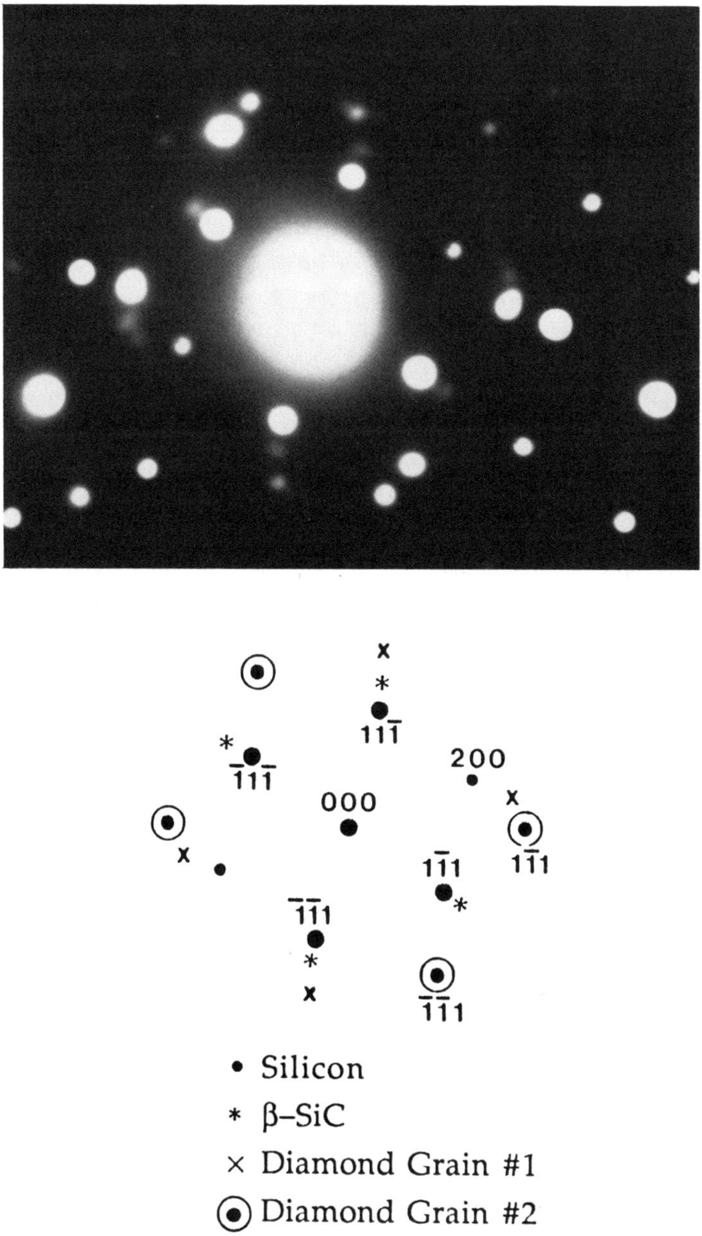

Figure 25. (a) An electron SAD pattern; and (b) its indexes from an interfacial region between a diamond film and Si substrate which indicates the epitaxial relationships between the diamond film, the β-SiC interface and Si substrate (53).

which gave arcs of stronger intensity to the graphite ring pattern. This indicated that graphite crystallites were preferentially oriented or textured about their basal planes in the diamond film. The length of the arc corresponded to the degree of orientational variation of graphite crystallites. It was noted that the graphite {00.2} diffraction arcs were aligned with the diamond <111> reciprocal lattice vectors with a spread of orientations within a cone angle of ~40°. It was thus suggested that graphite formed on or close to the {111} planes of diamond, and the hexagonal graphite axis was parallel with the trigonal diamond axis. Such a crystallographic relationship was also found between diamond and graphite by Evans (123) in a graphitization study of natural diamond. When the sample was tilted so as to have the diamond <111> zone axis parallel to the incident electron beam, there was an absence of graphite {10.0}-type diffraction, which suggested that graphite had a two-dimensional structure (biperiodical). This two dimensional structure was a result of graphite growing mainly in the basal plane, or ***a*** direction, and not in the direction perpendicular to the basal plane, or ***c*** direction. The textured graphite arc pattern was actually associated with a domain distribution of graphite crystallites in the diamond matrix as revealed by the dark field images which have been discussed previously in Sec. 4.1. When the methane concentration was increased to 5.0% in hydrogen, the resulting SAD pattern showed a strong and complete diffraction ring of graphite {00.2} planes (Fig. 26b) which indicated a loss of preferred orientation of graphite about their basal planes. Graphite crystallites were no longer textured and had a random distribution in the diamond matrix.

a

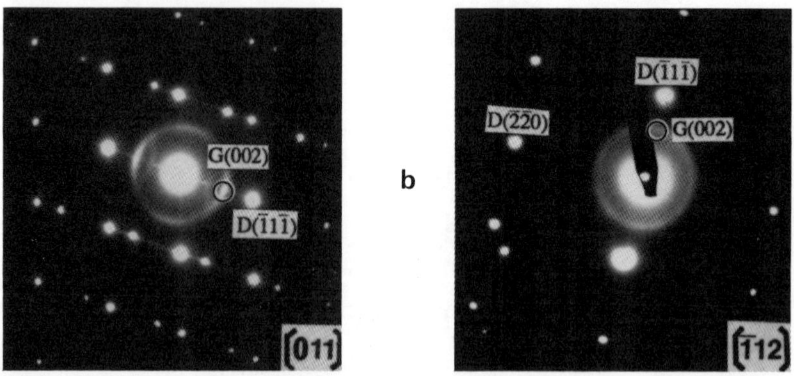

Figure 26. Electron SAD patterns of diamond films grown at (a) 2.0% CH_4/H_2; and (b) 5.0% CH_4/H_2, showing graphite coexisting with diamond (96).

It is worth noting that although twinning of crystals is very common in the <111> direction, the hexagonal sp^3 polymorph, Lonsdaleite, has not been reported in any electron diffraction study. Since a variety of carbon structures can be produced by CVD methods, it is remarkable that well crystallized Lonsdaleite or a mixture of Lonsdaleite and diamond has not been found. This is in sharp contrast to the fact that the analogous hexagonal and cubic polymorphs of SiC are readily obtained by the vapor phase deposition. Thus, as indicated by Yarbrough and Messier (13), any proposed mechanisms for diamond growth from the vapor phase must be consistent with the formation of cubic diamond, not Lonsdaleite.

It should be kept in mind that the sensitivity of electron SAD to crystalline phases (as opposed to amorphous materials) is sufficiently great to introduce the possibility that the diamond seen represents only a small fraction of the carbon present. Thus, films composed of both amorphous and crystalline components should be analyzed with various other techniques (i.e., Raman spectroscopy, EELS, etc.) to ensure that both regions are examined. Significant errors and uncertainties can also arise as a result of the extremely small volume of material being sampled and the strong possibility of texture or preferred orientation. Also, artifacts which could be introduced during the sample preparation should be treated cautiously during the interpretation of electron SAD patterns.

5.3 Low Energy Electron Diffraction

It was recognized very early in the history of diffraction studies that just as x-rays and relatively high energy electrons (say 50 keV) can give information about the bulk periodicity of a crystal, low energy electrons (~100 eV) whose penetrating power is only a few atomic diameters should give information about the surface structure of the solid (124)(125). In typical LEED "optics", an electron beam of 1 μA at energies of 20 - 300 eV from an electron gun is given an uniform acceleration, striking the crystal normal to its surface. The electrons may then be either elastically or inelastically scattered; only the former are used in the diffraction experiment. The diffraction pattern is simply a projection of the surface reciprocal net at a magnification determined by the incident electron energy. The pattern is primarily due to the first layer of atoms because of the small penetrating power of the low energy electrons. It is, therefore, an ideal technique to assess the translational symmetry of the surface, although not well suited to quantitative beam intensity measurements to determine atomic positions in addition to this translational symmetry. Furthermore, it is possible to modify LEED systems to conduct simple AES studies because their basic instrumental requirements are the same.

The usual object of a LEED study is the determination of the surface structure. This is not always easy because the LEED pattern is not simply a picture of the surface structure; rather, the pattern is one of the reciprocal lattice. Usually there will be more than one possible surface structure giving the same LEED pattern. In order to decide between possibilities, it is necessary to make calculations for not only each of the positions expected for the spots but also of the intensity of the spots. There can be problems in allowing for multiple scattering. Also, if a surface is nominally that of a high Miller index plane, what may actually be present is a series of steps and ledges made up of low Miller index planes, and this may show up as a splitting of the diffraction spots.

Most of the LEED studies for diamond have been made on natural diamond. Lander and Morrison (126) used LEED to study the diamond {111} surfaces. Clean surfaces obtained by heat treatment at 1300°C in UHV without surface graphitization occurring showed weak half-order diffraction features which could be removed by heating diamond in hydrogen. They interpreted these results in terms of an atomistic model involving the rearrangement of the outer atomic layer and the formation of surface vacancies. In addition, they studied the interactions of metals with diamond surfaces and found that aluminum and phosphorous formed ordered, two-dimensional structures on the {111} surface. Reactions with aluminum (at 700°C) produced a phase with "1/2-orders" which probably corresponded to a 1/2 monolayer coverage. Reactions with phosphorous (at 800°C, $<10^{-4}$ torr) produced a phase with (1/3n 1/3n) orders probably at a 1/3 monolayer coverage. Reactions with hydrogen at high temperatures (900 - 1000°C) and low pressures (10^{-4} torr) produced a simple phase without superstructure. Surface mobilities of carbon atoms (or vacancies) were detectable at 900°C. Based on this, they predicted that epitaxial growth of diamond on diamond was possible in the temperature range between 900°C and 1400°C if graphite nucleation was impeded. This is of great significance in the history of the development of diamond deposition from vapor phase at metastable conditions.

Lurie and Wilson (127) performed LEED studies on three low index faces of diamond: {111}, {110} and {100}. They found that the non-reconstructed surface atomic spacing was approximately the same (within ± 5%) as that of the bulk. Both {111} and {100} surfaces reconstructed upon heating (~1300°C), but the {110} did not. They extended Lander and Morrison's vacancy model (126) and suggested a two-step reconstruction process upon heating: *(i)* the removal of alternate rows of surface carbon atoms with their bonded hydrogen atoms plus another two atomic hydrogen atoms from adjacent remaining carbon atoms in the form of methane gas; and *(ii)* the two dangling bonds left on the surface then doubly bond to the second layer of carbon atoms (involving some bond bending). In this way, all the dangling bonds are saturated. They also examined gas adsorptions on diamond surfaces and found that diamond was inactive to common gases such as sulphur, oxygen, ammonia and nitrogen; but electron stimulated adsorption of sulphur could occur from an H_2S environment. Interactions with aluminum (a carbide former) and nickel (a carbon solvent) were also studied. Aluminum carbide (Al_4C_3) was formed on all three low index planes, but the epitaxial carbide growth was only observed on the {111} surface ([110]Al_4C_3 // [211]diamond and (111) Al_4C_3 // (111)diamond). In the case of Ni, epitaxial Ni island growth occurred on all three low index planes when the diamond was held between 500°C and 800°C during the Ni evaporation. Subsequent annealing at temperatures less than 1000°C improved ordering. Heating these diamond surfaces to 1000°C produced graphitic carbon at the surface, and a strong basal plane alignment of the graphite crystallites prevailed on the {100} surfaces.

Pate (35) found that the diamond {111} surfaces underwent reversible changes due to cycles of hydrogen adsorption and desorption. LEED indicated that the as-polished diamond {111} surface was hydrogen terminated and exhibited a (1x1) LEED pattern, but was otherwise relatively clean of contamination. Annealing to temperatures above 900°C desorbed the hydrogen and reconstructed the surface to a (2x2/2x1) LEED pattern (from the symmetry of the LEED pattern, it cannot be distinguished if one has a (2x2) reconstruction or three equivalent (2x1) domains), and corresponding surface state emissions near the valence band maximum were observed on the reconstructed surface by XPS. Adsorption of atomic hydrogen onto the reconstructed surface reverted the surface symmetry to the (1x1) structure. However, subsequent anneals did not change the LEED pattern from (1x1), and the LEED structure remained a (1x1) pattern thereafter, although surface state emissions were again detected. This persistent (1x1) LEED pattern, even when surface states were observed, was possibly related to etching and/or roughening of the surface by hydrogen adsorption/desorption cycles. Microscopic surface roughening might act to eliminate the long range order of the reconstructed surface (i.e., the lack of the (2x2/2x1) LEED pattern) while allowing for a local reconstruction which resulted in a surface electronic structure similar to that observed for the (2x2/2x1) reconstructed surface as probed by XPS.

In CVD processes, LEED has been used by Belton and Schmieg (42) to establish the surface order of diamond growth on Ni (100) as a function of time. The LEED experiments were used to address the question of when and why registry of the diamond film to the Ni (100) was lost. Prior to growth the Ni surface showed a p(2x2) LEED pattern due to approximately 0.25 monolayers of residual surface carbon which gave a pattern typical of low coverages of carbon on clean Ni (100). After only 2 min of growth, the LEED pattern changed to c(2x2) superimposed with a streaked 12 spot ring. These 12 streaked spots, as opposed to the six spot pattern expected for the hexagonal graphite, indicated that graphite was present primarily in two different domains that were related by a 90° rotation. The spots were streaked, almost into a ring, because there was considerable deviation from the two preferred orientations. If the azimuthal orientation was random, but the planes were parallel to the Ni (100) surface, the LEED pattern would be a continuous ring. The two preferred orientations were presumably related to the two symmetric axes of the substrate. The graphite layer was estimated from the Ni(2p) XPS intensity to be about 50 Å thick if it was uniform. However, considering the c(2x2) LEED pattern from the substrate, the graphite layer was probably not uniform over the entire surface. Instead islands of graphite were believed to be present on top of the c(2x2) carbon layer. After growing for 32 min, the c(2x2) pattern was not observed, and the 12 spots had joined to form a very faint six-lobed ring. This indicated that the microcrystalline graphite layer covered the entire Ni surface, and the graphite had almost no preferred azimuthal orientation. However, the observation of rings meant that the graphite planes must be largely parallel to the Ni surface. At longer growth times (>2 h), no LEED patterns were observed; therefore, if graphite was present, it was randomly oriented on the surface or in other words, the carbon surface was now more like amorphous or glassy carbon. The order of the graphite layer degraded smoothly as a function of thickness from the Ni surface. Thus, four distinct phases in the diamond growth on Ni substrate were identified: first, carbon was deposited on the Ni(100) surface in an ordered c(2x2) array that occupied 50% of the fourfold hollow sites; next, a graphite layer with poor azimuthal order grew in islands on top of the c(2x2) carbon deposit; then, the graphite gradually became very disordered appearing like glassy carbon; finally, diamonds nucleated and grew on the glassy carbon surface. These results were further confirmed by EELS, AES, and SEM studies. The intermediate layer of graphite and glassy carbon prevented the diamond film from "seeing" the order of the Ni (100) surface; thus, diamonds had no chance of growing in registry with the Ni substrate. However, as stated previously in Sec. 3.1, there is a recent report that diamond can be grown heteroepitaxially on Ni substrates (43). Confirmation and reproduction of this result is currently underway.

5.4 Reflection High Energy Electron Diffraction

In LEED, low energy electrons are used to ensure reasonably strong backscattering cross-sections and to minimize the inelastic scattering mean-free-path, thus enhancing the surface specificity of the technique. Reflection high energy electron diffraction (RHEED) represents an alternative solution to the same basic problem, although not quite as surface sensitive as LEED. Its surface specificity is realized at high energies (usually 30 - 100 keV) by using grazing incidence and emergence angles of electron beams with respect to the surface (<5°). A RHEED pattern of a good surface is usually streaked because of the large Ewald sphere relative to the spacing of the surface reciprocal net rods (cf. the LEED case) (128). Because there is a large energy difference between the elastically scattered electrons and the inelastically scattered background, and because these primary electrons are sufficiently energetic to produce fluorescence on a phosphor, careful energy filtering and post-acceleration are far less necessary than in LEED.

RHEED can be operated using conventional electron microscope optics with energies from as low as 3 - 5 keV up to about 100 keV. On the other hand, if a suitable electron gun is already fitted to a chamber for other purposes, the additional resources needed to exploit the technique are minimal. The RHEED geometry provides improved accessibility to the sample relative to LEED due to the space it leaves in front of the sample. In particular, the use of grazing incidence angles makes the technique sensitive to the quality of the surface which means that the technique can also identify changes in surface morphology. This technique has found an increasing use in the study of epitaxial growth and other multilayer surface processes. For CVD diamond films, RHEED has mainly been used to routinely examine and verify the surface or near-surface structure (129)-(131). The most significant advantage of RHEED relative to the high energy electron SAD utilized in transmission electron microscopy is that there is virtually no need for sample preparation other than keeping the surface somewhat clean of contamination. The time-consuming preparation for a sufficiently thin TEM sample in the case of transmission electron diffraction is not necessary for making a RHEED measurement. However, there is an absence of a complete quantitative theory of RHEED such as the multiple scattering theory of LEED, and a comparable experimental intensity database is also unavailable. Therefore, most applications of RHEED are concerned only with qualitative structure assessments and not with quantitative diffracted beam intensity analysis.

6.0 OTHER ANALYTICAL TECHNIQUES

In addition to the main characterization techniques discussed above, there are still many others which have been utilized for the analysis of CVD diamond. They often provide equally important, interesting and useful information about the structures and compositions of CVD diamond films. However, since each of these techniques and the interpretation of results constitute a field of specialty by itself, they cannot be covered in detail within the scope of this chapter. Therefore, they will only be described briefly in this section. Interested readers should consult various references given therein for technical or instrumental details. These techniques are not categorized in terms of their physical processes; rather, the choice and arrangement of these techniques are based on the extent of use and the usefulness of the yielded information in the understanding of CVD diamond to date.

6.1 Secondary Ion Mass Spectrometry

SIMS is an incident ion technique based on high energy (1 - 20 keV) ion beam sputtering (usually O^-, O_2^+ or Cs^+) of a material resulting in the ejection of neutral atoms and molecules along with a small fraction of ionized atoms (about 0.1 - 10%) (132). These secondary ions, either positive or negative, can be detected and analyzed by a mass spectrometer to provide elemental identification and chemical concentration data. Typical SIMS detection limits range from 1 ppm to 1 ppb. SIMS can detect hydrogen and helium, although the evaluation may not be very reliable, and is inherently capable of depth profiling. Moreover, ion beams (like electron beams in AES) can be readily focused onto a sample so that chemical composition imaging is possible. SIMS as a method of bulk composition analysis is a well-established technique which, with proper calibration and the use of a simple semi-empirical theory, is capable of providing quite good quantitative analysis with considerable sensitivity to some species. As a technique for surface studies, however, the understanding of the processes leading to ion emission is inadequate to provide the same kind of broad analytical capabilities. Nevertheless, SIMS spectral fingerprints do provide a means of characterizing different adsorption states and, particularly by using low incident ion energies, can prove extremely surface specific. Of course, a potential disadvantage of SIMS is that it is intrinsically destructive.

SIMS has been very useful in the compositional analysis of CVD diamond films. Although the quantification of SIMS data is not possible without implanted standards, a rough order of magnitude approximation of impurity concentrations can be obtained if matrix effects are ignored and published relative ion yield data are utilized to determine concentrations relative to carbon ion yield (133). In a compositional study of CVD diamond films, Williams et al. (53) obtained bar graphs (ion counts/sec vs. atomic weight), depth profiles and chemical maps using an O^- primary ion beam. The films they characterized were determined to be mostly carbon, and the various amounts of impurity elements identified throughout the films included hydrogen, oxygen, aluminum, silicon, calcium and copper. None of the impurities were present at concentrations greater than 1.0%, and all but H and O were less than 0.1%. Chemical maps of silicon, aluminium and oxygen with a spatial resolution of 5 - 10 μm and an image field of 400 μm indicated that these impurities were segregated in regions ranging from 5 to 20 μm in diameter. The depth profiles showed that the concentration changes of various elements across the interface were very smooth, and the change in carbon concentration across the interface was more rapid than the change in silicon, implying more silicon diffusion into the film than carbon diffusion into the silicon substrate.

SIMS has also been applied by Pickrell et al. (102) to investigate the composition of the distinctive intermediate layers formed between diamond films and silica substrates. Analysis of an as-received substrate showed metallic contaminants of Al, K, Na and Ca in addition to Si and O on the surface of the SiO_2 glass. Depth profiling of these contaminants revealed that they were restricted to the substrate surface. After polishing by diamond paste and subsequent methanol cleaning, the level of carbon on the substrate surface was increased. The authors attributed this result to the residual diamond crystals which were left embedded in the substrate after polishing. However, enhanced oxygenated carbon or hydrocarbon contamination of the polished surface would also contribute to the increase of the carbon level. In fact, this latter explanation has become more favorable recently (134). Analysis from the interfaces (the surface against the substrate) of two peeled free-standing films deposited at 0.5% and 5.0% methane concentrations in hydrogen revealed the presence of impurities picked up from contamination of the glass substrate. The concentrations of these impurities dropped with sputtering depth into the films. There was a peak for carbon in the spectra which indicated that initially the carbon was sputtered very quickly but then fell off to a constant level. This is expected when sputtering through a weakly bonded layer of carbon (the intermediate layer) comprised of disordered carbon and graphite, then into a more strongly bonded diamond film.

Although SIMS has been a basic tool for trace impurity analysis for CVD diamond films, there are several other techniques which have been reported to detect impurities in diamond films. For example, neutron activation analysis (NAA) (135) and Laser Microprobe Mass Analysis (LAMMA) (136) have revealed silicon, oxygen and metallic impurities (Na, Al, Ca, W, etc.) in diamond films which presumably result from the contamination of the growth environment or the etching of the substrate materials. Unfortunately, the quantification is again difficult (implanted standards are necessary) and was not attempted in either of these studies. As electronic applications are realized, such trace chemical analyses will become more important, and quantification of such data will be essential.

6.2 Luminescence Spectroscopy

Luminescence can be viewed simply as absorption followed by photon emission. The process involves incident photon excitation (for photoluminescence) or electron excitation (for cathodoluminescence, CL). The incident beam excites electrons in CL sample to higher energy levels, and subsequent radiative decay of electrons from the excited high energy states to some lower energy states results in the emission of photons. The frequency of these emitted photons is determined by the particular electronic excitation process. Comparing with Raman scattering which is an instantaneous process, there is a characteristic decay time associated with the excitation and emission (137).

There are characteristic luminescence features associated with CVD diamond films, possibly related to various impurities introduced during deposition. Several reports (66)(138)(139) on photoluminescence spectra excited by the same laser as that used to stimulate Raman scattering showed that almost all CVD diamond films exhibit a strong luminescence peak at 1.678 eV (738.4 nm) (Fig. 27a). This line is very close to the GR1 line at 1.674 eV common in natural diamond attributed to neutral vacancies (16)-(18)(140). There exists also a small satellite peak on the lower energy side of this strong peak at about 1.637 eV (757.4 nm). Hartnett (138) suggested that the 1.678 eV line was related to the GR1 center under stress, and that the 1.637 eV peak was due to impurities gathered near the 1.678

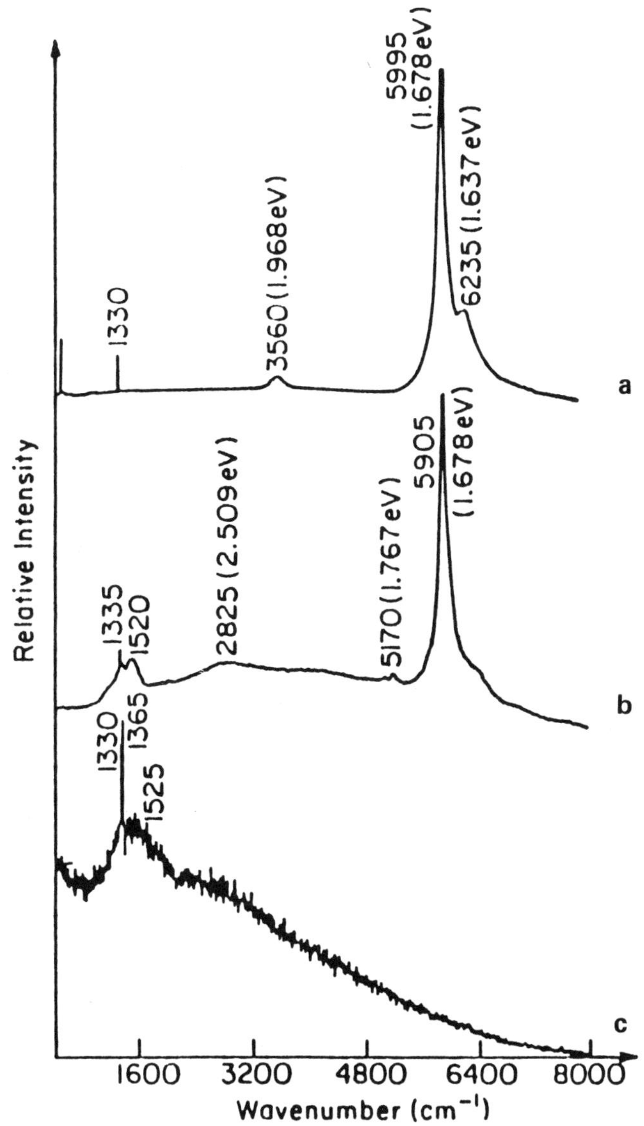

Figure 27. Photoluminescence spectra of diamond films deposited at (a) 1.0% CH_4/H_2 on Si by microwave plasma CVD; (b) 4.5% CH_4/H_2 on Si by microwave plasma CVD; and (c) 1.0% CH_4/H_2 on hexagonal BN by hot filament CVD (66).

eV center, similar to the broadening of the GR1 line to lower energies due to the nitrogen incorporation in natural diamond. But the relative intensities of the 1.637 eV and 1.678 eV lines can vary with the deposition conditions and the structural quality of the films. Therefore, they are considered as more probably related to two different defect centers. In addition, the 1.637 peak is always related to another small peak at 1.968 eV (630.3 nm). The 1.678 eV peak can become weak, and both the 1.637 and 1.968 peaks can disappear when the films contain a relatively large amount of graphitic phases (Fig. 27b). In such cases, additional broad luminescence features around 2.059 eV (602.0 nm) and a small peak at 1.767 eV (701.0 nm) can appear. Furthermore, as Hartnett reported (138), the Si incorporation increased the 1.678 eV peak, whereas B dopants eliminated this peak (Fig. 27c). Thus, it is suspected that B atoms incorporated in the diamond lattice effectively passivate the defects responsible for the 1.678 eV peak.

Cathodoluminescence (CL) spectroscopy can show a wider range of defects than radiation excited methods. This is because electron beam excitation can more easily provide the required high energies for facilitating luminescence detection in wide band gap materials such as diamond and can also allow for imaging the spatial distribution of the emitting centers. The depth of penetration of the electron beam can be selected, for example, to be from 1.4 to 23 μm by choosing the beam energy from 10 to 50 keV, respectively (141). In addition, electron beams produce no stray light in the UV and visible spectra, a problem that accompanies most sources of photons with energies greater than 5 eV. A special requirement is that ultra-high vacuum below 10^{-10} torr is necessary for CL measurements because electron beams can possibly convert condensed organic vacuum contaminants to diamond-like materials.

There is a large body of CL spectra for natural and HPHT synthetic diamond crystals (142). For CVD diamond films, Kawarada et al. (143) obtained visible cathodoluminescence spectra. The peaks occurred at different energies between 2.4 - 2.8 eV (green to purple-blue) and were attributed to donor-acceptor pair recombinations. The {111} and {100} growth sectors could be differentiated by the luminescence intensity. Other well-known diamond bands in the 1.4 - 3.5 eV range due to defects and impurities and an intrinsic diamond excitation emission near the 5.3 eV band gap have also been reported (144)-(146). In addition, the strong 1.678 eV line has also been identified in CL spectra by Ruan et al. (147). They found that this feature could sustain annealing temperatures up to 1200°C. This suggested that the 1.678 eV peak was not due to vacancies (i.e., GR1 center) as most people assumed, since the vacancies would be expected to become mobile and annihilate themselves or diffuse to a surface at these high annealing temperatures.

In general, the origins of the luminescence features in CVD diamond are still largely unknown. Because these luminescence features consistently appear over a wide range of deposition conditions and deposition methods, they are most probably due to commonly occurring impurities and defects which are inherent to the low pressure CVD processes. Badzian et al. (148) have ascribed the 1.678 eV luminescence peak to the incorporation of Si in octahedral holes in the diamond lattice. The most recent silicon ion implantation experiments on bulk diamond and an investigation of the 1.678 eV center under the effects of electron irradiation and subsequent heat treatment conducted by Clark and Dickerson (149) appear to support this assignment. They suggested that the defect center was due to a substitutional silicon atom associated with a vacancy in a nearest neighbor lattice site. However, as Badzian et al. (94) recently suggested, the impurities may not be the only sources of luminescence for CVD diamond. They found that the luminescence was stronger on the <111> growth sector than on the <100> growth sector in homoepitaxial diamond films. This was attributed to the different densities and configurations of planar defects on the different planes. The precise assignment of these features to specific sources requires improved luminescence spectroscopic techniques such as low temperature (liquid helium) excitation and decay time measurements. In addition, other factors such as the polycrystalline nature and graphitic phase inclusions which can complicate the study due to internal reflections, extra absorptions and strain should be considered more carefully. As electronic and electro-optic applications become possible, the luminescence measurements which supply information both about band structure and defects will become increasingly important.

6.3 Rutherford Backscattering Spectrometry

Since almost all of the CVD films are grown in a hydrogen diluted gas mixture, various amounts of hydrogen are incorporated in the diamond lattice. However, accurate measurement of the hydrogen concentration in the films presents a formidable problem because common tools for compositional analysis such as AES and x-ray microprobe are inherently unable to probe hydrogen. Other techniques which are, in principle, capable of detecting hydrogen such as SIMS, IR spectroscopy and mass spectrometry after combustion suffer from the incapability of providing accurate quantitative data without standards. RBS is one of the two techniques (another is nuclear reaction analysis (NRA)) which can supply quantitative information about hydrogen without standard samples; therefore, both RBS and NRA often serve as standards for other techniques that are much more sensitive but require calibration (150).

NRA utilizes nuclear reactions, for example, $^{1}H(^{19}F, \alpha\gamma)^{16}O$, to generate radioactive emissions (γ rays):

$$^{19}F + {}^{1}H \rightarrow ({}^{20}Ne) \rightarrow {}^{4}He\ (\alpha\ \text{particle}) + {}^{16}O(\text{excited state})$$
$$\rightarrow \gamma + {}^{16}O\ (\text{relaxed state})$$

and by measuring the energy and intensity of the γ rays, the hydrogen concentration can be determined (151) (the reaction $^{1}H\ (^{15}N, \alpha\gamma)^{12}C$ has also been reported) (138). This technique has good sensitivity (~1.0 ppm) and is capable of depth profiles. However, resolutions (a spatial resolution of 1 mm^2 and a depth resolution of 20 nm) are relatively poor, and the samples can be damaged by the energetic ion beams.

RBS utilizes a beam of ions (usually He^+) which is accelerated (about 2 MeV) until it collides with the sample of interest. The basic concept is that in an elastic collision, such as backscattering, energy transfer from a projectile to a target atom can be calculated from collision kinematics (150)(152). The mass of the target atom can be determined by measuring the energy transfer due to collision or by measuring the scattered energy of the backscattered particles. In addition, the probability of an elastic collision between the projectile and target atoms is highly predictable, and the scattering probability or scattering cross-section enables quantitative analysis of atomic composition. Thus, by detecting and analyzing the backscattered ions from the sample using a solid-state detector, quantitative compositional information about the sample where scattering events occur can be obtained. It is widely used for quantitative compositional analysis of thin film or bulk samples and measurements of heavy element surface impurities on substrates of lighter elements. Depth profiles of defect distribution can also be obtained, although TEM may be better for defining the nature of the defects. RBS typically has a depth resolution in the range of 5 - 15 nm, and the sensitivity is about 0.1% for heavy mass impurities and 10 % for light mass impurities. However, RBS does not yield any chemical state (i.e., bonding) information, and both the mass resolution for heavy element impurities and the sensitivity for light element impurities are poor.

By these two techniques, the measured typical atomic concentration of hydrogen in diamond films produced by microwave plasma CVD has been reported to be about 0.1 - 5.0% (138)(151)(153)(154). By contrast, a few percent to more than 30% hydrogen is common in DLC materials, although hydrogen-free DLC films composed of an amorphous, sp^3 bonded carbon structure have been reported (155). The incorporated hydrogen has been used to explain the behavior of electrical resistivities in CVD diamond films (26).

6.4 Spectroscopic Ellipsometry

Reflection based visible-near UV optical techniques such as ellipsometry yield information about materials and interfaces from the component of incident flux that is reflected as a result of the optical impedance mismatch between sample and ambient (156)(157). Owing to the high optical contrast between sp^2 and sp^3 bonded carbon in diamond films, spectroscopic ellipsometry (SE) is highly sensitive to the volume fraction of sp^2 bonded carbon, and thus can be used as a more quantitative probe for graphitic or sp^2carbon. In addition, bulk and surface roughness layer thickness and void volume fraction are outcome of an SE analysis. Therefore, spectroscopic and real time ellipsometry has been used to gain insights into growth mechanisms and film structure that would have been difficult to elicit by other methods.

An SE measurement can be conducted on a rotating analyzer spectroscopic ellipsometer, and spectra in the ellipsometry angles $\{\psi(h\nu), \Delta(h\nu)\}$ are obtained as a function of incident photon energy. These optical data $\{\psi, \Delta\}$ are then converted to the real and imaginary parts of the pseudomorphic function, $\{<\varepsilon_1>, <\varepsilon_2>\}$, using the Fresnel equations for a single interface (157). These raw spectra are interpreted by linear regression analysis (LRA) and multilayer optical techniques to deduce a number of photon energy independent structural parameters (layer thickness and volume fraction). However, the dielectric function of the anticipated material components of the sample and its structural model must be known in advance. For CVD diamond films, this includes diamond, disordered graphite and voids. Other phases may also be present in the films, such as diamond-like carbon or amorphous diamond, but these are expected to be simulated by composites of the above three basic components. The final dielectric function and structural model of the films are determined as those which, when used in the simulation, minimize the deviation between the experiment and the calculation.

By this method, Cong et al. (158) have been able to deduce the volume fraction of graphitic carbon in the bulk diamond films and the thickness of the roughness layer on the surface. For an optical quality film (fine grained microstructure) prepared to 1000 - 4000 Å under optimal conditions by microwave plasma CVD, a surface roughness of ~100 Å and bulk graphitic carbon of about 3 vol% were determined. SE measurements have also been applied to understand the structural evolution of the substrates in the initial stages of nucleation and growth of diamond films (159). The results showed that a stable carbide formed in the first few minutes of plasma exposure of the substrate. However, this carbide forming process was considered not a limiting factor for diamond nucleation which occurred on a much later time scale. It was further found that damage incurred by the substrate in the initial minutes of exposure was annealed as the surface temperature rose to reach equilibrium with the plasma. Thus, the single crystal silicon substrate was converted to a polycrystalline structure before diamond actually nucleated and grew. The role of these processes in the ultimate nucleation of diamond is not yet clear.

Another use of SE is that it can be applied in real time (in situ) in adverse environments for process and materials characterization. Real time ellipsometry has successfully been used to monitor and control the deposition of hydrogenated amorphous carbon films (a-C:H) (159). The measurements during the a-C:H deposition provided the bulk optical function, thickness, instantaneous growth rate, and the corresponding parameters for the film/substrate interfacial region. Recently real time SE has also been applied to monitor surfaces of optical quality diamond films during exposure to low energy Ar^+ and H^+ ion beams and atomic hydrogen (50). The SE experiments demonstrated the monolayer-level sensitivity to the conversion of diamond to optically absorbing sp^2 bonded carbon upon ion impact even in the presence of sub-microscopic surface roughness. For Ar^+ ions, a beam voltage of 50 V was sufficient to convert the surface to the equivalent of a single monolayer of sp^2 bonded carbon, whereas the topmost 12 - 16 Å layer was converted at the threshold for sputter etching, between 100 and 150 V. Atomic hydrogen and low energy H^+ ion treatments (<50 V) were found to eliminate optically absorbing defects within 15 Å of the surface. These results demonstrated that real time SE will be a highly sensitive in situ probe of diamond surface and bulk microstructure in the reactive CVD environments.

It should be noted that, to transform the optical data to sample properties (film thickness, compositions, void fractions, densities, and other structural details), a model or an idealized mathematical representation of a given sample that allows its optical properties to be calculated analytically is always required. The science (or art) lies in constructing a model that accurately represents the measured optical properties of a given sample over as many different conditions as possible. Considerable physical insights are needed to construct such a model. Potential difficulties include the possible existence of different but apparently equally valid models, overlooked models, inadequate data, convergence to a wrong solution, etc. In addition, optical analysis deals with macroscopic properties and, therefore, returns parameters that are average or effective quantities. Therefore, SE can only be applied to films having grain sizes smaller than the incident wavelengths. The model may not be sufficiently detailed, and some of the salient features may be lost. Also it should always be kept in mind that a model is an approximation, and as such, can always be pushed beyond its range of validity. Since data interpretation of SE measurements relies on an effective medium theory to predict the optical properties of the diamond films, structural measurements that can reveal physical features remain necessary to ensure homogeneity on the scale of the optical wavelength. Thus, SE is not expected to replace the routine characterization techniques. For many problems there is no alternative but to turn to techniques such as TEM that may be destructive, but nonetheless have the spatial resolution necessary to directly access the microstructure.

7.0 SUMMARY

This chapter dealt with a number of aspects of diamond analysis by various methods. It is rarely satisfactory to use only one of these techniques, since no single technique can yield all of the information necessary for a complete understanding of such a complex process as the CVD of diamond. Some information in each of the areas discussed above (e.g., composition, chemical bonding, interface, defects, etc.) is often needed simultaneously. Such a need is clearly illustrated in the proposed working definition for CVD diamond films which requires a combination of techniques (160). That is, a crystalline morphology discernible by electron microscopy; a single phase crystalline structure identified by X-ray or electron diffraction; and a Raman spectrum typical for diamond with a single, narrow line at 1332 cm^{-1}. In other words, observing a crystalline habit in a deposit by SEM is not adequate proof for the presence of diamond; but when combined with Raman and x-ray or electron diffraction measurements, microscopic observations can be a reliable indicator of the presence of diamond. Another example which reflects such a need is that although AES can conveniently perform fast routine chemical analysis with a sensitivity of 0.1% for most elements (except H which is virtually undetectable), it will not yield any information about trace impurities due to the sensitivity limitations of the technique. More sensitive techniques such as SIMS have to be utilized for such purposes. For these reasons, the multi-technique approach is desirable to understand the fundamental characteristics of CVD diamond.

Concerning comparisons of results from different techniques and investigations, special aspects of each technique must be considered. One technique may be dominated by signals from minority species or particular components of the structure and be insensitive to other components, while another probes the average properties. Attempts to identify these special aspects are essential to make specific and meaningful comparisons. For instance, x-ray and electron diffractions suffer from being sensitive only to the crystalline phases and give little or no information on any non-crystalline carbon that might be present. Hence, it is extremely difficult to determine whether the observed crystalline phase constitutes a minor or major component in the system. Partly because of this difficulty, Raman spectroscopy is popular. It offers the advantages of being sensitive not only to the crystalline phase, but also to the various non-crystalline phases which are present. However, the Raman scattering measurements can be affected by the differences in optical absorption of the different components in the diamond films. The different optical absorption coefficients of the different structures (sp^3 vs. sp^2 phases) lead to large differences in the Raman sensitivity to diamond and graphite. As a result, quantification is extremely difficult.

New analysis techniques are constantly being applied to the characterization of diamond films. They supply unique and complementary information on various aspects of diamond analysis. Some recent examples are: *(i)* electron channeling microscopy which has been used to measure the strain of diamond films with large crystal sizes (>100 µm) and smooth surfaces (161); *(ii)* nuclear magnetic resonance (NMR) which has been employed for quantitative determination of carbon bonding ratios (sp^2/sp^3) in CVD diamond (162); *(iii)* analytical color fluorescence electron microscopy (ACFEM) which has been applied to sensitively distinguish the types, sizes and distribution of impurities and defects (163); *(iv)* extended x-ray absorption fine structure (EXAFS) and extended electron energy loss fine structure (EEELFS) (164) which are beginning to be utilized for chemical bonding analysis and may help achieve the quantification necessary in this area; and finally, *(v)* scanning tunneling microscopy (STM) which is being used to provide atomic resolution images of the diamond surface structure (165)(166). With such extensive efforts utilizing a combination of characterization techniques, a more complete understanding of the growth, structures and properties of diamond films will continue to develop rapidly in the future.

ACKNOWLEDGEMENTS

This chapter was made possible by the various diamond research projects in which the authors are involved at North Carolina State University. The authors would like to acknowledge the many researchers, especially the graduate students, who have contributed to these projects. In particular, useful discussions and support from R. F. Davis, R. J. Nemanich, J. Prater, G. H. M. Ma, M. McClure, F. R. Sivazlian, B. R. Stoner, T. Tachibana, D. Tucker, M. Veil, X. H. Wang, B. E. Williams, J. von Windheim and S. Wolter as well as others are appreciated. Financial support from various sponsors is also gratefully acknowledged including: Strategic Defense Initiative Organization/Innovative Science and Technology Branch through the Office of Naval Research, National Science Foundation, Kobe Steel, Ltd., the Oak Ridge Associated Universities SHaRE program at Oak Ridge National Laboratory, MITI/NEDO and the Army Research Office.

REFERENCES

1. Tobin, M. C., *Laser Raman Spectroscopy*, Wiley (Interscience), NY (1971)
2. Colthup, N. B., Daley, L. H. and Wiberley, S. E., *Introduction to Infrared and Raman Spectroscopy*, Academic Press, NY (1975)
3. Raman, C. V. and Krishnan, K. S., *Nature* 121:501-502 (1928)
4. Nemanich, R. J., Glass, J. T., Lucovsky, G. and Shroder, R. E., *J. Vac. Sci. Tech. A* 6:1783-1787 (1988)
5. Knight, D. S. and White, W. B., *J. Mater. Res.* 4:385-393 (1989)
6. Solin, S. A. and Ramdas, A. K., *Phys. Rev. B* 1:1687-1698 (1970)
7. Nemanich, R. J. and Solin, S. A., *Phys. Rev. B* 20:392-401 (1979)
8. Tuinstra, F. and Koenig, J. L., *J. Chem. Phys.* 53:1126-1130 (1970)
9. Tsai, H. C. and Bogy, D. B., *J. Vac. Sci. Tech. A* 5:3287-3312 (1987)

10. Yoshikawa, M., Katagiri, G., Ishida, H., Ishitani, A. and Akamatsu, T., *J. Appl. Phys.* 64:6464-6468 (1988)

11. Wada, N. and Solin, S. A., *Physica* 105B:353-356 (1981)

12. Shroder, R. E., Nemanich, R. J. and Glass, J. T., *Phys. Rev. B* 41:3738-3745 (1990)

13. Yarbrough, W. A. and Messier, R., *Science* 247:688-696 (1990)

14. Griffiths, P. R., *Chemical Infrared Fourier Transform Spectroscopy*, John Wiley and Sons, NY (1975)

15. Lax, M. and Burstein, E., *Phys. Rev.* 97:39-52 (1955)

16. Davies, G., in: *Chemistry and Physics of Carbon, Vol. 13*, (P. L. Walker, Jr. and P. A. Thrower, eds.), pp.1-143, Marcel Dekker Inc., NY (1977)

17. Clark, C. D., in: *Physical Properties of Diamond*, (R. Berman, ed.), pp. 295-324, Clarendon Press, Oxford (1965)

18. Clark, C. D., Mitchell, E. W. J. and Parsons, B. J., in: *The Properties of Diamond*, (J. E. Field, ed.), pp. 23-77, Academic Press, London (1979)

19. Herzberg, G., *Molecular Spectra and Molecular Structure. II Infrared and Raman Spectra of Polyatomic Molecules*, pp. 193-195, D. Van Nostrand Company, Inc., Princeton, NJ (1945)

20. Couderc, P. and Catherine, Y., *Thin Solid Films* 146:93-107 (1987)

21. Yarbrough, W. A., Rosen, N. D., Pilione, L. J. and Drawl, W. R., *Proceedings of SPIE on Diamond Optics II* 1146:2-11 (1989)

22. Wild, C., Herres, N., Wagner, J., Koidl, P. and Anthony, T. R., in: *Proceedings of the First International Symposium on Diamond and Diamond-Like Films, Vol. 89-12* (J. P. Dismukes, et al., eds), pp. 283-295, The Electrochemical Society, Pennington, NJ (1989)

23. Bi, X. X., Eklund, P. C., Zhang, J. G., Rao, A. M., Perry, T. A. and Beetz, C. P. Jr, *J. Mater. Res.* 5:811-817 (1990)

24. Hartnett, T. M. and Miller, R. P., *Proceedings of SPIE on Electroptical Materials for Switches, Coatings, Sensor Optics and Detectors* 1307:474-484 (1990)

25. Cody, G. D., Abeles, B., Wronski, C. R. and Brooks, B., *J. Non-Cryst. Solids* 35/36:463-468 (1980)

26. Landstrass, M. I. and Ravi, K. V., *Appl. Phys. Lett.* 55:1391-1393 (1989)

27. Snail, K., *SPIE Proceedings on Optical Surfaces Resistant to Severe Environments* 1330:46-64 (1990)

28. Taborek, P., *Proceedings of SPIE on Window and Dome Technologies and Materials* 1112:205-211 (1989)

29. Yoshikawa, M., *Proceedings of SPIE on Diamond Optics III* 1325:210-221 (1990)

30. Zhao, T. J., Grogan, D. F., Bovard, B. G. and Macleod, H. A., *Proceedings of SPIE on Diamond Optics III* 1325:142-151 (1990)

31. Wang, X. H., Pilione, L., Zhu, W., Yarbrough, W. A., Drawl, W. and Messier, R., *J. Mater. Res.* 5:2345-2350 (1990)

32. Meilunas, R., Wong, M. S., Sheng, K. C., Cheng, R. P. H. and Van Duyne, R. P., *Appl. Phys. Lett.* 54:2204-2206 (1989)

33. *Practical Surface Analysis by Auger and X-ray Photoelectron Spectroscopy*, (D. Briggs and M. P. Seah, eds.), John Wiley and Sons, NY (1983)

34. Pate, B. B., Oshima, M., Silberman, J. A., Rossi, G., Lindau, I. and Spicer, W. E., *J. Vac. Sci. Tech. A* 2:957-960 (1984)

35. Pate, B. B., *Surf. Sci.* 165:83-142 (1986)

36. McFeely, F. R., Kowalczyk, S. P., Ley, L., Cavell, R. G., Pollak, R. A. and Shirley, D. A., *Phys. Rev. B* 9:5268-5278 (1974)

37. Morar, J. F., Himpsel, F. J., Hollinger, G., Jordan, J. L., Hughes, G. and McFeely, F. R., *Phys. Rev. B* 33:1340-1345 (1986)

38. Belton, D. N. and Schmieg, S. J., *J. Vac. Sci. Technol. A* 8:2353-2362 (1990)

39. Belton, D. N., Harris, S. J., Schmieg, S. J., Weiner, A. M. and Perry, T. A., *Appl. Phys. Lett.* 54:416-417 (1989)

40. Williams, B. E., Stoner, B. R., Asbury, D. A. and Glass, J. T., presented at the *NATO Advanced Study Institute on Diamond and Diamond-like Films and Coatings*, Castelvecchio Pascoli, Italy (1990)

41. Williams, B. E., "Surface Analysis and Microstructural Characterization of Diamond Thin Films," Ph. D. Thesis in Materials Science and Engineering, North Carolina State University, Raleigh, NC, August 1991

42. Belton, D. N. and Schmieg, S. J., *J. Appl. Phys.* 66:4223-4229 (1989)

43. Fujita, O. H., Ando, T., Kamo, M., Tanaka, T. and Sato, Y., in: *Extended Abstracts of the Fourth Diamond Symposium of Japan New Diamond Forum*, pp. 13-14, Nagaoka, Japan (1990)

44. Auger, P., *Le Journal de Physique et le Radium* 6:205-208 (1925)

45. Chang, C. C., *Surf. Sci.* 25:53-79 (1971)

46. Feldman, L. C. and Mayer, J. W., *Fundamentals of Surface and Thin Film Analysis*, pp. 257-279, North Holland, NY (1986)

47. Lurie, P. G. and Wilson, J. M., *Surf. Sci.* 65:476-498 (1977)

48. Madden, H. H., *J. Vac. Sci. Tech. A* 18:677-689 (1981)

49. Painter, G. S., Ellis, D. E. and Lubinsky, A. R., *Phys. Rev. B* 4:3610-3622 (1971)

50. Cong, Y., Collins, R. W., Messier, R., Vedam, K., Epps, G. F. and Windischmann, H., *J. Vac. Sci. Technol. A* 9:1123-1128 (1991)

51. Egerton, R. F., *Electron Energy Loss Spectroscopy in the Electron Microscope*, Plenum Press, NY (1986)

52. Egerton, R. F. and Whelan, M. J., *Phil. Mag.* 30:739-749 (1974)

53. Williams, B. E. and Glass, J. T., *J. Mater. Res.* 4:373-384 (1989)

54. Wang, Y., Chen, H., Hoffman, R. W. and Angus, J. C., *J. Mater. Res.* 5:2378-2386 (1990)

55. Pepper, S. V., *Appl. Phys. Lett.* 38:344-346 (1981)

56. Pepper, S. V., *Surf. Sci.* 123:47-60 (1982)

57. Pandy, K. C., *Phys. Rev. B* 25:4338-4341 (1982)

58. Heidenreich, R. D., *Fundamentals of Transmission Electron Microscopy*, Wiley (Interscience), NY (1964)

59. Grundy, P. J. and Jones, G. A., *Electron Microscopy in the Study of Materials*, Edward Arnold Publication, London (1976)

60. Hirsch, P. B., Howie, A., Nicholson, R. B., Pashley, D. W. and Whelan, M. J., *Electron Microscopy of Thin Crystals*, Butterworths, London (1971)

61. Thomas, G. and Goringe, M., *Transmission Electron Microscopy of Materials*, Wiley, NY (1979)

62. Thompson-Russell, K. C. and Edington, J. W., *Electron Microscope Specimen Preparation Techniques in Materials Science*, MacMillan, London (1977)

63. Goodhew, P. J., in: *Practical Methods in Electron Microscopy, Vol. 1*, (A. M. Glauert, ed.), North-Holland, Amsterdam and American Elsevier, NY (1973)

64. Goodhew, P. J., in: *Practical Methods in Electron Microscopy, Vol. 11*, (A. M. Glauert, ed.), Elsevier, Amsterdam (1988)

65. Evans, T., in: *Physical Properties of Diamond*, (R. Berman, ed.), pp. 120-124, Clarendon Press, Oxford, England (1965)

66. Zhu, W., "Microwave Plasma Enhanced Chemical Vapor Deposition and Structural Characterization of Diamond Films," Ph. D. Thesis in Solid State Science, pp. 123, The Pennsylvania State University, University Park, P.A. (1990)

67. Phaal, C. and Zuidema, G., *Phil. Mag.* 14:79-86 (1966)

68. Trueb, L. F., *J. Appl. Phys.* 39:4707-4716 (1968)

69. Woods, G. S., *Phil. Mag.* 23:473-484 (1971)

70. Woods, G. S., *Phil. Mag.* 34:993-1012 (1976)

71. Walmsley, J. C. and Lang, A. R., *J. Mater. Sci. Lett.* 2:785-788 (1983)

72. Moriyoshi, Y., Kamo, M., Setaka, N. and Sato, Y., *J. Mater. Sci.* 18:217-224 (1983)

73. Evans, T. and Phaal, C., *Phil. Mag.* 7:843-846 (1962)

74. Evans, T. and Wild, R. K., *Phil. Mag.* 12:479-489 (1965)

75. Humble, P., *Proc. Roy. Soc. Lond. A* 381:65-81 (1982)

76. Humble, P., Mackenzie, J. K. and Olsen, A., *Phil. Mag. A* 52:605-621 (1985)

77. Humble, P., Lynch, D. F. and Olsen, A., *Phil. Mag. A* 52:623-641 (1985)

78. Matsumoto, S. and Matsui, Y., *J. Mater. Sci.* 18:1785-1793 (1983)

79. Kawarada, H., Mar, K. S., Suzuki, J., Ito, Y., Mori, H., Fujita, H. and Hiraki, A., *Jpn. J. Appl. Phys.* 26:L1903-1906 (1987)

80. Zhu, W., Badzian, A. R. and Messier, R., *J. Mater. Res.* 4:659-663 (1989)

81. Clausing, R. E., Heatherly, L. and More, K. L., *Surf. Coat. Technol.* 39/40:199-210 (1989)

82. Kaae, J. L., Gantzel, P. K., Chin, J. and West, W. P., *J. Mater. Res.* 5:1480-1489(1990)

83. Hetherington, A. V., Wort, C. J. H. and Southworth, P., *J. Mater. Res.* 5:1591-1594 (1990)

84. Williams, B. E., Kong, H. S. and Glass, J. T., *J. Mater. Res.* 5:801-810 (1990)

85. Tamou, Y., Ohsawa, Y. and Kikuchi, N., in: *Program and Abstracts of the First International Conference on the New Diamond Science and Technology,* (Japan New Diamond Forum, ed.), pp. 30-31, Tokyo, Japan (1988)

86. Narayan, J., *J. Mater. Res.* 5:2414-2423 (1990)

87. Ma, G. H. M., Lee, Y. H. and Glass, J. T., *J. Mater. Res.* 5:2367-2377 (1990)

88. Ma, G. H. M., Williams, B. E., Prater, J. T. and Glass, J. T., *Diamond and Related Mater.* 1:25-32 (1991)

89. Meakin, D., Stoemenos, J., Migliorato, P. and Economou, N. A., *J. Appl. Phys.* 61:5031-5037 (1987)

90. Clausing, R. E., Heatherly, L., Specht, E. D. and More, K. L., in: *Proceedings of the Second International Conference on New Diamond Science and Technology* (R. Messier, J. T. Glass, J. E. Butler and R. Roy, eds.), pp. 575-580, Materials Research Society, Pittsburgh, PA (1991)

91. Moore, M. and Lang, A. R., *Phil. Mag.* 26:1313-1325 (1972)

92. Fujimori, N., Imai, T., Nakahata H., Shiomi, H. and Nishibayashi, Y., in: *Mater. Res. Soc. Symp. Proc. on Diamond, Silicon Carbide and Related Wide Bandgap Semiconductors, Vol. 162* (J. T. Glass, R. Messier and N. Fujimori, eds.), pp. 23-33, Materials Research Society, Pittsburgh, PA (1990)

93. Geis, M. W., in: *Mater. Res. Soc. Symp. Proc. on Diamond, Silicon Carbide and Related Wide Bandgap Semiconductors, Vol. 162* (J. T. Glass, R. Messier and N. Fujimori, eds.), pp. 15-22, Materials Research Society, Pittsburgh, PA (1990)

94. Badzian, A. R., Badzian, T. and Pilione, L., *Int. J. Refract. Metals and Hard Mater.* 9:92-95 (1990)

95. Zhu, W., Wang, X. H., Badzian, A. R. and Messier, R., in: *Proceedings of the Second International Conference on New Diamond Science and Technology* (R. Messier, J. T. Glass, J. E. Butler and R. Roy, eds.), pp. 821-826, Materials Research Society, Pittsburgh, PA (1991)

96. Zhu, W., Randall, C., Badzian, A. R. and Messier, R., *J. Vac. Sci. Technol. A* 7:2315-2324 (1989)

97. Ono, A., Bada, T., Funamoto, H. and Nishikawa, A., *Jpn. J. Appl. Phys.* 25:L808-810 (1986)

98. Gildenblat, G.S., Grot, S. A., Wronski, C. R., Badzian, A. R., Badzian, T. and Messier, R., E., *Appl. Phys. Lett.* 53:586-588 (1988)

99. Spence, J. C. H., *Experimental High Resolution Electron Microscopy*, Clarendon Press, Oxford (1980)

100. Narayan, J., Srivatsa, A. R., Peters, M., Yokota, S. and Ravi, K. V., *Appl. Phys. Lett.* 53:1823-1825 (1988)

101. Goldstein, J. I., Newbury, D. E., Echlin, P., Joy, D. C., Fiori, C. and Lifshin, E., *Scanning Electron Microscopy and X-ray Microanalysis*, Plenum Press, NY (1981)

102. Pickrell, D., Zhu, W., Badzian, A. R., Messier, R. and Newnham, R. E., *J. Mater. Res.* 6:1264-1277 (1991)

103. Zhu, W., Messier, R. and Badzian, A. R., in: *Proceedings of the First International Symposium on Diamond and Diamond-Like Films*, Vol. 89-12 (J.P. Dismukes, et al., eds), pp. 61-79, The Electrochemical Society, Pennington, NJ (1989)

104. Spitsyn, B. V., Bouilov, L. L. and Deryagin, B. V., *J. Cryst. Growth* 52:219-226 (1981)

105. Kobashi, K., Nishimura, K., Kawate, Y. and Horiuchi, T., *Phys. Rev. B* 38:4067-4084 (1988)

106. Haubner, R. and Lux, B., *Int. J. Refract. Hard Metals* 6:210-215 (1987)

107. Zhu, W., Badzian, A. R. and Messier, R., *Proceeding of SPIE on Diamond Optics III* 1325:187-201 (1990)

108. Komoda, T., *Jpn. J. Appl. Phys.* 7:27-30 (1968)

109. Iijima, S., *Jpn. J. Appl. Phys.* 26:357-364 (1987)

110. Bragg, W. L., in: *The Crystalline State, Vol. I*, (W. H. Bragg and W. L. Bragg, eds.), pp. 12-21, George Bell and Sons Ltd., London (1933)

111. Azaroff, L. V., *Elements of X-ray Crystallography*, McGraw-Hill, NY (1968)

112. Wild, C., Herres, N. and Koidl, P., *J. Appl. Phys.* 68:973-978 (1990)

113. Kobashi, K., Nishimura, K., Miyata, K., Kumagai, K. and Kawate, Y., in: *Proceedings of the First International Symposium on Diamond and Diamond-Like Films, Vol. 89-12* (J. P. Dismukes, et al. eds.), pp. 296-305, The Electrochemical Society, Pennington, NJ (1989)

114. Specht, E. D., Clausing, R. E. and Heatherly, L., *J. Mater. Res.* 5:2351-2355 (1990)

115. Van der Drift, A., *Philips Res. Rept.* 22:267-288 (1967)

116. Geis, M. W., *Appl. Phys. Lett.* 55:550-552 (1989)

117. Sato, Y. and Kamo, M., *Surf. Coat. Technol.* 39/40:183-198 (1989)

118. Moore, M., *Ind. Dia. Rev.* 48:59-64 (1988)

119. Edington, J. W., *Electron Diffraction in the Electron Microscope*, MacMillan, London (1975)

120. Pashley, D. W. and Stowell, M. J., *Phil. Mag.* 8:1605-1632 (1963)

121. Liaw, H. P. and Davis, R. F., *J. Electrochem. Soc.* 131:3014-3018 (1984)

122. Williams, B. E., Glass, J. T., Davis, R. F., Kobashi, K. and Kawate, Y., in: *MRS Extended Abstracts on Diamond and Diamond-like Materials Synthesis, Vol. EA-15* (G. H. Johnson, A. R. Badzian and M. W. Geis, eds.), pp. 59-62, The Materials Research Society, Pittsburgh, PA (1988)

123. Evans, T., in: *The Properties of Diamond*, (J. E. Field, ed.), pp. 403-424, Academic Press, London (1979)

124. Van Hove, M. A. and Tong, S. Y., *Surface Crystallography by LEED*, Springer Series in Chemical Physics, Vol. 2, Springer-Verlag, Berlin (1979)

125. Pendry, J. B., *Low Energy Electron Diffraction: The Theory and Its Application to Determination of Surface Structure*, Academic Press, London (1974)

126. Lander, J. J. and Morrison, J., *Surf. Sci.* 4:241-246 (1966)

127. Lurie, P. G. and Wilson, J. M., *Surf. Sci.* 65:453-475 (1977)

128. Woodruff, D. P. and Delchar, T. A., *Modern Techniques of Surface Science*, pp.71-75, Cambridge University Press, Cambridge, 1986

129. Matsumoto, S., Sato, Y., Kamo, M. and Setaka, N., *Jpn. J. Appl. Phys.* 21:L183-185 (1982)

130. Sawabe, A. and Inuzuka, T., *Appl. Phys. Lett.* 46:146-147 (1985)

131. Kawarada, H., Mar, K. S. and Hiraki, A., *Jpn. J. Appl. Phys.* 26:L1032-1034 (1987)

132. *SIMS Principles and Applications*, (J. C. Vickerman, A. Brown and N. M. Reed, eds.), Clarendon Press, Oxford (1989)

133. Storms, H. A., Brown, K. F. and Stein, J. D., *Anal. Chem.* 49:2023-2030 (1977)

134. Glass, J. T., presented at the *Workshop on Diamond Films Science and Technology,* Stanford University, Stanford, CA (February 1, 1991)

135. Sato, Y., Kamo, M. and Setaka, N., in: *Proceedings 8th. Int. Symp. on Plasma Chem. Vol. 1,* (K. Akashi and A. Kinbara, eds.), pp. 2446-2451, International Union of Pure and Applied Chemistry, Oxford, England (1987)

136. Matsumoto, O., Toshima, H. and Kanzaki, Y., *Thin Solid Films* 128:341-351 (1985)

137. Guilbault, G. G., in: *Practical Fluorescence,* (G. G. Guilbault, ed.), Marcel Dekker, Inc., NY (1990)

138. Hartnett, T. M., "Characterization of Diamond Deposition in a Microwave Plasma," M. S. Thesis in Solid State Science, The Pennsylvania State University, University Park, PA (1988)

139. Knight, D. S. and White, W. B., *Proceedings of SPIE on Raman Scattering, Luminescence and Spectroscopic Instrumentation in Technology* 1055:144-151(1989)

140. Walker, J., *Rep. Prog. Phys.* 42:1605-1659 (1979)

141. Yacobi, B. G. and Holt, D. B., *J. Appl. Phys.* 59:R1-R24 (1986)

142. Davies, G., in: *The Properties of Diamond,* (J. E. Field, ed.), pp. 165-181, Academic Press, London (1979)

143. Kawarada, H., Nishimura, K., Ito, T., Suzuki, J., Mar, K. S., Yokota, Y. and Hiraki, A., *Jpn. J. Appl. Phys.* 27:L683-L686 (1988)

144. Partlow, W. D., Ruan, J., Witkowski, R. E., Choyke, W. J. and Knight, D. S., *J. Appl. Phys.* 67:7019-7025 (1990)

145. Robins, L. H., Cook, L. P., Farabaugh, E. N. and Feldman, A., *Phys. Rev. B* 39:13367-13377 (1989)

146. Collins, A. T., Kamo, M. and Sato, Y., *J. Phys. D: Appl. Phys.* 22:1402-1405 (1989)

147. Ruan, J., Choyke, W. . and Partlow, W. D., in: *Proceedings of the Second International Conference on New Diamond Science and Technology* (R. Messier, J. T. Glass, J. E. Butler and R. Roy, eds.), pp. 703-708, Materials Research Society, Pittsburgh, PA (1991)

148. Badzian, A. R., Badzian, T., Roy, R., Messier, R. and Spear, K. E., *Mater. Res. Bull.* 23:531-548 (1988)

149. Clark, C. D. and Dickerson, C. B., *Surf. Coat. Technol.* 47:336-343 (1991)

150. Chu, W. K., in *Metals Handbook, Vol. 10: Materials Characterization,* pp.628-636, American Society for Metals, Metals Park, OH (1986)

151. Celii, F. G., Purdes, A. J., Gnade, B. E. and Weathers, D. L., in: *Proceedings of the Second International Conference on New Diamond Science and Technology,* (R. Messier, J. T. Glass, J. E. Butler and R. Roy, eds.), pp. 631-636, Materials Research Society, Pittsburgh, PA (1991)

152. Chu, W. K., Mayer, J. W. and Nicolet, M. A., *Backscattering Spectrometry,* Academic Press, NY (1978)

153. Lanford, W. A., presented at: *Second Annual Diamond Technology Initiative Seminar,* Durham, NC (July 7-8, 1987)

154. Chu, W. K., Parikh, N. R., King, M. and Swanson, M. L., presented at: *Second Annual Diamond Technology Initiative Seminar,* Durham, NC (July 7-8, 1987)

155. Angus, J. C., Koidl, P. and Domitz, S., in: *Plasma Deposited Thin Films,* (J. Mort and F. Jansen, eds.), pp. 89-127, CRC Press, Inc., Boca Raton, FL (1986)

156. Vedam, K., Knausenberger, W. and Lukes, F., *J. Opt. Soc. Amer.* 59:64-71 (1969)

157. Azzam, R. M. A. and Bashara, N. M., *Ellipsometry and Polarized Light,* North-Holland, Amsterdam (1977)

158. Cong, Y., Collins, R. W., Epps, G. F. and Windischmann, H., *Appl. Phys. Lett.* 58:819-821 (1991)

159. Collins, R. W., Cong, Y., Kim, Y. T., Vedam, K., Liou, Y., Inspektor, A. and Messier, R., *Thin Solid Films* 181:565-578 (1989)

160. Messier, R., Badzian, A., Badzian, T., Spear, K. E., Bachmann, P. and Roy, R., *Thin Solid Films* 153:1-9 (1987)

161. Vardiman, R. G., Vold, C. L., Snail, K. A., Butler, J. E. and Pande, C. S., *Mater. Lett.* 8:468-471 (1989)

162. McNamara, K. M., Gleason, K. K. and Geis, M. W., in: *MRS Proceedings of Diamond, SiC and Related Wide Bandgap Semiconductors, Vol. 162* (J. T. Glass, R. Messier and N. Fujimori, eds.), pp. 207-212, Materials Research Society, Pittsburgh, PA (1990)

163. Nishimura, K., Nakano, T., Koike, H., Tomimori, H., Kawarada, H., Hiraki, A. and Ogawa, K., *Proceedings of SPIE on Diamond Optics III* 1325:267-282 (1990)

164. Capehart, T. W., Perry, T. A., Beetz, C. B., Belton, D. N., Fisher, G. B., Beall, C. E., Yates, B. N. and Taylor, J. W., *Appl. Phys. Lett.* 55:957-959 (1989)

165. Turner, K. F., Stoner, B. R., Bergman, L., Glass, J. T. and Nemanich, R. J., in: *Proceedings of the Second International Conference on New Diamond Science and Technology,* (R. Messier, J.T. Glass, J. E. Butler and R. Roy, eds.), pp. 607-612, Materials Research Society, Pittsburgh, PA (1991)

166. Yugo, S., Izumi, A., Kanai, T., Muto, T. and Kimura, T., in: *Proceedings of the Second International Conference on New Diamond Science and Technology,* (R. Messier, J.T. Glass, J. E. Butler and R. Roy, eds.), pp. 385-389, Materials Research Society, Pittsburgh, PA (1991)

7

Non-Electronic Applications of Diamond Films

Linda Plano and Michael Pinneo

1.0 INTRODUCTION

The growing interest in diamond films is dramatically illustrated by the explosion of diamond-related papers and patents generated primarily by Japan and the United States. Several joint ventures and major research programs have recently been initiated by companies such as General Electric, Asahi, Kobe Steel, and Norton. Industry is recognizing the potential of diamond films as a technologically useful material, and is accelerating basic and applied research in the field. This chapter discusses the research needed to support development of non-electronic applications for CVD diamond materials. The reader is referred to Ch. 8 for an exposition of electronic applications.

Initially, studies of diamond film focused on basic parametric experiments. More recent fundamental research into gas phase and surface chemistry has employed increasingly sophisticated diagnostic and analytic methods, combined with theoretical kinetic and thermodynamic analysis and modeling. A third, rapidly growing segment of research in CVD diamond is being driven by application-specific requirements, with investigations designed to yield data useful for specific products rather than for increased fundamental insight into the deposition process or material properties. This phenomenon is a natural consequence of increased interest in moving diamond films out of the laboratory into industrial use. This class of research is performed most often by industrial laboratories and usually includes the elements of the cycle illustrated in Fig. 1: *(a)* diamond properties of interest are identified; *(b)* diamond is then deposited and characterized; *(c)* characterization data are then assessed in light of known application requirements; *(d)* further development work proceeds to optimize diamond properties, or attention shifts to more attractive application opportunities, and the cycle repeats.

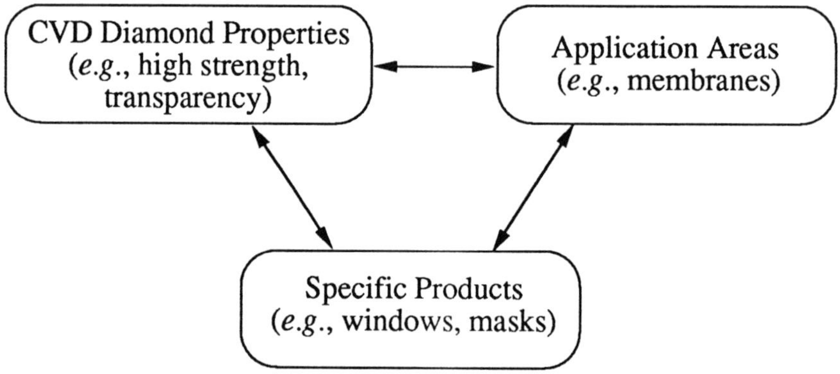

Figure 1. CVD diamond applications research and development cycle.

The cycle can be entered at any point: for example, unusual characterization data may identify an application, rather than the reverse. Once it has been established that the material exhibits useful properties, promising general application areas can be selected (e.g., membranes). Selection of application areas can suggest directions for process improvement or changes needed in the film's materials properties. Specific prototype products can then be developed (e.g., x-ray detector windows, x-ray lithography masks, etc.). Experience gained in prototype development and use can then be fed back to accelerate process and film property optimization, which then opens up a broader spectrum of potential applications, and so on.

Referring again to the cycle in Fig. 1, many potentially useful natural diamond properties have been identified and, to some extent, prioritized. Many application areas have likewise been identified. What remains is to reproduce those properties in diamond films, characterize them with methods appropriate to the eventual product, and enhance those properties such that useful products can be manufactured. In the next two sections, the development process is discussed, followed by a chronological forecast of future products.

2.0 GOALS OF APPLIED CVD DIAMOND RESEARCH

2.1 Identification of Relevant Properties

The properties of diamond have been discussed earlier (Ch. 1). In developing applications using CVD diamond, it is necessary to identify those properties which will be relevant to the application. It is often the case that several properties are critical and that optimizing all simultaneously is not possible. In this section, a variety of applications are discussed with regard to critical properties. Table 1 shows the relationships among five diamond film application areas and corresponding critical properties and characterization requirements. These application examples are selected from a wide range of potential products. A more complete summary can be found in Sec. 4.

Table 1. CVD Diamond Applications and Associated Critical Film Properties

Application Area	Critical Material Properties and Characterization Needs
X-ray Detector Windows	Tensile strength, non-carbon impurity content, thickness, gas permeability, nucleation density, pinhole defect density
Abrasives	Hardness, morphology, fracture mechanics, size dispersion, surface chemistry, non-carbon impurity content
Electronic Packaging	Thermal conductivity, expansion coefficient, internal stress, adhesion to substrate, resistivity, dielectric constant & loss factor, surface chemistry, film flatness, surface roughness, machinability
IR Optical Coatings and Elements	Optical transparency, thickness, hardness, modulus, thermal expansion coefficient, internal stress, surface roughness, crystallite size and structure, resistivity, deposition temperature, conformal coating
Machine Tool Coatings	Adhesion to substrate, thickness, workpiece material compatibility, oxidation threshold, graphitization threshold, thermal conductivity, dynamic friction coefficient, wear rate and homogeneity, deposition contour coverage, surface finish

X-ray Detector Windows. X-ray detector windows for scanning electron microscope x-ray spectrometers must be transmissive to x-rays of all energies, impermeable to gases, opaque to visible light, and capable of withstanding slightly over an atmosphere differential pressure. A typical window is shown in Fig. 2. For this application, the most important of diamond's properties are its very high strength and low atomic number (Z). Its strength allows use of much thinner films (< 0.5 µm compared to 8 - 12 µm for Be windows) while still supporting the required pressure differential. Thin section and low atomic number combine to allow very low energy x-rays from low-Z materials to reach the detector, as illustrated by the graph in Fig. 3. Very high nucleation density is, of course, essential to obtaining a coalesced film (impermeable to gases) at thicknesses of less than a micron. Gas permeability at grain boundaries is an important variable in determining required film thickness because vacuum cryogenic conditions must be maintained for proper detector operation. Substrate adhesion is not critical since the film must be free of its substrate to transmit x-rays. Finally, due to diamond's optical transparency, an opaque coating material must be applied to prevent transmission of visible light which would cause signal artifacts from the x-ray detector.

Figure 2. PECVD diamond window used for transmitting characteristic x-rays down to low energies to a detector.

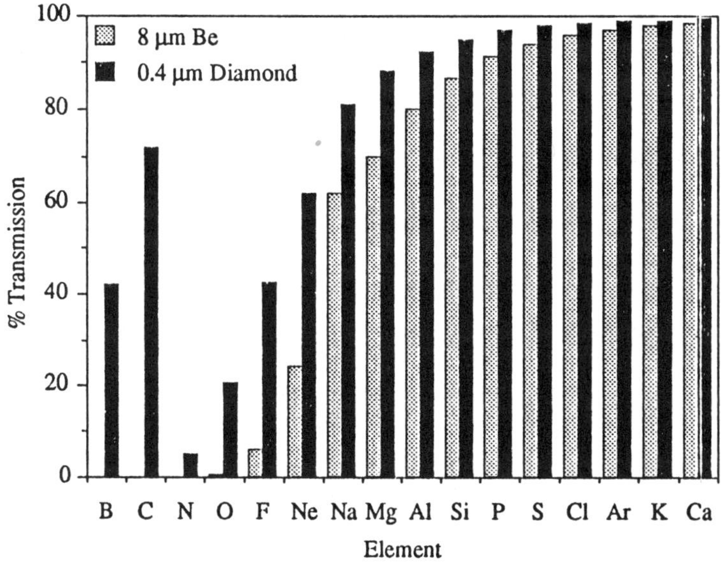

Figure 3. Transmission data for beryllium and diamond windows for low energy characteristic x-rays.

Diamond Abrasives. Satisfactory performance of diamond particles in abrasive applications clearly requires attention to hardness, the basic diamond property on which efficient material removal depends. Material removal rate and workpiece surface finish are affected strongly by particle morphology and fracture mechanics, the latter parameter determining the rapidity with which new, sharp surfaces are exposed as diamond grit wears away from grinding component surfaces. Surface chemistry is important in assuring adequate bonding between particles and their metallic or epoxy binders. Non-carbon impurities are usually unimportant in many abrasive diamond applications. Finally, particle size distribution must be carefully controlled to prevent scarring of surfaces by an occasional oversize particle during finish polishing operations, the stage at which the workpiece value is nearing a maximum. Diamond single-point tools, which might be considered a special case of diamond abrasives, require excellent crystal quality to support fabrication of cutting edge surfaces with closely controlled dimensions, angles, edge radii, and edge finish.

Electronic Heat Sinks. The prime diamond property of interest for all electronic heat sink applications is thermal conductivity. Electrical resistivity is important to the extent that it is often necessary that a diamond heat spreader provide electrical isolation. Some electronic packaging applications require good interface bond strength between diamond and an underlying ceramic or metal substrate. Diamond surface chemistry can be important in those applications which require metallization for die attach processes. Diamond's thermal expansion coefficient is important in that it is much lower than common electronic materials like silicon and GaAs, and devices attached to diamond heat sinks can experience significant mechanical stress due to thermal coefficient mismatch. Dielectric constant and loss tangent can strongly affect design of packaging for fast devices. Surface roughness and flatness are among other specifications required for certain applications. For those electronic heat dissipation applications in which it is required that a diamond film be deposited directly on an active device, deposition process temperature and radiation damage are only two of the important process integration challenges which must be met.

Infrared Optical Coatings. Many available IR optical materials exhibit excellent IR transmission, but have relatively poor mechanical properties and are soft and easily damaged by environmental insults like dust, rain, and industrial process chemicals. A bright potential applications area for diamond films is as IR-transparent armor for soft optical materials like ZnS and ZnSe. Diamond's IR transmission losses are of prime interest, and include absorption, reflection, and complex scattering from surfaces and within the bulk from grain boundaries. Deposition temperature, adhesion to IR materials, and resistance to abrasion and impact damage are equally important in this application. Selected military uses may also require tailored electrical resistivity and controlled IR emissivity at elevated temperatures.

Machine Tool Coatings. Conformal, uniform coating of machine tools (see Fig. 4) may require extensive development since many tools have complex topologies (e.g., drill bits, gear hobs, etc.) which induce local film nonuniformities in several parameters. Inhomogeneities in thickness or microstructure can lead to sudden failure of the coating during use. To avoid failure, tool coatings must adhere strongly to the substrate and the wear rate should be uniform and predictable. Adhesion issues may be complicated by materials incompatibilities: for instance, diamond does not easily nucleate on untreated stainless steel. Film morphology is an important consideration since different tools may need different surfaces: a grinding tool may be better served by a very rough film surface while a cutting tool may require as smooth a surface as possible to minimize tearing. Control of surface roughness and other friction-related characteristics leads to the ability to tailor the dynamic coefficient of friction to the application. Most machine tool applications produce considerable heat during operation, and high thermal conductivity of the diamond film is likely to be useful in reducing local temperatures. This is important not only for overall tool life, but also because diamond films oxidize in air at temperatures above 600°C and graphitize above 1300°C (in the absence of oxygen) (1).

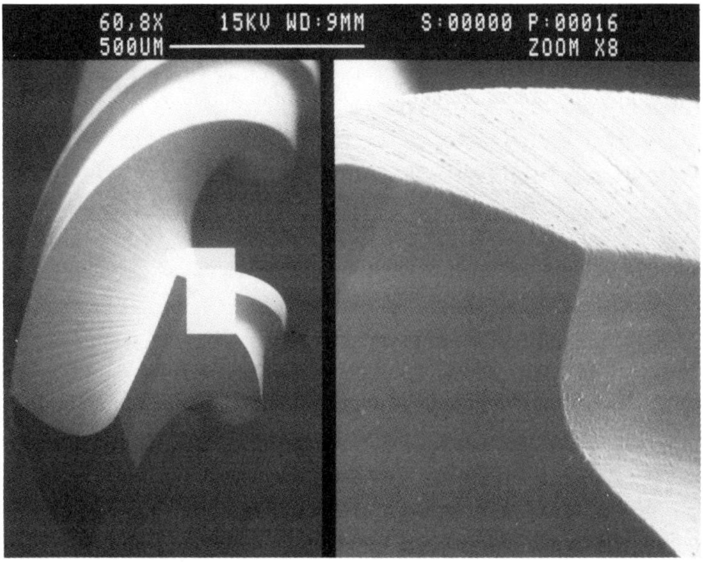

Figure 4. Diamond coated twist drill. Note conformal coating.

2.2 Applications-Related Characterization of Diamond Films

Characterization of polycrystalline diamond films for use in applications is more complex than characterization of single-crystal materials like silicon. Complexities include presence of nondiamond carbon, grain boundaries, varying grain size and morphology, surface roughness, grain orientation, etc. Characterization of the material for applications development requires use of a variety of techniques to achieve satisfactory results. Some commonly employed techniques are enumerated in Table 2.

Thermal Conductivity. Thermal conductivity in diamond films is sensitive to the presence of nondiamond carbon in small concentrations. Characterization of a diamond film for graphitic content is often impossible using classical diffraction methods, which fail to detect small proportions of nondiamond carbon (graphite, disordered sp^2 and/ or sp^3). Neither x-ray nor electron diffraction is sensitive to graphitic phases at less than ~0.5 atomic percent (2), and a film which appears to be entirely polycrystalline diamond according to diffraction tests can be essentially useless for heat transfer applications due to impaired heat conductivity. Neither ESCA nor Auger is useful for bulk carbon bonding determination in part because of their very shallow analytic depth. NMR can detect graphitic phases, but may require more material than is available in many thin film applications. Raman spectroscopy is currently the most commonly employed method for semi-quantitatively determining the relative amounts of diamond, graphite, and other types of bonding in carbon films (3). Additional information is available in the Raman spectrum about film stress state and temperature (sp^3 peak shift) (4), as well as thermal conductivity (sp^3 peak FWHM) and film defect structures (photoluminescence features).

Table 2. Major characterization methods and relationship to applications areas.

Characterization Technique	Information	Applications Areas
Raman Spectroscopy	bonding state(s)	All
SEM	morphology, grain size, thickness, growth rate	All
Profilometer	surface roughness	Optical Coatings, Wear Coatings
Thermal Conductivity	thermal conductivity,	Electronic Packaging
Nanoindenter, Ball Indenter	hardness, adhesion, fracture toughness	Wear Coatings
Pin-on-Disk Reciprocal/Sliding	friction and wear	Wear Coatings, Bearings
Nuclear Reaction Analysis	quantitative hydrogen content	Membranes, possibly Friction and Wear Coatings
SIMS, EDX	impurity identification and profile	All
XRD	preferred orientation detection	Nonscattering Light Pipes

The thermal conductivity of CVD diamond was first established by direct measurement (5) and correlated with the full width at half maximum intensity (FWHM) of the diamond Raman peak (6) as illustrated by the graph in Fig. 5. The methodology in these studies employed optical measurement of thermal gradients along free-standing diamond films using carefully controlled heat inputs and radiation backgrounds, and is somewhat cumbersome and slow due to sample preparation requirements. However, it is now possible to infer thermal conductivity of a diamond film rapidly from its Raman spectrum with sufficient accuracy for most heat transfer applications development.

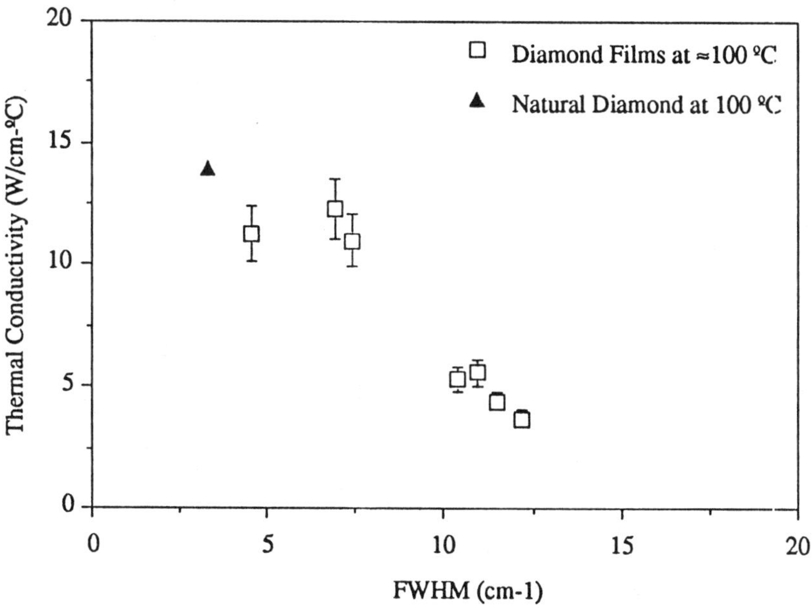

Figure 5. Diamond film thermal conductivity as a function of diamond Raman peak (1332 cm^{-1}) FWHM.

Mechanical Properties. Film hardness, important for wear applications and protective coatings, is challenging to quantify. Since diamond is the hardest known material, indentation measurements of hardness which employ a diamond indenter are somewhat suspect. Indentation hardness measurements of diamond films can require replacement of the diamond indenter on a regular, frequent basis. In addition, morphology effects such as roughness can affect measurements (7). Measurement of thin film hardness can be further complicated by substrate deformation effects. At worst, it may not be possible to obtain meaningful data, except to say that the film is at least as hard as the indenter material. Scratch hardness tests can yield substantial errors because under some conditions, hard crystals can be worn by materials with considerably lower indentation hardness values (8).

Friction and wear behavior can also be difficult to quantify in diamond films, because, as in the case of natural diamond, the material is so hard that the stylus used for tribological testing wears at least as quickly as the diamond, complicating data analysis. Analysis of diamond films presents the additional complications of grain boundary effects, surface roughness and morphology, and the presence of nondiamond carbon. Surface roughness has a strong effect on friction and wear (9). Diamond-on-diamond reciprocating sliding tests have shown that the coefficient of friction observed for natural diamond (~0.05) can be attained by diamond films (10) and that this behavior appears to be strongly dependent on film surface roughness and less so on the ratio of diamond to nondiamond carbon. This may arise from weaker nondiamond material being turned into debris particles which act as minute roller bearings at the interface, reducing the apparent dynamic friction coefficient. Wear of either the diamond film (due to breaking off of asperities) or the opposing stylus also generates debris. Further complexities are introduced when using nondiamond styli or diamond films of different surface morphologies (e.g., 111 vs. 100; euhedral vs. faceted). Deconvolution of the various effects is difficult (11). Prediction of sliding friction coefficients for diamond films in bearing applications is under study.

Thickness. Film thickness is a critical parameter for nearly every diamond film application and does not yet enjoy a simple, economic measurement method suitable for most circumstances. Film thickness uniformity strongly affects deposition economics for most applications. The best diamond deposition uniformities reported to date (12)(13) show variations of ~10% across a 4" diameter, which is quite primitive by the standards for silicon, for which 1% uniformity can be maintained over an 8" diameter. Two direct measurement methods, profilometry of a step created by masking during deposition and SEM examination of broken edges, have severe limitations. Deposition masking cannot be used for applications which require uniform film thickness over the entire substrate. SEM inspection requires breakage of the film to expose a cross-section. For thicker films, direct measurement with calipers is sometimes feasible.

The more desirable optical methods in common use in the semiconductor industry have not yet been successfully adapted to diamond films. Surface roughness, internal scattering, and nonuniform optical properties caused by graphitic inclusions combine to preclude use of ordinary ellipsometry. For example, the refractive index of a diamond film can vary as much as 50% among different areas of a 4" diameter film. Multispectral ellipsometry (14) offers promise, but has not been fully developed, nor is it widely available. Most desirable would be an in situ monitor which could be used as a deposition thickness probe. IR interferometry, implemented using IR laser diodes, may be useful in this regard.

Transparency. CVD diamond films are being developed for use as transparent windows in several regions of the electromagnetic spectrum. IR, UV, and x-ray windows are of particular interest for near-term products (see Fig. 6). Successful development requires characterization of the material's optical properties. Characterization methods employed include conventional UV/VIS transmission spectroscopy, FTIR transmission spectroscopy, microwave attenuation and loss measurements at various frequencies, and low energy x-ray transmission tests. Optical scattering is an important phenomenon in CVD diamond (15). Separate assessments of surface vs. bulk scattering have not yet appeared, and will be very important for applications in imaging optics.

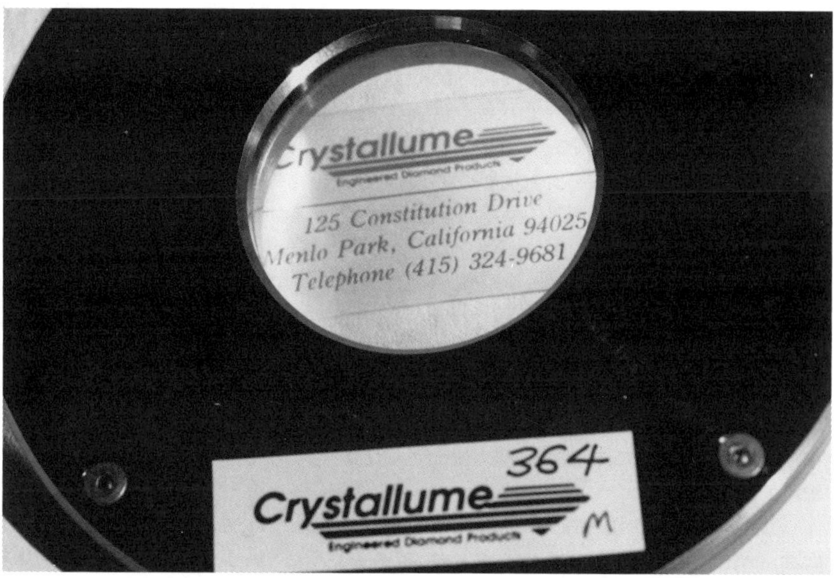

Figure 6. Prototype x-ray lithography mask. Note optical transparency.

Functional Characterization and Tests. As indicated in the preceding discussion, some important diamond film properties cannot be satisfactorily characterized for applications use with currently available fundamental methods. In addition, it is often difficult to predict film behavior in a complex operating environment from data gathered from several individual tests. Thus, for many applications, it is most useful to design functional tests which closely simulate the expected product operation environment. One example may be drawn from product testing evolved during development of x-ray detector windows.

Two failure mechanisms for x-ray detector windows are development of leaks or outright rupture under pressure changes which occur as the SEM chamber is vented and evacuated during sample changeout. As yet, we have found no fundamental characterization property or combination thereof which predicts probability of these pressure cycle failures. Similarly, correlations between gross observations of diamond films and their tendency to fail when fabricated into windows have been weak at best. To help assure that customers' windows did not fail, a functional test was implemented in which fully fabricated windows were subjected to rapid cyclic pressurization of 1.25 atmospheres differential pressure for 5,000 cycles prior to shipment. This test was followed by a leak test in which windows were mounted on the inlet to a mass spectrometer with one atmosphere of helium on the other side of the window. Detection of helium above the instrument blankoff background level (equivalent to a leak rate of $<10^{-10}$ sccs He) was cause for rejection of the window.

Records of window rupture events disclosed that nearly all units which failed did so within a few dozen cycles after test start. Windows which survived past a few hundred cycles went on to survive thousands of cycles. Selected windows have been cycled over 100,000 times without bursting or developing detectable leaks, suggesting that a fatigue failure mechanism, if it exists, lies beyond these limits. An additional benefit of this functional testing is that when window breakage occurs in the field, engineering resources can be assigned to investigate more probable failure mechanisms than fatigue fracture (e.g., handling, particle impact, etc.). This assists in rapidly identifying and suppressing failure modes, and provides more reliable performance, a critical factor in developing product acceptance.

The degree of fundamental insight into diamond film properties provided by such functional tests is arguably nil. However, the extent to which such tests facilitate the successful development of diamond films for applications use is enormous and provides a good example of the essential difference between fundamental research and applications-oriented activities. The point here is not to diminish the value of fundamental research in diamond films, but to illustrate both its necessity and insufficiency in meeting applications and production challenges. The most efficient, cost-effective development programs recognize the complementarity inherent in these two types of work and properly balance the two.

3.0 RESEARCH APPROACHES FOR DIAMOND PROPERTY ENHANCEMENT

Diamond exhibits a host of desirable properties. These were described in detail in Ch. 1. In the previous section, we identified several product areas and the properties each would require, then discussed methods of characterizing each. In this section, we discuss research directions for enhancement and optimization of those properties in light of applications requirements.

Optimization of CVD diamond properties for applications takes place at three levels. The first is simple establishment of deposition capability, i.e., the ability to grow a film which consists mostly of diamond rather than other forms of carbon. The second involves exploration of deposition parameters and film properties sufficient to provide an existence proof that diamond can be made to work in a specific application. The third consists of the considerable engineering effort needed to make production profitable rather than merely possible. Many diamond deposition techniques have been developed in the last decade, so the first level is well-established. The two remaining levels, proof of principle and bringing the product out of the laboratory and into production, compose the essence of applications-oriented research. Often, product development is an iterative process between these two levels, as described below.

3.1 Proof of Principle

In order to establish that CVD diamond can be used for a specific application, it is necessary to deposit a sample of diamond film with the required properties. An existence proof is a critical milestone in application development research, and no amount of "favorable trends" or "promising research results" can sidestep the need to make that first single unit which works as desired. Doing so usually requires a good deal of exploration of deposition parameters and correlation of deposition conditions with resulting film properties. This exploration can be guided by theoretical models to some extent, but necessarily involves substantial empirical work, often with significant serendipity. Statistical experiment design can be of use in reducing the large number of experiments needed to optimize deposition processes for a particular film property (16).

In developing the x-ray window prototype, several film characteristics were paramount: high tensile strength was essential because of the repeated application of an atmosphere differential pressure across the window; low impurity content and ≤2 μm thickness to maximize x-ray transmission; and absence of pinholes and other avenues for gas leakage through the film were the major issues to be resolved. Calculations of strength (using values obtained for natural material) and x-ray transmission indicated the possibility of making windows sufficiently thin to pass light element x-rays, while preserving the detector vacuum environment. Gas diffusion data were not available, and there remained the possibility that diffusion rates through the grain boundaries would be unacceptable.

Iteration between customer engineering design requirements and film deposition process development occurs often as the prototype is developed. This process is essential since, particularly with a "new" material like diamond film, the customer is not always sure of his needs. We will draw again on the x-ray window as an example: The customer originally specified a 2.5 μm thick film with a helium leak rate of 10^{-6} sccs. After prototype testing, it was determined that 2.5 μm was too thick to allow sufficient low energy x-ray transmission and that 10^{-6} sccs He leak rate was three orders of magnitude too high. This required a return to process development to enhance film nucleation in order to produce thinner films that were capable of reducing the helium leak rate to 10^{-9} sccs. Only after a prototype that satisfied the customer's requirements had been developed (a real proof of principle) could full attention be focused on converting the process into commercial production. Before discussing commercialization of this process, we will present an overview of the research needed to resolve issues for several types of film characteristics in the prototype stage. The relation between product-related development issues and research oriented towards resolving those issues is summarized in Table 3. Selected issues from the Table will be discussed in more detail. Fabrication and testing procedures are highly product-specific and will not be discussed further.

Substrate Adhesion. Adhesion of diamond film to most carbide formers (e.g., silicon) is often quite good without any preparation outside of normal nucleation procedures (typically scratching with diamond dust). Evidence of silicon carbide interlayers formed during diamond deposition on silicon has been revealed through TEM work (17). Diamond films tend to delaminate from many non-carbide forming materials (e.g., copper), although diamond does nucleate and grow. A final category is those substrate materials which do not tend to nucleate diamond nor to foster good adhesion, the most important representative of which (for products) is stainless steel, and other nickel or ferrous alloys. Deposition on ferrous material has been accomplished with very high growth rate techniques (18), suggesting that the species relevant to diamond growth diffuse immediately into the bulk of the substrate, rather than remaining on the substrate surface and fostering diamond formation.

Table 3. Diamond Film Product-Related Development Issues and Suggested Research Directions for Their Resolution.

Development Issue	Affected Product(s)	Research Direction
Substrate Adhesion	Machine Tool Coatings	Predeposition Surface Modification
Film Morphology	Wear Coatings, Optical Coatings	Surface Modification, Growth Rate, Substrate Temperature
Grain Orientation	Thermal Management, Optical Coatings	Predeposition Surface Modification
Extent of Diamond Bonding	All Products	Substrate Temp., Plasma Excitation Mode, Precursor Mix
Complex Substrates	Machine Tool Coatings	Reactor Design
Reduced Substrate Temperature	Plastic Substrates, IR Substrates	Reactor Design, Growth Rate
Mechanical Properties	Machine Tool Coatings, Wear Coatings	Impurities, Surface Orientation
Thermal Conductivity	Thermal Management	Diamond Bonding Content
Oxidation Resistance	Turbine Engine Coatings	Film Microstructure and Diamond Bonding Content

For substrate materials with adhesion problems, one promising route is the use of an interlayer of material compatible with and adherent to both the substrate material and diamond. An adequate material for stainless steel substrates has not been published at this writing. Alternatively, it may be economic to redesign the product such that the substrate material is compatible with adherent diamond growth. This approach is being taken in tooling where ceramic tool materials such as silicon carbide, SIALON, and silicon nitride are replacing carbide tools in some applications.

Film Morphology and Grain Orientation. The importance of film morphology and grain orientation is easily illustrated by analyzing the requirements for optical coatings. Surface scattering can drastically reduce the total light entering an optical coating. The very rough coatings typical of diamond films present a serious obstacle to the adoption of diamond films in this application. Reduced grain size (i.e., increased nucleation density) and less faceted films may provide short term solutions. In the long run, in situ planarization techniques must be developed. The classic process of film etch and regrowth is a logical place to begin. Preliminary research indicates that in situ hydrogen plasma etching of a diamond film followed by renucleation and growth has little effect on surface roughness (19), but more work remains to be done.

Once light has entered an optical coating, it can be absorbed by impurities and/ or scattered by grain boundaries and crystal defects. If the grains are preferentially oriented, grain boundary scattering may be reduced. Other film properties, such as thermal conductivity, may eventually be found to be anisotropic, or highly dependent on crystal perfection. In these cases, preferential orientation combined with intelligent engineering can maximize material efficiency in performing the application.

Diamond Bonding Content. Most research efforts strive to deposit diamond with the highest degree of perfection possible with their deposition technique. In general, "perfection" is defined by Raman spectral features: a very intense, narrow peak at approximately 1332 cm^{-1} with a flat background (i.e., no background photoluminescence, no graphitic Raman peaks (1550 - 1600 cm^{-1}, 1340 - 1370 cm^{-1}), and no other peaks associated with nondiamond bonded carbon (e.g., 1470 - 1500 cm^{-1}). This type of spectrum does indicate that the film approximates natural diamond in structure, but in the development of products, reproduction of natural diamond may not be necessary, or even desirable. Applications requiring smooth surfaces may be difficult to satisfy with purely diamond bonded material since these films tend to be heavily faceted and, as a result, quite rough. Their extreme surface hardness can make polishing processes lengthy and expensive. The euhedral films commonly observed when high carbon precursor concentrations are employed during deposition may not exhibit impressive Raman spectra but may be satisfactory or even preferable in some applications.

Some applications do require the purest diamond bonding possible. Thermal conductivity and electronic properties, for instance, are strongly affected by even small amounts of nondiamond bonded carbon. The most common approach for maximizing diamond bonding is to reduce the carbon precursor concentration in the reactant gas mix. In addition, high power densities and low growth rates appear to aid the development of diamond bonding. Thus, enhancement of diamond bonding is almost completely dependent on deposition parameters.

Low Temperature Deposition. Diamond films have been deposited at temperatures below 300°C (Fig. 7 shows the Raman spectrum from a film deposited below 380°C) (20). Low deposition temperatures open up an enormous range of important applications, from scratch-proof eyeglass and microscope lenses to wear coatings on plastics. The difficulties in low temperature deposition include both engineering and fundamental issues. Rapid growth of essentially purely diamond-bonded material generally requires high power densities, which in turn raise substrate temperatures to the 800 - 1000°C range. Thus, the substrate must be independently cooled for low temperature deposition. With cooler substrate temperatures, lower growth rates are typical, thus, research to improve film quality at low temperatures must focus on growth rate enhancement. In addition to independent control of substrate temperature, attention should be paid to surface preparation and gas phase chemistry to enhance nucleation and growth.

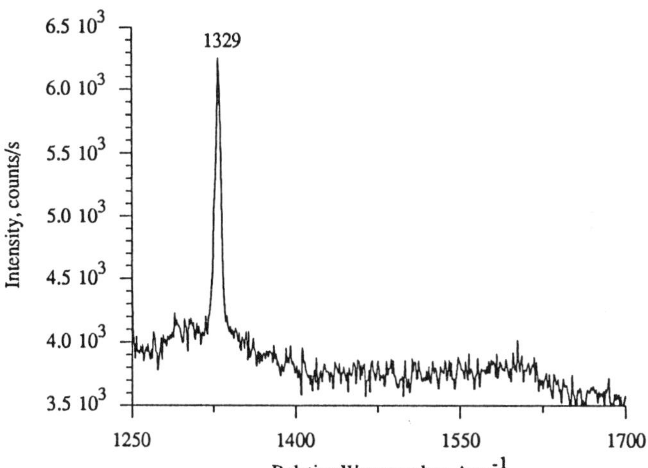

Figure 7. Raman spectrum of a diamond film grown on a gold-silicon eutectic (melting temperature 380°C). Nominal temperature was 270°C. Note absence of non-diamond bonded features.

Mechanical Properties. The hardest films are likely to be those which contain the least nondiamond bonded material and the least defects, although one may expect nitrogen impurities to enhance diamond film hardness as they do in natural and high-pressure single-crystal diamonds. Reciprocating sliding tests of natural diamond on natural diamond can produce dynamic coefficients of friction as low as 0.05 (21), indicating that properly engineered diamond films could be useful in wear applications, such as ball bearing coatings. Finally, the speed of sound in diamond is also the highest known, a fact of significance for both advanced armor and hypervelocity kinetic penetrators, as well as potential acousto-optical and acousto-electronic devices.

Diamond's superlative mechanical properties make it attractive for many tribological applications but, at the same time, make analysis of these properties difficult. Preliminary research has shown that diamond films can exhibit approximately the same hardness (22) and coefficient of friction (23)(24). Sonic velocity measurements are not available at this writing.

Research for improving film mechanical properties should involve study of substrate adhesion, grain size and orientation, and diamond bonding. Substrate adhesion is vital for most applications exploiting diamond's mechanical properties since these are often protective coatings subjected to mechanical and thermal stresses. Grain size and orientation are important in that they determine surface roughness. Preliminary tribological results indicate that surface roughness plays a more important role in determining friction behavior than does extent of diamond bonding. Surface roughness can also affect hardness measurements. While extent of diamond bonding may play a secondary role to that of surface roughness, it is likely to be of central importance to many mechanical properties, so fundamental process research should be continued in the development of tribological applications.

Thermal Conductivity. Maximum thermal conductivity is essential for thermal management applications. It is a secondary, though still important, property for some wear coatings because of strong localized heating which occurs during tool use. Since natural diamond has the highest room temperature thermal conductivity of known materials, films containing nondiamond bonded material might be expected to exhibit reduced thermal conductivity. Data from several investigators (25)(26) have shown that the less nondiamond bonded material, the higher the thermal conductivity (see Fig. 5). Thus, the first step in developing a thermally conducting film is to determine the combination of parameters that result in maximum diamond bonding. This clearly requires formation of void-free material as well as maximum sp^3/sp^2 Raman ratios and minimum sp^3 Raman peak FWHM. Substrate adhesion may or may not play a role in diamond film thermal applications, depending on whether a free-standing film or one deposited directly on a substrate is required.

Oxidation Resistance. The oxidation behavior of diamond films has been shown to be similar to that of natural diamond (27)(28). Oxidation under atmospheric conditions begins at temperatures above 600°C and increases in rate with increasing temperature. This temperature is sufficiently high to make diamond films of interest as an oxidation resistant coating in many applications, such as jet engine bearings and seals.

Interestingly, films with a greater extent of non-diamond bonded material appear to oxidize more slowly than do essentially purely diamond bonded films (29). Apparently, the non-diamond bonded material is selectively etched revealing a dendritic structure. As oxidation continues, increasingly greater surface area becomes exposed to the oxidizing atmosphere. Thermodynamic equilibrium is achieved and the oxidation rate slows. In contrast, primarily diamond bonded material etches preferentially along structural defects and the entire film oxidizes at a higher rate than does natural diamond. Thus, for enhanced oxidation resistance, a complex microstructure of mixed bonding types appears to be superior to pure diamond bonding. Research areas of importance include substrate adhesion, grain size, orientation, and morphology, and extent of diamond bonding. Surface chemistry modifications produced by termination of diamond deposition with thin layers of more oxidation-resistant material (e.g., SiC) offer good potential in those applications for which diamond's bulk, rather than surface, properties are important.

Optical Transparency. Hard, transparent coatings for infrared detectors are a very promising application for diamond films. Infrared optical materials such as ZnSe are very soft and exhibit rapid degradation in optical performance (total transmission and modulation transfer function) when exposed to erosion by high-speed rain and dust, or by abrasive process slurries in contact with IR analysis windows. Diamond films should provide an excellent anti-erosion coating for IR optics across a broad range of wavelengths. Low temperature deposition, enhanced adhesion, reduced damage to the substrate by excited species, and excellent transmission are all essential to a satisfactory product.

Direct deposition of diamond on ZnS has been tried at relatively low temperatures without success, apparently due to severe etching by atomic hydrogen (30). Damage to the fragile infrared material can be circumvented through alternative manufacturing techniques. Adhesion and transmission are still important issues. Good adhesion may require a thin interlayer such as boron phosphide or Ge:C. Nucleation enhancement by scratching is impractical at all but the longest wavelengths due to unacceptable scattering increases. Diamond surface morphology modification needs additional work because the deposition parameters which currently give relatively smooth film surfaces also result in either unacceptable IR absorption (due to C-H and C=C structures) or intolerably low growth rates. An efficient, economical method of film planarization needs to be developed: a potentially fruitful direction lies in modification of deposition chemistry to provide simultaneous etch and deposition activity. In the absence of such methods, post-deposition polishing, whether by mechanical means or ion milling, will have to be employed for most CVD diamond optical applications.

3.2 Commercialization of Prototype Diamond Film Products

Once a prototype product has been developed to the point that the potential customer(s) are satisfied with its performance, the production process must usually be made faster, cheaper, more reliable, and operate on a larger scale. New requirements of fabrication, testing, quality assurance and control, and shipping come into play. Directions for improving the reproducibility, deposition area, and film growth rate will be discussed in this section. Other requirements are so product specific as to be impossible to treat adequately; their development will be left to the entrepreneur.

Deposition Reproducibility and Scale-up. The ability repeatedly to deposit diamond with the desired characteristics over an adequate substrate area is essential to the success of any diamond product development effort. Currently, the value in final product of a 4" diamond-coated silicon wafer can easily exceed by one order of magnitude the value of a 6" wafer of finished 80486 or 68040 microprocessors. This places a very high premium on enhancing deposition run reliability and minimizing variation among runs and reactors, and provides economic justification for such techniques as continuous monitoring of on-line feed gas purity by multichannel mass spectroscopy rather than cheaper batch sampling methods.

Reactor automation is essential for process reliability. Computerized data acquisition and control systems must be integrated into reactors at the earliest design stages. Attention to reactor instrumentation will be repaid with enhanced repeatability. For example, because substrate temperature is a critical process variable, it is desirable that it be measured with optical methods which more closely indicate true substrate temperature than do thermocouples. Optical methods are more expensive and must avoid artifacts due to plasma radiation, surface emissivity changes, and deposition on in situ optics. Acquired machine performance data, correlated with resulting film properties, form the basis for statistical process control of operations, and automated reactor control reduces process variation related to human factors. Resources must be allocated to the routine review of performance data and maintenance of the database, so that comparison standards will be available when the inevitable "loss of recipe" process excursion occurs. Very high volume applications (e.g., coated cutting tool inserts) require integration of robotic materials handling technology into the diamond deposition process to meet the needs of both reliability and economics.

Deposition area scale-up can be a particularly difficult development barrier: it is common that deposition processes which work well in a 1-inch tube reactor go awry when carried out in a 6" reactor, requiring repetition of parameter variation experiments and reconfiguration of deposition chambers. Furthermore, a diamond deposition reactor may perform quite differently when loaded with five hundred printed circuit microdrills rather than the one or two drills used for development runs, even if chamber size, shape, and other known critical parameters are unperturbed. In addition, each of the different variations in diamond deposition technology offers its own idiosyncratic scale-up engineering problems. For example, scale-up of plasma torch systems (without reduction of growth rate) requires close attention to the problem of removing as much as a megawatt of heat from a 6" diameter substrate. In contrast, microwave systems may require additional substrate heating as substrate diameters increase.

Finally, ongoing fundamental diamond deposition research forms the solid foundation on which to build development of reliable production processes. Continued investigation of the fundamentals will disclose additional unrecognized parameters important in maintaining reliability of developed processes and in extending known process capabilities for new application requirements.

Large Area Uniform Deposition. Large area diamond film deposition of uniform quality and thickness is a key element of almost any successful diamond application. Few deposition parameters more strongly affect production deposition economics than deposition area. Large area deposition will be taken here to mean > 70 cm^2 of uniform material, since a large area with wildly differing morphology, for instance, is of limited use to a production line. "Uniform" indicates thickness and property variations within ± 10 % of the mean. Note that this is quite primitive compared with the state of the art for more mature materials like silicon.

Uniform diamond film deposition over more than 40 cm^2 has been reported by only a few research groups (e.g., Fig. 8 and Refs. 31, 32) although it is likely that better results exist and remain unpublished because of their commercial sensitivity. Diamond deposition nonuniformities arise from two major sources, choice of gas excitation technology, and reactor configuration.

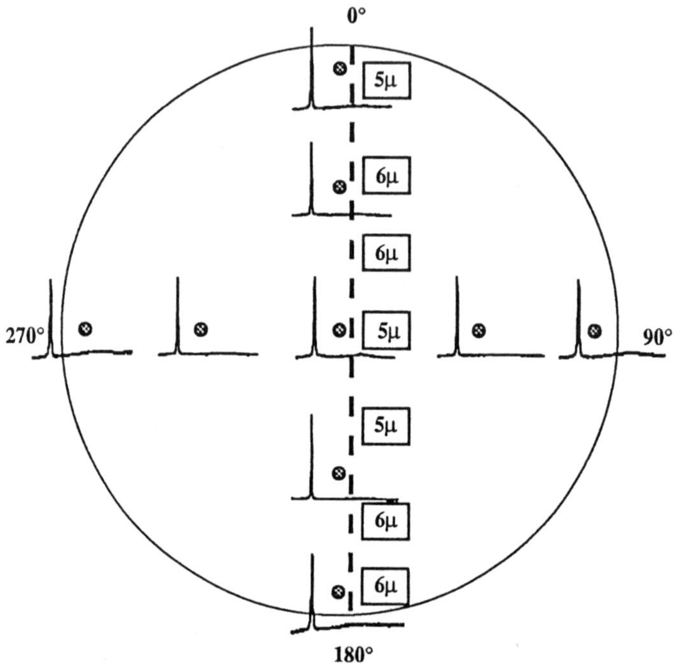

Figure 8. Film thickness and bonding uniformity of a diamond film across a 75 mm silicon wafer (32).

Most excitation methods (hot filament, high-pressure microwave, plasma torch) provide highly localized generation of excited species, which then must move by diffusion or bulk flow to the deposition substrate. This in itself is sufficient to create substantial spatial variations in growth rate and film composition at the substrate. However, since most excitation methods also significantly heat the substrate in nonuniform ways, further film property variations occur due to large substrate thermal gradients. Some production-related strengths and weaknesses of different deposition techniques with large area deposition are discussed below.

Two common thermally-driven diamond deposition techniques employ either a heated filament (33) or combustion of a fuel/oxidizer mixture (34). Hot filament systems depend on thermal dissociation of hydrogen and carbon source gases by passage near a refractory metal wire typically heated to > 2000 K.

Benefits of this system include low capital and operating cost (at least at the research reactor level) and ease of operation. Disadvantages of this method for production use include limited filament lifetime, loss of filament material and possible contamination thereby of the growing diamond film, and strong heating of the substrate by radiation, a particularly difficult problem with nonplanar substrates. A further disadvantage of hot filament systems involves limitations on the types of precursor gases which can be used without adversely affecting filament lifetime. Some of the more interesting oxygen-bearing precursor mixes (CO/O_2) rapidly oxidize filaments under some conditions. Finally, maintaining high growth rates over large areas using hot filament methods requires large power inputs which can degrade process economics and further complicate substrate temperature control.

Thermal combustion deposition systems like the oxyacetylene torch offer even lower capital costs (one's local welding shop is now a potential diamond merchant!) and ease of operation due to the simplicity of their equipment and ability to operate in air. Use of flame deposition in production environments presents significant challenges, however. One of these, as with the hot filament method, is management of large heat inputs to the substrate. Operation at reduced pressure can assist in substrate temperature control, but this shifts flame chemistry substantially. Adhesion of flame-grown diamond to many substrates is poor relative to other techniques (e.g., microwave plasma), and film surface roughness tends to be high. Again, coating of three-dimensional substrates can be difficult due to overheating and large thermal gradients. Flame systems may not be suitable for high-volume production needs due to the inability to recycle gases for process cost reduction.

A third class of diamond thermal CVD systems which employ CF_4/F_2 mixtures is attracting investigation (35), but too little data exist at this writing to offer any commentary.

Plasma torch diamond deposition proceeds by both thermal excitation and ionization of feed gases. Deposition rates can be extremely high (~ 1mm/hour), with typical reported deposition areas of approximately 1 cm^2 (36). Commercial plasma torches are available which can supply high-enthalpy plasma streams of over 6" diameter (37) but which require nearly

2 MW_e for excitation, of which about 1 MW_t appears as heat at the substrate. This creates tremendous substrate temperature control problems and may not be a viable scaling means except for very limited applications. Substrate scanning has been successfully used to expand plasma torch deposition area, but with average growth rate reduction in proportion to the ratio of unscanned deposition area to the total scanned substrate area. Internal film stress buildup may limit the utility of substrate scanning. As with flame systems, poor adhesion and high surface roughness are common characteristics of plasma torch diamond films. Unlike flame systems, gas recirculation offers potential process economic benefits.

The best apparent near-term deposition area increase possibilities appear to lie in DC glow discharge and conventional cavity-mode microwave systems. In the longer term, ECR microwave plasmas may eventually be extremely attractive for growth of uniform diamond films over large areas under relatively mild conditions. To date, published reports of diamond deposition in ECR systems have not been at pressures sufficiently low to satisfy the ECR condition, i.e., in which gas collision frequency is substantially less than the electron cyclotron resonance frequency (38). For typical ECR systems (F_0 = 2.45 GHz, B_0 = ~850 Gauss), gas pressures must be below ~10^{-4} torr. Plasma uniformity in ECR systems far exceeds that in high-pressure microwave systems, but diamond deposition chemistry is known to be pressure dependent, with a transition to diamond-like or graphitic carbon below about 1 torr, depending on substrate bias conditions and the presence or absence of magnetic assist (39). ECR may be the most promising large area deposition technique in the long run, due to the ability to control ion fluxes and energies with extraction grids and substrate bias. Substantial research remains to be done to develop the method for diamond deposition.

Growth Rate. Film growth rate is a key issue in diamond deposition applications development. Deposition rate affects film structure and properties, and fixes production economics. Early work in diamond CVD emphasized growth rate "hero experiments", with little consideration of the resulting variations in material properties. The realization that many of the desirable properties of CVD diamond (e.g., thermal conductivity, adhesion strength, and optical transparency) degrade when growth rates are pushed to their upper limits has caused reassessment of the importance of high growth rate as a desirable end in itself by entities involved in commercial CVD diamond development. Low growth rates do not, however, solely determine the economic viability of a given product.

Process modifications known to enhance deposition rates include use of appropriate precursors, with oxygen-bearing materials such as methanol, acetone, CO, and CO_2 apparently giving enhanced material deposition rates under appropriate conditions (40)(41). Excitation technology strongly influences deposition rate, with the plasma torch system currently at the top (~1000 µm / hr) (42), with combustion flames next (~40 - 100 µm / hr) (43)(44), followed by hot filament (~10 µm/hr) (45), with microwave (~3 µm / hr) (46) and DC glow discharge (<1 µm / hr) among the slower methods. Promising avenues for enhanced growth rates, beyond simply putting more and more power into plasma or thermal excitation, may include alternate deposition chemistries (e.g., CF_4/F_2). Substantial improvements in growth rates without degradation of film properties probably await a detailed understanding of the chemistry specific to each deposition technology. This, in turn, will require continued commitment to fundamental diamond deposition chemistry research.

4.0 FUTURE NON-ELECTRONIC DIAMOND APPLICATIONS

We believe it is difficult for even the most energetic of CVD diamond partisans to overestimate the breadth of potential applications for this material. In this section, we will try to impart a sense of the enthusiasm we feel in contemplating the development opportunities already identified for CVD diamond by enumerating a few drawn from a wide variety of non-electronic applications areas. If some of these applications seem impractical in light of today's primitive diamond CVD capabilities, we would ask the reader to remember the explosive progress in modern materials science and technology, and to consider CVD diamond as an undeveloped material with perhaps more potential than most.

Our view of the likely order of appearance of several major diamond applications is shown in Table 4, along with a few of the major challenges to be overcome during development. This type of prediction is always hazardous, in that the forecasts are bound to be either too optimistic or too conservative, depending on the application. The large range cited for diamond semiconductors indicates our estimate that a few very simple diamond semiconductor devices (thermistors, etc.) may appear relatively quickly, while transistors and more complex devices will take much longer.

We will group by application area, and within areas, loosely by expected order of development. We believe these represent only a small fraction of the eventual applications to be developed.

Table 4. Predicted Timing for Development of Various Diamond Film Based Products.

Application	Year(s) Expected	Development Challenges
X-ray detector windows	1989	pinhole defects, very thin films, high nucleation density, gas diffusion
X-ray lithography membranes	1991-1992	stress control, surface smoothness, pinhole defects
Electronic packaging	1992	deposition economics
Coated cutting tools	1993-1994	adhesion to substrate
Diamond-coated IR optics	1992-1994	adhesion, optical scattering
Diamond ceramics and fibers	1993-1995	particle consolidation & production economics
Diamond abrasives	1992-1995	uniform particle size, economics
Diamond semiconductors	1992-2000	heteroepitaxy, doping, packaging

4.1 Mechanical Applications

Near Term Mechanical Applications. Diamond films are under consideration as a replacement for other hard materials (such as sapphire and SiC) in a number of applications. Limitations on CVD diamond's interface bond strength to certain commonly used structural materials suggest that the first simple wear resistance applications will avoid high dynamic environments in which the coated material is flexed or shocked. One important example is that of paper *guide surfaces* in papermills. Papermills are currently subject to expensive mill downtime caused by guide wearout. Paper contains silica spicules which originate in wood, and these silica particles are extremely abrasive when imbedded in paper sheets passing at high speed over guide surfaces. Current hard guide surface materials, such as single-crystal sapphire or silicon carbide, wear rapidly in this service. Such guides, which are not subject to impact or flexure, are logical candidates for early diamond wear coatings. Similar guide wearout problems exist in fiber and fabric manufacturing operations in which downtime is extremely costly. Existing diamond deposition technology can serve many of these applications, with the principal need being for larger deposition areas.

Extrusion dies used to form highly abrasive materials such as SiC whisker-reinforced aluminum would gain extended useful life through diamond coatings. In this application, extrusion fluid dynamics result in substantial pressure gradients along the die, but little die deformation. Strain rates are relatively low (except for rapid impact extrusion processes), and spot temperatures for many processes are within diamond's thermal limits. Today's diamond film interface adhesion strength is probably adequate for many smaller extrusion die applications. Larger dies will require some extension of current technologies to larger deposition areas. The ability to coat within small holes has been demonstrated to some degree already, as shown in Fig. 9. This indicates the potential for replacement of single-crystal diamond or PCD (diamond particles in metallic binder) wire *drawing dies* with CVD diamond-coated parts. Surface finish is an important parameter in most extrusion applications, indicating more needed work on surface orientation and morphology control, and on post-deposition polishing methods.

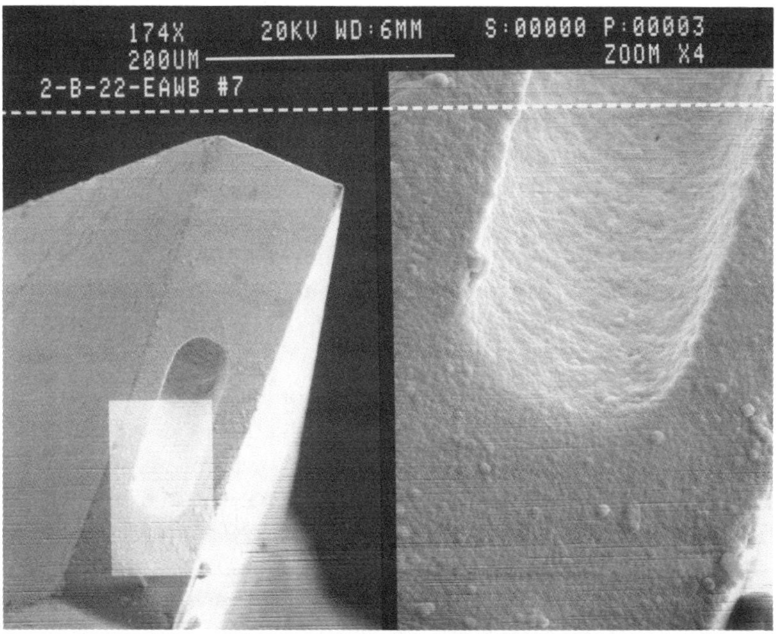

Figure 9. Micrograph of a diamond film coated wirebond tool.

Medium Term Mechanical Applications. Diamond-coated sliding seals and bearings are in development and may be expected to appear by 1993 in selected applications. Mud pumps used in rock drilling applications use impellers and seals of Stellite, Hastelloy, silicon carbide, and other hard materials. These parts wear rapidly during exposure to fast flows of rock chips suspended in Bentonite mud. Pump failure causes expensive downtime because pump repair requires removal and replacement of a segmented drill string which may be as much as two miles long. These components operate under substantial vibration and point stress concentration, but with relatively low flexural loads. Similar applications appear in *jet engines* in areas which operate below 600°C. The potential exists to replace selected lubricated turbine components with unlubricated diamond coatings, achieving significant engine weight reduction and reliability enhancement through elimination of lubrication system components. Interface adhesion, deposition uniformity, and surface finish are areas which need further development to generate suitable diamond coatings. The U.S. Air Force has begun to sponsor applications development of diamond coatings for turbine engine applications (47).

Selected diamond-coated cutting tools can be expected to appear along with coated sliding seals. These tools will probably be coated inserts used for long cuts (e.g., turning operations) on nonferrous metals such as 390 aluminum, a high-silicon alloy used extensively in the automotive industry. Turning operations usually place lower stresses on tools than interrupted cutting operations (fly cutting, broaching, etc.) and will therefore be among the first operations to use diamond-coated cutting tools. Enhanced interface adhesion suggests the early use of ceramics (e.g., SIALON and silicon nitride) as substrates for diamond-coated cutting tools, but conservative attitudes in some sectors of the machining industry toward adoption of new materials technology will provide an incentive to solve the problems of diamond adhesion on more familiar tungsten carbide tools. Perhaps the most obvious of CVD diamond cutting edge applications, which we believe will await reductions in deposition costs as well as adhesion engineering, will be diamond-coated knives for consumer use.

A different category of medium-term CVD diamond applications will arise from the adaptation of CVD deposition to the high-volume production of diamond abrasive particles. Deposition of diamond particles in fluidized-bed systems offers the potential for extremely high mass deposition rates due to the very large surface areas available in fluidized particle beds. For example, assuming a 2" diameter plasma deposition volume, over 4×10^4 cm^2 of deposition area is available given a cloud of seed particles of average (spherical) diameter of 10 μm, packed to occupy only 10% of the 2" spherical growth zone at the beginning of growth. This contrasts with only about 8 cm^2 of deposition surface area for ten seed particles occupying 10% of the area within a 2" diameter planar deposition zone. Preliminary plasma ignition and fluidization tests have been carried out with promising results, but production development will require substantial resources. The control over morphology and purity afforded by the CVD process and high mass deposition rates are expected to provide diamond abrasives at dramatically reduced production costs compared with current high-pressure grown particles or natural sources. Reduced diamond abrasive prices should stimulate greater use of the material, as the market for diamond abrasives is still price-constrained. A diamond particle grown by CVD is shown in Fig. 10.

Figure 10. Micrograph of large, PECVD diamond particle.

Long Term Mechanical Applications. Availability of low-cost, high-purity diamond particles with controllable morphology and size distribution is expected to lead to production of diamond ceramics, by which we mean solid materials built by consolidation of diamond or diamond composite particles. Examples include solid diamond cutting tools and large mechanical parts, such as gauge blocks. The first of these materials to appear will be formed by consolidation of composite particles which consist of diamond core particles (probably grown in fluidized-bed systems) overcoated with thin layers of conventional ceramics such as silicon carbide or silicon nitride. This composite powder feedstock can then be consolidated using conventional methods already developed, and will produce a ceramic material whose properties approach those one would expect of a pure diamond ceramic material. Production of all-diamond ceramics (no non-diamond material) awaits the development of appropriate consolidation technology, perhaps by modifying the CVD process to function within the interstices of compacted diamond powder preforms. Either of these technologies is expected to make possible fabrication of large ceramic pieces having many of the properties of diamond. Advantages of such ceramics should include high thermal conductivity, resistance to thermal shock, controllable electronic properties, and excellent dimensional stability. This technology is currently in development (48).

Diamond-coated reinforcing fibers may allow manufacture of fiber-reinforced composite materials with enhanced properties resulting from diamond's high modulus and other properties. Growth of CVD diamond in fiber or whisker morphology has been reported (49)(50), possibly by VLS growth. It is more likely that the first industrially useful diamond fibers will consist of non-diamond fibers which have been coated with CVD diamond. Fibers of silicon carbide, silicon nitride, and graphite have been successfully coated with thick, adherent CVD diamond layers. Surface finish can be varied from smooth to rough, an important parameter in determining fiber-matrix interface behavior (51). Short fibers (whiskers) may be economically grown in fluidized-bed systems, while continuous feed methods are under consideration for production of long, unbroken lengths of diamond-coated fibers. Fig. 11 shows a fracture cross-section of a hollow silicon nitride fiber (Dow Corning HPZ) coated with CVD diamond.

Development of diamond films with excellent interface adhesion may bring diamond-coated ceramic rolling bearings into use. A diamond-coated silicon nitride ceramic ball bearing is shown in Fig. 12. Ceramics are suggested as likely substrates for diamond-coated rolling bearings because of their high modulus and consequent minimum flexure under load, because of their relatively low thermal expansion coefficients which may be tailored to some degree, and because of their general tendency to form strong interface bonds with diamond films. Coated carbide or steel bearings may eventually be available if very thick diamond coatings are used to reduce flexure of the lower modulus substrate. Successful development of dia-

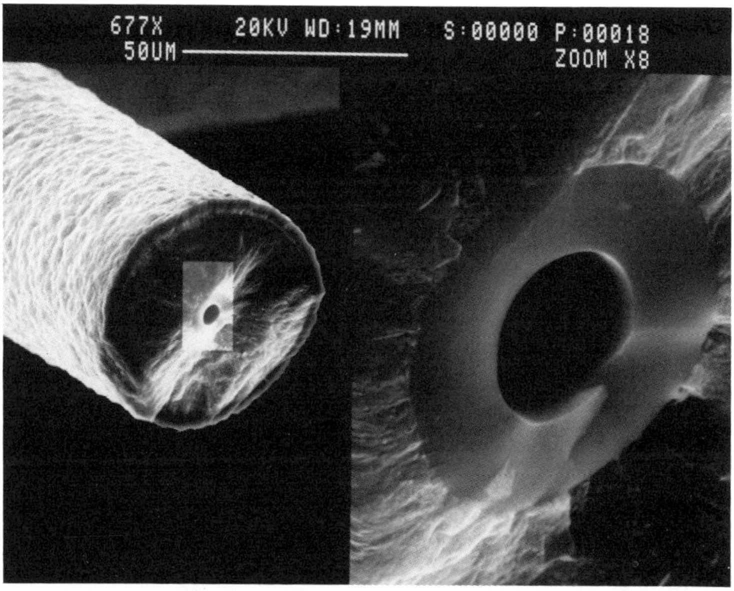

Figure 11. Micrograph of a fracture cross-section of a hollow silicon nitride fiber (Dow Corning HPZ) coated with CVD diamond. Right half of photo is an 8x magnification of left half.

mond-coated rolling bearings will require that race surfaces which carry rolling elements also be coated. The technology which makes possible these bearing applications should be directly transferable to placing diamond coatings on compressor blades in jet engines, which are subject to foreign object damage and dust erosion. Compressor stage blades run cool, but undergo considerable flexure and stretch due to centrifugal, aerodynamic, and vibrational loads. High interface adhesion strength will be essential, and may require development of hydrogen barrier interlayers to prevent hydrogen embrittlement of blades during diamond deposition. Achieving economic volume production will be a definite challenge.

A second category of longer-term applications for diamond wear coatings may arise from the need for enhanced wear life for prosthetic implants such as hip and knee joints. Corrosion by body electrolytes and wear during use forces frequent major surgery for replacement of the prosthesis. Advances in prosthetic technology have resulted in implants being used in younger patients whose higher physical activity levels impose greater mechanical loads and decreased wear life. Diamond's superior corrosion resistance and presumed biocompatibility suggest its use in this

a

b

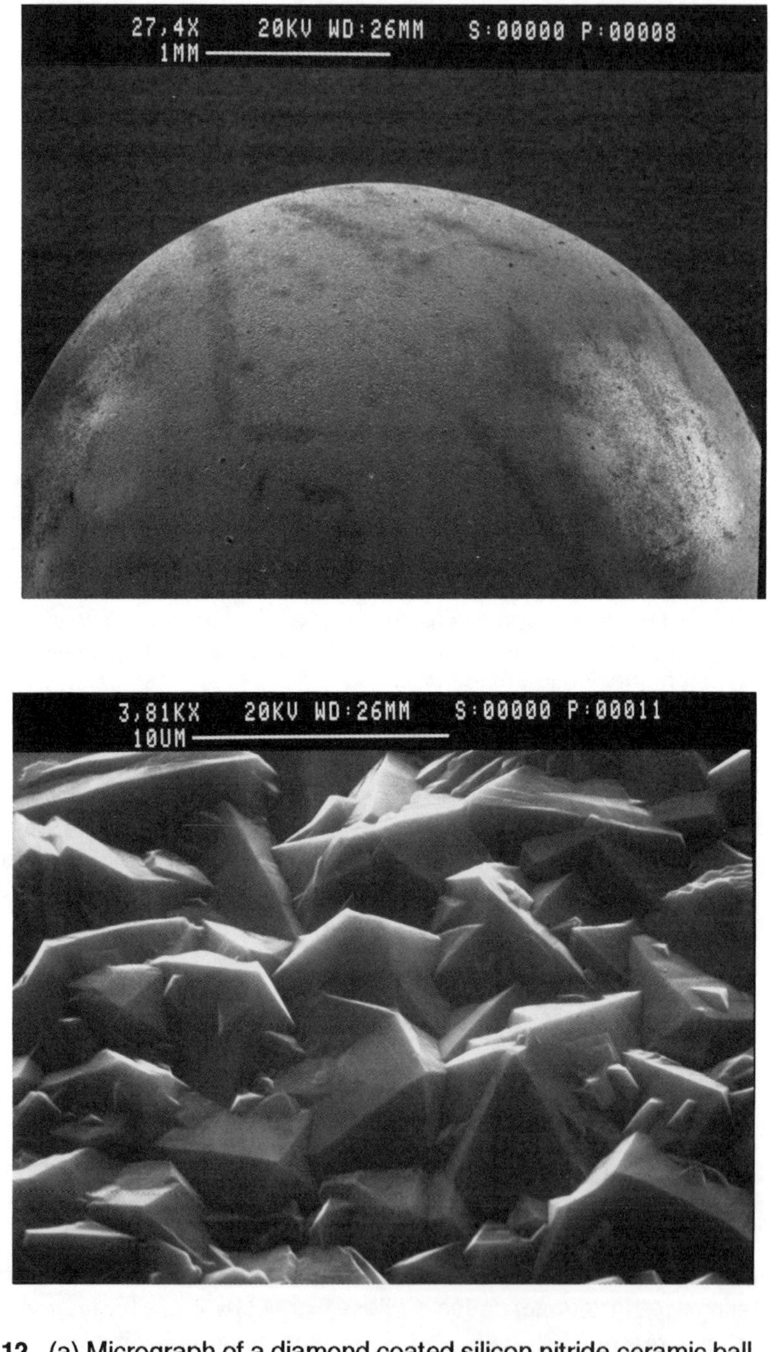

Figure 12. (a) Micrograph of a diamond coated silicon nitride ceramic ball bearing. (b) Detail of diamond morphology on ball bearing.

application, and initial tests of CVD diamond coating on an implant alloy have confirmed expectations. The basic technical requirements of this application are probably less stringent than those for applications which we expect to appear sooner. However, we believe development time for diamond-coated implants will be fairly long because of nontechnical issues, such as the need for extensive testing and FDA approval. Product liability issues could further delay introduction of diamond-coated implants.

4.2 Passive Thermal Applications

Near Term Passive Thermal Applications. The advantages of small single-crystal diamond as heat spreaders for high power density devices are well-known. Due to the unavailability and prohibitive costs of large single-crystal substrates, few potential users have yet seriously considered using large-area CVD diamond films in electronic packaging. We expect this to change rapidly in the early 1990s.

The earliest products which take advantage of diamond's high thermal conductivity will probably be substitutes for beryllium oxide heat spreaders in various high-dissipation discrete devices. The use of BeO, with a thermal conductivity of about 2.5 W/cm-K (*cf.* copper @ 4.0 W/cm-K) is under increasing review due to fears about toxicity of the material. Aluminum nitride, while an inferior thermal conductor (~1.7 W/cm-K), is nonetheless being adopted on grounds of price and nontoxicity. CVD diamond, with a measured thermal conductivity of up to 15 W/cm-K @ 100°C (52)(53), is attractive now as a BeO substitute in some low-volume, cost insensitive markets like airborne electronics with high packing densities. Fujitsu has successfully tested a InGaAsP laser diode on a CVD diamond heat spreader grown by a plasma torch process (54).

Microwave power devices comprise another early market for CVD diamond heat spreaders. Control over film dielectric constant will be necessary for many devices, and package redesign may be needed because package transmission line behavior varies with dielectric substrate properties. Production techniques which remain to be developed before commercial introduction of diamond heat spreaders include methods of cutting and polishing, and metallization for device attachment. Substantial progress has been made at the single-unit demonstration level for each of these fabrication technologies. Deposition cost reduction is another requirement for widespread introduction of large-area diamond heat spreaders. In general, deposition process parameters which enhance thermal conductivity reduce growth rate and increase costs. It is likely that the technologies which have been used to grow material for prototype and small volume production runs (e.g., hot filament) will give way to large-scale deposition technologies with significantly better process economics (e.g., plasma torch).

A second near-term CVD diamond film application which takes advantage of diamond's thermal properties may be high-power microwave transmission windows. Microwave power tubes often employ dielectric windows to provide vacuum isolation between tube and waveguide, and to couple microwave radiation out of the tube. These windows are adversely heated by microwave losses and often must be actively cooled, leading to complex, expensive windows which can fail due to thermal shock. CVD diamond should provide a new material choice for such windows. Its dielectric constant and loss tangent have been found to be ~5.2 and ~1×10^{-3}, respectively, @ 94 GHz (55). These data compare favorably with other dielectric materials, and diamond's thermal conductivity and low thermal expansion coefficient render it quite resistant to thermal shock failure mechanisms. Technology for integrating large diamond plates into high temperature window structures must be developed.

Medium Term Passive Thermal Applications. With declining deposition and fabrication costs, large CVD diamond plates may see service as packaging substrates for very fast microprocessors and multichip modules whose densely packed chips present severe thermal management problems. This will require great reductions in deposition cost and attention to mechanical stress concentration problems at interfaces with large chips due to expansion coefficient mismatch effects. Advanced cooling technology (56) may be significantly helpful in this type of application. Multilayer metallization methods with *via* connections will be required as well. Extension of passive thermal spreading technology to the device level is possible using silicon-on-insulator (SOI) structures in which the insulator is CVD diamond. Figure 13 shows a thin (~0.5 µm) layer of single-crystal silicon over a thicker polycrystalline CVD diamond layer. In addition to its electronic advantages, this structure offers significantly better thermal performance than silicon-on-sapphire (SOS) or SIMOX materials.

Diamond coatings may make it possible to enhance the brightness of color video display tubes. Most tubes employ color phosphor triads (RGB dots or stripes) which are excited by scanned, modulated electron beams. Interposed between the phosphors and the electron beams is a thin metal sheet (shadow mask) of minute apertures which are in register with the phosphor pattern and which prevent unwanted electron beam excitation of phosphors adjacent to the intended target. In doing so, the shadow mask intercepts considerable electron current (up to 70% of the beam current) and is heated. Although continuous beam scanning prevents local overheating, the need to prevent thermal distortion and consequent misregistration with phosphor groups sets a limit on the allowable energy input to the mask. This in turn limits beam current and display brightness. It may be possible to diamond-coat the shadow mask structure, enhancing both heat conduction and dimensional stability. Increased allowable electron beam current would lead to substantially higher display brightness. The coating would probably have to be electrically conductive to prevent beam deflection due to charging effects. Termination of the deposition with a layer of graphitic material could provide this function and also increase emissivity to enhance radiative heat rejection. This is an attractive application for CVD diamond because of its very high unit volume. Development research would focus on economic deposition uniformity over large areas, and on the adverse tradeoff between growth rate and thermal conductivity.

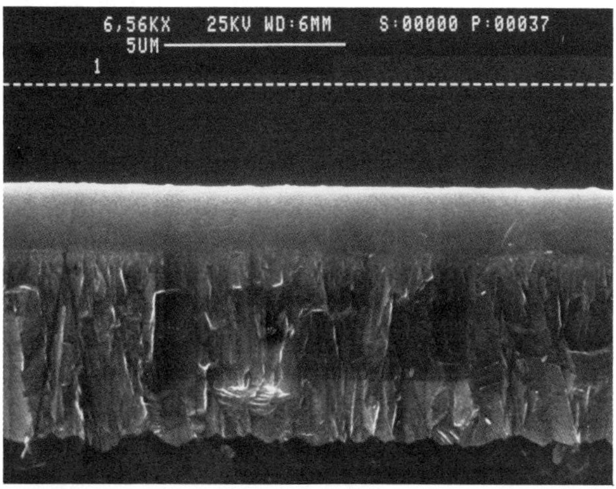

Figure 13. Micrograph showing silicon-on-diamond structure. Eventual utilization in SOI applications is promising.

Long Term Passive Thermal Applications. Longer-term thermal management applications for CVD diamond products may include advanced combustion engine components. In the past two decades, metallic materials technology has been vigorously pressed to achieve ever higher combustion chamber temperatures in jet engines for better Carnot efficiency and power output. Substantial development funds have been spent to gain as little as 30°C in hot section operating temperatures. Oriented polycrystalline blades have given way to actively cooled, oriented single-crystal superalloy blades which can operate at slightly higher temperatures due to suppression of grain boundary corrosion. Nevertheless, conventional metallic blade technology is probably close to its thermal limit, and attention has turned to ceramics and intermetallics (e.g., nickel aluminide).

Diamond ceramics may play a part in high-temperature advanced turbine engines by providing blades which can be more efficiently cooled, allowing use of increased combustion temperatures. Such blades would be actively cooled by gas flow through internal passages, as are today's metal blades. Oxidation barrier coatings will be necessary on both internal and external surfaces. Advantages of diamond ceramic blades would include: reduced thermal gradients and blade temperature due to high thermal conductivity and consequent reduction in thermomechanical stress; better dimensional stability; and greatly reduced weight and rotational inertia by virtue of diamond's low density compared with current superalloys. The last advantage may benefit other engine subsystems by allowing lighter support structures, enhanced bearing lifetimes, etc. An intermediate step in achieving diamond hot section turbine blades may be the use of diamond-coated fibers as reinforcers for intermetallic or ceramic blades. Increased thermal conductivity and dimensional stability should result.

4.3 CVD Diamond Optical Applications

Near Term Optical Applications. Development of early optical applications for CVD diamond is hampered by the general problem of scattering due to the material's polycrystallinity and rough surface morphology. Optical transmission, if not yet as broadband as the best natural material, seems rapidly improving as a result of deposition chemistry development. Arguably, the x-ray detector window now available (57) qualifies as an optical product within the meaning of transparency to electromagnetic radiation. X-ray lithography mask membranes (see Fig. 6) may more readily qualify as an optical application in the usual sense. Scattering at x-ray lithography wavelengths is not an issue, but mechanical requirements for these membranes are extreme. Diamond membranes in tensile stress, a key requirement for a mask membrane, have been fabricated (58), and one company has reportedly successfully made a functioning circuit with x-ray exposure through a diamond mask. Widespread use of diamond lithography masks will probably be paced more by the timing of acceptance of x-ray technology as a production fabrication method than by any aspect of diamond technology development.

Early opportunities for more conventional CVD diamond optical applications appear in infrared optical systems, principally less cost-sensitive military devices. Early laser damage threshold (59) and x-ray pulse damage measurements (60) indicate diamond's use as a damage-resistant antireflection layer on silicon optical system components. Figure 14 shows the IR transmission spectrum of a 1 µm diamond film on silicon, compared to the (normalized) uncoated silicon transmission, and demonstrates the antireflection function expected given a refractive index of 2.4 for diamond (61). The ability to control CVD diamond's electrical conductivity is important in that it provides reduced radar cross-section (stealth) capabilities as well as optical functionality.

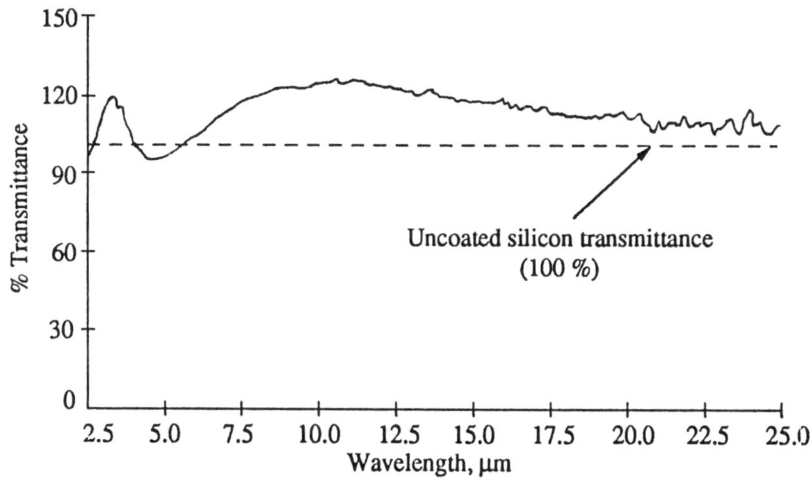

Figure 14. IR transmission spectrum of a 1 µm diamond film on silicon, compared to the (normalized) uncoated silicon transmission. Antireflective properties of the film are apparent.

Medium Term Optical Applications. CVD diamond films may find use in high-energy reflective optics. Liquid-cooled metal mirrors are currently used as high power laser optical elements. Diamond's thermal conductivity and dimensional stability may be very helpful in making more robust optics in which a thin mirror layer (or optical multilayer stack) is deposited on a smooth, thick diamond film which is fluid-cooled on the other side. Such an optic should operate under higher power densities with reduced thermally-induced beam distortion. Table 5 shows a number of mirror materials and their properties in comparison with CVD diamond. Note diamond's large performance advantage implied by the equilibrium and transient thermal distortion figures of merit in the last two columns.

Table 5. Mirror Material Comparisons.

Material	ρ g/cm³	E GPa	E/ρ	α ppm/K	κ W/cm-K	D cm²/sec	α/κ	α/D
Diamond	3.54	1054	297	~1	~18	28.6	0.06	0.03
Beryllium[a]	1.85	287	100	11.4	1.93	0.57	5.9	20
Aluminum	2.7	68	16	23.6	1.67	0.69	14.1	34
Fused Silica	2.19	72	21	0.5	0.014	<0.01	36	59
ULE	2.21	67	19	<0.02	0.013	<0.01	1.2	1.9
Zerodur	2.53	92	23	-0.09	0.016	<0.01	5.7	12
Silicon Carbide	3.15	403	103	3.4	0.88	0.42	3.9	8.2
Silicon	2.33	137	59	3.0	0.84	0.9	3.6	3.3

[a]Non-diamond data from Brush Wellman, Inc.
ρ = density
E = Young's modulus
E/ρ = normalized stiffness
α = thermal expansion coefficient
κ = thermal conductivity
D = thermal diffusivity
α/κ = equilibrium thermal distortion figure of merit (lower is better)
α/D = transient thermal distortion figure of merit (lower is better)

Long Term Optical Applications. Finally, a longer-term optical application for CVD diamond may be as a laser host material. Natural diamond containing nitrogen (H3 color center) has been optically pumped to coherent emission at 540 nm using only the Fresnel reflections at the air interfaces of a 2 mm thick natural diamond (62). Furthermore, there is at least one report of near-band gap emission (~5.3 eV) from a natural diamond in response to electrical input (63). Diamond's potential as a laser host material is very attractive: its thermal conductivity, low thermal expansion coefficient, and broadband transparency indicate the possibility of high-power solid-state lasers with greatly reduced thermal beam distortion effects. Splitting of diamond's electronic levels by phonon levels may indicate tunability of coherent emission from some color centers. Substantial work remains before the potential of CVD diamond micro- and macro-lasers can be realized. Transparent CVD diamond can now be routinely produced, but creation of appropriate color centers and solution to scattering problems will require time and significant resources.

REFERENCES

1. Plano, L. S., Yokota, S. and Ravi, K. V., *Proc. First International Symposium on Diamond and Diamond-Like Films* (J. P. Dismukes, et al., eds.), p. 380, Electrochemical Society, Pennington, NJ (1989)
2. Plano, L. and Adar, F., *Int'l Conf. on Raman and Luminescence Spectroscopy in Technology* (J. E. Griffiths, F. Adar, eds.), Proc. SPIE 822, pp. 52 - 56 (1987)
3. Bou, P. and Vandenbulcke, L., *Proc. First International Symposium on Diamond and Diamond-Like Films* (J. P. Dismukes, et al., eds.), p. 610, Electrochemical Society, Pennington, NJ (1989)
4. Krishnan, R. S., *Proceedings of the Indian Academy of Sciences*, 25A:45-57 (1946)
5. Ono, A., Baba, T., Funamoto, H. and Nishikawa, A., *Jap. J. App. Physics* 25(10):L808 (1986)
6. Herb, J. A., Bailey, C., Ravi, K. V. and Dennig, P. A. *Proc. First International Symposium on Diamond and Diamond-Like Films* (J. P. Dismukes, et al., eds.), p. 366, Electrochemical Society, Pennington, NJ (1989)
7. Buckle, I. H., *Metallurgical Reviews* 4(13):49 (1959)
8. Brookes, C. A., in: *The Properties of Diamond* (J. E. Field, ed.) p. 384, Academic Press, NY (1979)
9. Plano, L. S., "CVD Diamond Films for Tribological Applications." SDIO/IST Phase I SBIR Final Report for Contract No. N00014-89-C-0151 (1990)
10. Personal communication, John Wegand, Geo-Centers (currently at Naval Research Laboratories); data on Crystallume diamond films (1990)
11. Plano, L. S., Hayward, I. and Wegand, J., "CVD Diamond Films for Tribological Applications." Final Report for Office of Naval Research SBIR Contract No. N00014-89-0151 (1990)
12. Conner, L. C., Presented at *Gorham Advanced Material Institute International Conference on High Performance Inorganic Thin Film Coatings*, Monterey, CA (1988)
13. Ishibori, K., and Ohira, Y., *Abst. of First Int'l. Conf. on the New Diamond Sci. & Tech.*, p. 104, Japan New Diamond Forum (1988)
14. Collins, R. W., Cong, Y., Kim, Y. T., Vedam, K., Liou, Y., Inspektor, A. and Messier, R, *Thin Solid Films*, 181:565 (1989)
15. Bi, X. X., Eklund, P. C., Zhang, J. G. and Rao, A. M., *J. Mater. Res.*, 5(4):811 (1990)

16. There are several companies marketing statistical experiment design software and related seminars. For example: Edgework, Inc. Seattle, WA.

17. Williams, B. E. and Glass, J. T., *J. Mater. Res.*, 4(2):373 (1989)

18. Ohtake, N., Tokura, H., Kuriyama, Y., Mashimo, Y., and Yoshikawa, M., *Proc. First International Symposium on Diamond and Diamond-Like Films* (J. P. Dismukes, et al., eds.), p. 93, Electrochemical Society, Pennington, NJ (1989).

19. Plano, L. S., Hayward, I. and Wegand, J., Office of Naval Research SBIR Contract No. N00014-89-0151 (1990)

20. Pinneo, J. M., "PECVD Diamond Thin Films on GaAs Substrates." Final Report for U.S. Army Strat. Def. Command SBIR Contract No. DASG60-87-C-0109 (1988)

21. Tabor, D., in *The Properties of Diamond* (J. E. Field, ed.) p. 336, Academic Press, NY (1979)

22. Data from Crystallume films, Atlas Testing Laboratories, Commerce, California, U.S.A. (1989)

23. Gardos, M. N., and Ravi, K. V., *Proc. First International Symposium on Diamond and Diamond-Like Films* (J. P. Dismukes, et al., eds.), p. 475, Electrochemical Society, Pennington, NJ (1989)

24. Wong, M. S., Meilunas, R., Ong, T. P. and Chang, R. P. H., *Appl. Phys. Lett.*, 54(20):2006 (1989)

25. Ono, A., Baba, T., Funamoto, H. and Nishikawa, A., *Jap. J. App. Physics* 25(10):L808 (1986)

26. Herb, J. A., Bailey, C., Ravi, K. V. and Dennig, P. A. *Proc. First International Symposium on Diamond and Diamond-Like Films* (J. P. Dismukes, et al., eds.), p. 366, Electrochemical Society, Pennington, NJ (1989)

27. Plano, L. S., Yokota, S. and Ravi, K. V., *Proc. First International Symposium on Diamond and Diamond-Like Films* (J. P. Dismukes, et al., eds.), p. 380, Electrochemical Society, Pennington, NJ (1989)

28. Weimer, W. A., Johnson, C. E. and Harris, D. C., "Optical and Thermal Properties of PACVD Diamond Films." Presented at SDIO/IST - ONR *Diamond Technology Initiative Symposium* (1989)

29. Plano, L. S., Yokota, S. and Ravi, K. V., *Proc. First International Symposium on Diamond and Diamond-Like Films* (J. P. Dismukes, et al., eds.), p. 380, Electrochemical Society, Pennington, NJ (1989)

30. Pinneo, J. M., "CVD Diamond Protective Coatings for Alpha Measuring Survey Instruments." Final Report for Defense Nuclear Agency SBIR Contract No. DNA001-88-C-0161 (1989)

31. Kimock, F. M., Hoover, D. S., Garg, D., Dyer, P. N., Iampietro, R. L., Monk, V. A. and Tsai, W., *Proc. First International Symposium on Diamond and Diamond-Like Films* (J. P. Dismukes, et al., eds.), p. 536, Electrochemical Society, Pennington, NJ (1989)

32. Conner, L. C., Presented at *Gorham Advanced Material Institute International Conference on High Performance Inorganic Thin Film Coatings*, Monterey, CA (1988)

33. Hiroyuki, A. and Kondo, K., *Jap. J. of Appl. Phys.* 28(9):L1631 (1989).

34. Cappelli, M. A. and Paul, P. H., *J. Appl. Phys.* 67(5):2596 (1990).

35. Markunas, R. J., Rudder, R. A., Posthill, J. B. and Thomas, R. E., "Semiconductor Diamond Technology." SDIO/IST Contract No. N00014-86-C-0460 Quarterly Report, p. 23 (1989)

36. Ohtake, N., Tokura, H., Kuriyama, Y., Mashimo, Y., and Yoshikawa, M., *Proc. First International Symposium on Diamond and Diamond-Like Films* (J. P. Dismukes, et al., eds.), p. 93, Electrochemical Society, Pennington, NJ (1989).

37. TAFA, Inc. model 69 high power torch

38. Kawarada, H., Mar, K. S. and Hiraki, A., *Jap. J. Appl. Phys.* 26(6):L1032 (1987)

39. TAFA, Inc. model 69 high power torch

40. Harris, S. J. and Weiner, A. M., *Appl. Phys. Lett.*, 55(21):2179 (1989)

41. Hirose, Y. and Terasawa, Y., *Jap. J. Appl. Phys.* 25(6):L519 (1986)

42. Ohtake, N., Tokura, H., Kuriyama, Y., Mashimo, Y., and Yoshikawa, M., *Proc. First International Symposium on Diamond and Diamond-Like Films* (J. P. Dismukes, et al., eds.), p. 93, Electrochemical Society, Pennington, NJ (1989).

43. Tzeng, Y., Cutshaw, C., Phillips, R., Srivinyunon, T., Ibrahim, A. and Loo, B. H., *Appl. Phys., Lett.* 56(2):134 (1990)

44. Hirose, Y., *Abst. of First Int'l. Conf. on the New Diamond Sci. & Tech.*, p. 39, Japan New Diamond Forum (1988)

45. Harris, S. J. and Weiner, A. M., *Appl. Phys. Lett.*, 55(21):2179 (1989)

46. Kamo, M., Sayo, Y., Matsumoto, S. and Setaka, N., *J. Crystal Growth*, 62:642 (1983)

47. "Enhanced Oxidation Resistant Diamond Film Coatings," SBIR Phase II Contract No. F33615-90-C-5913 (August 1991)

48. U.S. Patent # 4,882,138

49. Deryagin, B. V., Fedoseev, D. V., Luk'yanovich, V. M., Spitsyn, B. V., Ryabov, V. A. and Lavrent'ev, A. V., *Soviet Physics-Doklady*, 13(8):783 (1969) and *Doklady Akademii Nauk SSSR*, 181(5):1094 (1969)

50. Deryagin, B. V., Builov, L. L., Zubkov, V. M., Kochergina, A. A. and Fedoseev, D. V., *Soviet Physics-Crystallography*, 14(3):449 and *Kristallografiya*, 14(3):535 (1969)

51. Pinneo, J. M., "CVD Diamond Coating for Fibers and Particulates in High Temperature Composites." DARPA SBIR Contract No. DAAH01-89-C-0782, DARPA, 1400 Wilson Blvd., Suite 111, Arlington, VA 22209-2308 (1990)

52. Ono, A., Baba, T., Funamoto, H. and Nishikawa, A., *Jap. J. App. Physics* 25(10):L808 (1986)

53. Herb, J. A., Bailey, C., Ravi, K. V. and Dennig, P. A. *Proc. First International Symposium on Diamond and Diamond-Like Films* (J. P. Dismukes, et al., eds.), p. 366, Electrochemical Society, Pennington, NJ (1989)

54. Kurihara, K., Sasaki, K. and Kawarada, M., *Fujitsu Scientific and Technical Journal*, 25(1):44 (1989)

55. Personal communication, data from Crystallume films, J. C. duBois, Thomson-CSF Laboratoire Central de Recherches, Orsay, France (1990)

56. Mundinger, D., Beach, R., Benett, W., Solarz, R., Krupke, W., Staver, R. and Tuckerman, D., *Appl. Phys. Lett.*, 53(12):1030 (1988).

57. Conner, L. C., Presented at *Gorham Advanced Material Institute International Conference on High Performance Inorganic Thin Film Coatings*, Monterey, CA (1988)

58. Herb, J. A. and Peters, M. G., "Diamond Membrane X-Ray Lithography Masks." Final Report for Naval Research Labs SBIR Contract No. N00014-89-C-2256 (1990) (to be submitted)

59. Albin, S., Cropper, A., Watkins, L., Byvik, C. E., Buoncristiani, A. M., Ravi, K. V. and Yokota, S., *Proceedings SPIE* (1988)

60. Personal communication, G. A. Gurtman, S-Cubed, San Diego, California (1989)

61. Lin, A. L., Stapelbroek, M. G., Ravi, K. V. and Peters, M. G., *Proc. First International Symposium on Diamond and Diamond-Like Films* (J. P. Dismukes, et al., eds.), p. 261, Electrochemical Society, Pennington, NJ (1989)

62. Rand, S. C. and DeShazer, L. G., *Proc. Int'l. Conf. on Lasers '84, Society for Optical & Quantum Electronics*, p. 543, STS Press, McLean, VA (1985)

63. Male, J. C. and Prior, J. R., *Nature* 186(4730):1037 (1960)

8

Electronic Applications of Diamond Films and Coatings

Kalyan Kumar Das

1.0 INTRODUCTION

Diamond, by virtue of its wide band gap, predicted high carrier mobilities, high breakdown fields, saturation velocity and thermal conductivity, is almost ideally suited as the material for devices which operate at high temperatures, voltages, power-levels, frequencies and high-radiation environments. The potential of diamond as a material for solid state devices has been the subject of a few excellent reviews (1)-(7). These reviews discuss the electronic material parameters of diamond and the simulated characteristics that can be obtained.

Simple basic experimental devices in diamond have been demonstrated, primarily in monocrystalline natural or high pressure synthetic wafers or homoepitaxial layers grown by chemical vapor deposition (CVD) on these wafers. Photodetectors, light emitting diodes, nuclear radiation detectors, thermistors, varistors and negative resistance devices in synthetic crystals have been demonstrated primarily by researchers in the Soviet Union (2). Several groups (8)-(12) have also demonstrated basic field effect transistor (FET) device operation in homoepitaxial diamond films and B doped layers on insulating single crystal diamond substrates. However, for wide application of diamond solid-state devices, high quality films on more commonly available substrates are essential. For certain applications insulating substrates would be a preferred choice. The device potential of polycrystalline films must also be further studied. So far only thermistors (13) and Schottky diodes (14)-(17) have been produced and characterized in the polycrystalline material.

Device fabrication involves a number of process steps. In addition to the growth of a high quality crystalline film, generally these steps include controlled doping of the films, as well as selected area doping with p and n type dopants, etching, formation of ohmic and rectifying contacts, either deposition or formation of dielectric films and passivation of the surface. In the following sections we review and discuss these processes as they have been employed by various researchers in their attempts to fabricate devices in diamond. The recent progress in device fabrication and the potential for additional structures are also described.

2.0 DOPING

The ability to introduce and activate dopants reproducibly and controllably, the achievements of understanding and characterization of the associated physical processes of compensation, carrier transport properties and the presence and activity of deep-level states, are of fundamental importance for an emerging device technology.

Boron has been identified as the dopant responsible for p-type behavior in naturally occurring semiconducting diamond (1). From geometric and energetic considerations B is probably the only element that can substitutionally dope diamond without generally distorting the lattice. The activation energy associated with B in diamond is 0.37 eV (1)(18). It has been incorporated into homoepitaxial and polycrystalline films during chemical vapor deposition (CVD) from either a B_2H_6 gas source or from various solid sources (17)(19)(20). For relatively small concentrations, the activation energy is close to the characteristic 0.37 eV, consequently only a partial activation of B is observed at room temperature. For highly doped (10^{19} - 10^{20} B cm^{-3}) films, almost total activation is achieved (17)(19)(20) with activation energies as low as 0.002 eV.

In addition to in situ doping of the CVD grown films, ion implantation offers an alternative technique for the introduction of electrically active impurities in a more precisely controlled manner. Experiments conducted by Vavilov et al. (21), Braunstein and Kalish (22), Prins (23), Sandhu and his coworkers (24), and de la Houssaye et al. (25) have shown promising results. Vavilov et al. (21) reported the observation of p-type electrical conductivity in B implanted diamond crystals annealed at temperatures between 800°C and 1200°C. They used ion doses of the order of 10^{14} cm^{-2} at

40 keV. Extensive radiation damage resulted in diamond implanted with a B dose exceeding ~10^{15}. This damage led to the formation of a surface layer of graphite upon annealing. Braunstein and Kalish (22) in their implantation doping experiments used a B dose of 1 x 10^{16}cm^{-2} at 40 keV. They annealed the implanted diamond substrate at a temperature of 1400°C. The resulting layer of graphite was removed by chemical etching. A portion of the implanted B distribution was retained in the substrate. This remaining distribution of B was sufficient to provide a highly conducting surface layer. They observed a positive Hall coefficient and a relatively low mobility of 0.5 cm^2 / Vs, and also that the Hall coefficient and the mobility were nearly independent of temperature. These observations suggested that the charge carriers were holes and the conduction mechanism was predominantly hopping. A different approach was used by Prins (23) and Sandhu et al. (24) to enhance the substitutional incorporation of B in diamond. These researchers introduced co-implantation of B and C into insulating single crystal diamond at liquid N_2 temperature. The implantation of C introduced vacancies at a depth that overlapped with the profile of the subsequently implanted (also at liquid N_2 temperature) B profile, thus the probability of substitutional incorporation of B through vacancy-interstitial recombination during subsequent annealing was enhanced. Prins (23) annealed his implanted samples initially at 500°C and subsequently at 1200°C in an argon ambient. Sandhu et al. (24) obtained annealing by heating directly from liquid N_2 temperature to 1100°C in a rapid thermal annealing system. Sandhu et al. observed a higher level of dopant activation by rapid thermal anneal as compared to a furnace anneal, although it is expected that a furnace anneal would be a more effective means for damage reduction. Implantation of B alone at liquid N_2 temperature has also been employed by de la Houssaye et al. (25) for producing doped semiconducting layers in insulating diamond crystals (25). De la Houssaye et al. used a multiple B implant scheme (25 keV, 1.5 x 10^{14}; 50 keV, 2.1 x 10^{14}; and 100 keV, 3.0 x 10^{14}), that was conducted at liquid N_2 temperature, to obtain an approximately uniformly doped active layer of ~2100 Å in thickness. A subsequent anneal at 990°C contributed to room temperature carrier concentration of ~5 x 10^{15}cm^{-3}, with a mobility of ~30 cm^2/Vs. The maximum allowable dose in these experiments is determined by the amount of damage which leads to graphitization.

For the flexibility in device design and operation, it is of great advantage to be able to introduce shallow n-type dopants in diamond, however n-type electrical activity has not been observed in natural or synthetic crystals.

Nitrogen, a commonly occurring impurity in diamond, contributes to donor-type centers with a low energy threshold for the photoconductivity at ~2 eV and a thermal activation energy between 1.6 and 1.7 eV (1). This level is electrically inactive, although it acts as a recombination center (1). Numerous efforts to dope diamond n-type by implanting various ions, such as Li, P, As, Sb, C, Kr, Xe, have been reported (26). In some cases observed n-type activity was attributed to residual damage. Following annealing at high temperatures the n-type behavior was not retained in these samples (21)(27)(28). A recent theoretical study by Kajihara et al. (29) suggests the potential of using P, Li and Na as shallow n-type dopants. Phosphorus is expected to occupy a substitutional site, while Li and Na should reside in interstitial positions. The solubility of these elements in diamond is low, therefore, kinetic trapping during ion implantation would be a preferred technique for introducing an excess concentration of these dopants in the diamond lattice. Phosphorus doped layers in diamond are expected to be stable due to its low diffusivity. Lithium diffuses rapidly in diamond even at room temperature and Na is expected to have an intermediate value of diffusivity and to be stable dopant to moderate temperatures.

Phosphorus doping of films during growth in a hot-filament deposition system has been reported recently by Okano et al. (30). They employed a solution of P_2O_5 in methanol which had been diluted in acetone. This solution was vaporized and used as the dopant source gas. They obtained a carrier concentration of 1×10^{15} cm^{-3} and a mobility of 50 cm^2/Vs. The activation energy of the conduction process was found to be 0.05 eV. Phosphorus doping was employed for the fabrication of p-n junction diodes (31). However, these authors were unable to control the film resistivity. They only obtained films with a resistivity of 100 Ω cm. This leaves some uncertainty regarding the controllability of n-type doping using P during thin film growth. Although ion implantation of P would be an attractive means for producing n-type activity, early results (21) indicate that P implanted n-type layers are stable only to 600°C.

Lithium ion implantation studies reported by Vavilov et al. (32) indicate that n-type activation was achieved. However, recent experiments (10)(33)(34) on Li doping studies by in-diffusion and in situ doping during growth have not produced desired results. In one such study (33) Li was introduced by in-diffusion into diamond films grown by the hot filament technique. The diffusion experiments were conducted in sealed quartz ampoules at temperatures between 400 - 900°C. In these experiments Li was not allowed to come in contact with the diamond films. Subsequently the presence of Li in the films was established by SIMS analysis. However, neither a major change in the electrical conductivity nor thermionic emission from the donor states was observed in these films. External photoemission experiments indicated that Li donors were compensated by a high density of acceptor states distributed over several electron volts in the band gap (33). The origin of these states has not been established.

Fountain et al. (10) reported Li doping of homoepitaxial films deposited in a plasma-enhanced CVD reactor with a solid dopant source of LiF in the growth environment. In some of the doping experiments they used the solid source to 'dose' the reactor prior to loading the substrate for diamond growth. The presence of both Li ~10^{21} cm^{-3} and B ~10^{18} cm^{-3} was established from SIMS analysis of these films. They found, from Hall effect measurements, the films to be p-type with a carrier concentrations between ~$2.0 \times 10^{16} cm^{-3}$ and $1.0 \times 10^{18} cm^{-3}$ and mobilities between 143 cm^2/Vs and 0.13 cm^2/Vs. The higher concentrations of Li in the films resulted in lower mobility values. Fountain et al. obtained an activation energy of 0.23 eV from the measurement of resistivity as a function of temperature. Schottky diodes fabricated on these films with an electron-beam evaporated Au film indicated good rectifying characteristics. The direction of current flow in these diodes also indicated a p-type conduction. The Li appeared to be electrically inactive in these films, although may have contributed to an improvement of the electrical characteristics. These improvements were generally higher mobilities for the lower concentrations of Li and the ease with which rectifying Au contacts could be fabricated. Rectifying contacts could not be fabricated on similarly grown B doped films without the presence of Li.

Lithium doping of diamond films was also reported by Beetz et al. (34). These researchers used Si substrates, an organometallic compound as the Li source and a filament assisted CVD system. Seebeck effect measurements indicated p-type conduction in these films. The presence of a large peak in photocurrent at 1.1 µm in the photoconductive excitation spectrum from these films was speculated to be a Li-vacancy complex.

Thus, the current state of understanding of n-type doping of diamond is far from clear, therefore the subject deserves a thorough and systematic study. In the case of implantation doping, donor activity of residual damage must be eliminated by suitable annealing. It is well known that a charge state associated with any defect can drastically modify electrical characteristics of a given semiconductor, including type conversion, by trapping carriers or contributing centers that could participate in the generation and recombination of carriers. In case of trapping or recombination, the carrier lifetime is reduced and the generation process would contribute to large reverse currents in the depletion region of a device. A high concentration of these defects is also expected to contribute to a rapid rise in the reverse current with temperature of a rectifying contact. These defects will also result in a general degradation in mobility. Both Hall effect and current-voltage (I-V) measurements as a function of temperature, capacitance-voltage (C-V), deep level transient spectroscopy (DLTS) and optical techniques should help in a complete characterization of the doped films.

3.0 CONTACTS

3.1 Effect of H Content in CVD Films on Contact Properties

Chemical vapor deposited diamond films invariably contain some amount of H since the deposition is conducted in a H plasma environment. Electrical effects of H and associated active-defects can significantly influence contact properties. Also, it is well known that the presence of H significantly modifies the resistivity of diamond films. As-deposited undoped films exhibit fairly low resistivities of the order of 10^6 Ω cm. Annealing of these films at temperatures as low as 400°C increases the resistivity by a few orders of magnitude (35)-(37). A subsequent exposure to an H_2 plasma reduces the resistivity close to the as-deposited value (35). A reduction in resistivity has also been observed in natural insulating (36) and semiconducting crystals (37) on exposure to a H plasma. The increase in resistivity in CVD films has been attributed to dehydrogenation resulting in the activation of deep traps with a consequent increase in resistivity (35)-(37). Using an elastic recoil detection (ERD) technique (38) small but measurable reduction in the H content was observed in annealed CVD samples (39). Presence of B in certain samples resulted in a reduction in H loss (39). The observed effects may be a result of complex interactions between H, other defects present in the film, the grain boundaries (40) and the intentionally introduced B during CVD growth.

Aluminum metal contacts on as-deposited B doped polycrystalline diamond films on Si substrates exhibit rectifying characteristics. Rectifying characteristics of these contacts can be improved by depositing a thin film of undoped diamond (8)(41) or a thin film of CVD SiO_2 (42) prior to metallization (see Sec. 3.4 on rectifying contacts). However, rectifying contacts cannot be formed, using direct metallization on the same films, following anneal at 950°C and a post-anneal treatment in a hot solution of $CrO_3 + H_2SO_4$ (see Sec. 3.2). Contacts at this stage exhibit highly resistive and only slightly asymmetric I-V characteristics. This observation is in conformity with earlier reported observations (35)-(37) of increase in resistivity of CVD grown diamond films with anneal. It is, therefore, believed that such an anneal treatment of as-deposited films is necessary in order to obtain stable electrical characteristics as demonstrated by measured high resistance values. Such 'stabilization' annealed films may permit the fabrication of contacts with reproducible characteristics and may also prove to be a suitable material for device fabrication. It is also speculated that an optimum concentration of B would contribute towards the passivation of some of the electrically active defects as well as providing carriers for the transportation process. There probably is a bulk, as well as a surface, component associated with the observed contact characteristics. A better understanding of the effects of H in the films and that of subsequent anneal treatment would certainly contribute to a better control of electrical parameters of CVD grown films and enhance their device potential.

3.2 Effect of Surface Preparation on Metal Contacts

Grot et al. (43) have shown that surface preparation plays an important role in determining the characteristics of metal contacts on the diamond surface. They observed that Au contacts on as-deposited homoepitaxial films were ohmic regardless of the B concentration and that the resistance measured by the four-point probe technique was not representative of the doped film. They attributed this to a nondiamond component present on the surface of the films. When these films were treated in order to remove any nondiamond carbon in a solution of CrO_3+H_2SO_4 at 170°C for 3 min, followed by a rinse in 1:1 boiling solution of H_2O_2 and 30% NH_4OH, rectifying characteristics were obtained. It was established that the exposure to the H plasma during deposition was responsible for the observed effect. Fabricated Au contact diodes on the treated films were rectifying to 580°C. These authors also found that both Al and Ti formed rectifying contacts when deposited on the as-deposited surface of both undoped and doped films. A better understanding of these observed effects must be obtained through the use of surface analytical techniques.

3.3 Ohmic Contacts

Traditionally, silver paint or a damaged surface (44) and W probe contacts modified by passing a high current (45) have been employed for establishing ohmic contacts on semiconducting diamond for performing electrical measurements. In some cases the painted contacts were annealed to reduce contact resistance (44). Indium has also been used (6) to make an ohmic contact on diamond; however this metal is unsuitable for high-temperature studies. Recently, photolithographically defined contacts of various high melting point metals have been investigated. Moazed et al. (46)(47) reported the formation of ohmic contacts on diamonds by using metals, e.g., Ti, Ta, Ni, and Mo, that form carbides upon annealing. In a typical application, 100 - 150 Å thick layer of Ti followed by a 1500 Å layer of Au was employed. The ohmic contact was obtained following an anneal at 885°C for 8 - 16 min in a hydrogen ambient. The surface roughness observed in the Au/Ta, Au/Ti, and Au/Mo contacts is indicative of carbide formation.

A Si/SiC/diamond multilayer structure has been investigated by Fang et al. (48) for its potential as an ohmic contact. These researchers deposited 500 Å of Si on cleaned diamond substrates. This was followed by the implantation of Kr^+ at an energy of 240 keV, dose and temperature of 2 x $10^{16}/cm^2$ and 700°C respectively, to achieve ion-beam mixing. The samples were then annealed at a temperature 1200°C for 1 hr in flowing forming gas. In some samples, they pre-implanted the diamond substrates with Kr^+ using the conditions of 150 keV, 2 x $10^{16}/cm^2$ and 700°C prior to the deposition of the 500 Å of Si. This implantation step caused the formation of an n-type layer. Following the Si deposition, the structure was annealed in the same manner as the previous case. The post-deposition ion-mixing contributed to a lower contact resistivity of 1 x 10^{-3} Ω cm^2 as compared to 5 x 10^{-3} Ω cm^2 for the pre-implanted structure.

Geis et al. (49) have established ohmic contacts to p-type diamond using surface modification induced by intense argon-fluoride excimer laser radiation. The modified layer was ~500 Å thick and composed of diamond and graphite-like material. It was also stable at high temperatures to 1800°C. Geis (6) also observed that ohmic contacts can be obtained on heavily B doped ~10^{21} cm^{-3} diamond with a deposited layer of each of several metals. Similar results have also been reported for polycrystalline diamond films (50).

Surface modification using other techniques has also been employed for the formation ohmic contacts on diamond surfaces. Childs, Sandhu and Chu (51)-(53) reported the formation of ohmic contacts by pre-deposition exposure to an Ar sputtering environment prior to the deposition of Au on semiconducting substrates. The pre-sputtering also enhanced the adhesion of the Au film. Normally Au forms rectifying contacts on B doped semiconducting diamond crystals.

Ohmic contact formation by high dose B implantation has been reported by Sandhu et al. (53) and Prins (54). Sandhu et al. conducted a selected-area B implantation at room temperature with 65 keV ions to a dose of 3 X 10^{16} cm^{-2} (53) to establish ohmic contact to previously implanted layers in insulating crystals (22)(23). This high-dose implantation step introduced extensive radiation damage leading to graphitization of the surface layer during a subsequent anneal treatment at 950°C in vacuum for 30 min. The graphitized surface was removed by boiling in a solution of $HClO_3$ + HNO_3 + H_2SO_4 (1:4:3). Tungsten probes on the remaining heavily doped film provided ohmic contact. It should be pointed out that Braunstein and Kalish (22), in their earlier work on B implantation-doping of diamond, were also able to establish ohmic contact to their highly conducing film with sharp W probes. Prins (54) obtained ohmic contact on semiconducting crystals using an implantation energy of 35 keV and a B ion dose of 3 X 10^{16} cm^{-2} at a substrate temperature of 200°C. Subsequent to annealing at 1200°C, the graphitized layer was removed and contact established with a film of metal, such as Ag, Cu, or Au. Metallization at a temperature of ~800°C with an alloy of Ag, Cu, and In provided a strong mechanical bond.

A similar study has also been conducted (55) on polycrystalline films deposited on Si and SiAlON. These films were polished (56) prior to ion implantation. In this study, a B dose of 3.5 x 10^{16} cm^{-2} at 60 keV was employed. An optimum substrate temperature of 100°C was determined empirically. Boron profiles, as determined by Secondary Ion Mass Spectroscopic (SIMS) analysis, in the as-implanted state and that obtained subsequent to anneal and graphite removal are shown in Fig. 1. A surface concentration of B of ~7 x 10^{20} cm^{-3} in the diamond filmwas achieved. Using a conventional TLM pattern a specific contact resistance of 3.7 x 10^{-3} Ω cm^2 was obtained with an as-deposited film of Au. With a Au/Ti metallization and subsequent anneal, at a temperature of 850°C and a pressure of 10^{-6} Torr for 30 min, a specific contact resistance of the order of 10^{-6} Ω cm^2 was obtained.

3.4 Rectifying Contacts

Rectifying contacts have been reported for naturally occurring semiconducting crystals (57), synthetic B doped crystals (18) and CVD grown homoepitaxial (43)(58) and polycrystalline films (14)-(17)(59). These contacts have been formed by using point probes (15)(59) or by evaporation of films of various metals (14)-(16)(18)(43)(58), e.g., Al and Au. Rectifying behavior with sputtered platinum on polycrystalline films has also reported by Nishimura et al (17). High breakdown voltages have been reported for polycrystalline films (16). Shiomi et al. (8) have demonstrated that by placing a rectifying contact on an undoped diamond film on a previously deposited doped film, the reverse leakage current is substantially reduced and the breakdown voltage improved. Similar observations have also been reported for polycrystalline films (41). Introduction of a thin (~20 Å) plasma enhanced CVD (PECVD) SiO_2 film has also been used for improving rectifying characteristics of metal contacts on both polycrystalline and homoepitaxial films (42). These improvements in characteristics are probably due to a modification in the barrier height and field distribution and/or to passivation of the surface. Shiomi et al. (8) employed an undoped diamond film deposited on a doped film for the fabrication of the gate structure in their MESFETs.

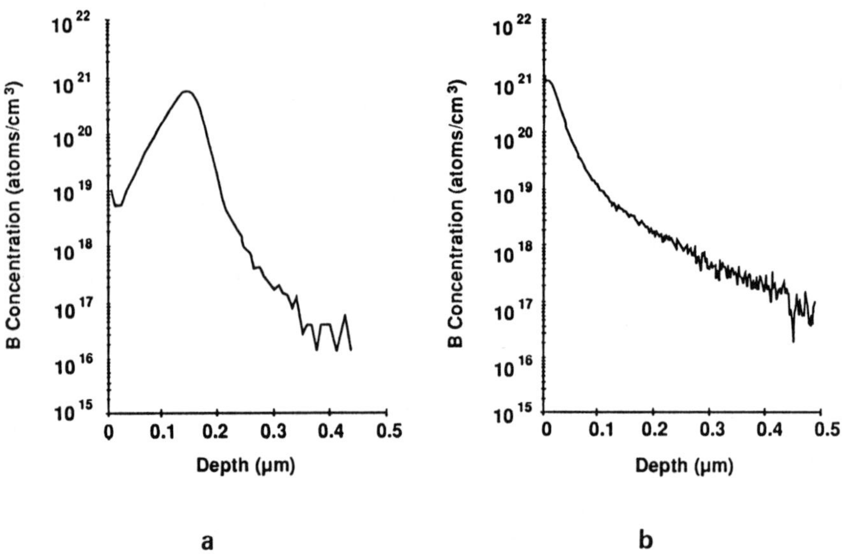

Figure 1. Boron profile as determined by SIMS. A B dose of 3.5×10^{16} cm^{-2} at 60 keV was used. A substrate temperature of 100°C was maintained during implantation. (a) As-implanted profile. (b) Profile following anneal at 1200°C and graphite removal.

More recently, Humphreys et al. (60) have demonstrated that deposited heteroepitaxial films of Ni on semiconducting single crystal diamond exhibit excellent rectifying contacts both at room temperature and 400°C, as shown in Fig. 2. With Ni films deposited at a temperature of 500°C, for a 625 μm diameter dot, they observed a reverse leakage current of 2 nA at a bias of 20 V for a room temperature measurement. At a measurement temperature of 400°C the leakage current increased only to 4 nA, which corresponds to a current density of 5 x 10^{-6} A/cm^2. Moreover, as a result of strong interface bonding, the Ni films exhibit excellent adhesion properties with the underlying diamond substrate. In contrast, metal films deposited at room temperature, e.g., Au, have a tendency to peel from the diamond surface during processing (61).

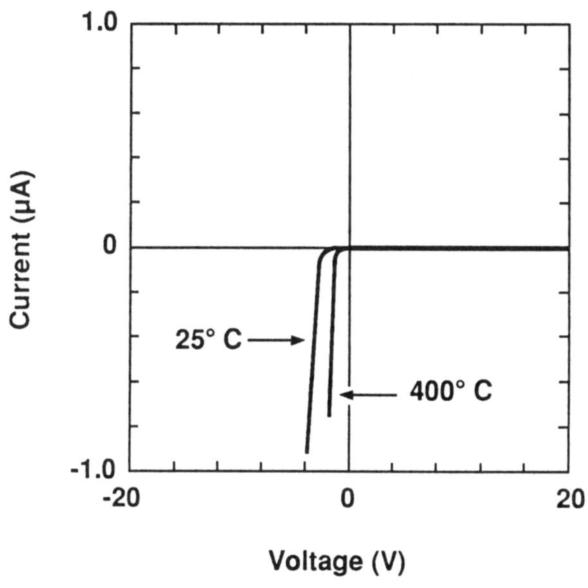

Figure 2. Current-voltage characteristics of Ni contacts on a semiconducting diamond crystal (Ref. 60).

Contacts formed with $TiSi_2$ deposited by co-evaporation of Ti and Si have also been studied by Humphreys et al. (62). These contacts were nominally rectifying in the as-deposited state. However, following an anneal treatment at a temperature of 1100°C the rectifying properties improved significantly. These diodes also showed rectification at a measurement temperature of 400°C. However, the leakage currents in these devices were significantly higher than that of the Ni diodes. Excellent rectifying contacts on semiconducting single crystal diamond have also been achieved using a composite film of co-sputtered Ta and Si (63). At room temperature these contacts showed a reverse leakage current two orders of magnitude lower than the Ni contacts, however, at a measurement temperature of 400°C these had approximately three orders of magnitude higher reverse leakage current. For both $TiSi_2$ and Ta/Si films excellent adhesion on the diamond substrate was reported.

Barrier height measurements for Au and Al contacts on single crystal diamond have been reported by a number of authors (18)(64)(65). Most of these studies were conducted on the (111) surface of diamond (63)(64). Both (C-V) and internal photoemission spectroscopy techniques were used. Glover (19) obtained from C-V measurements a barrier height of 1.7 eV for Au on the (100) surface of a chemically cleaned synthetic B doped diamond. Mead and McGill (64) obtained 1.7 - 2.0 eV for Au and 1.9 - 2.0 eV for Al on chemically cleaned semiconducting diamond. The lower values were obtained from C-V measurements whereas the higher values were from internal photoemission spectroscopy. These measurements probably were taken from the (111) surface of natural crystals. Barrier heights reported by Himpsel et al. (65) were significantly different from the above values. They reported barrier heights of 1.3 eV for Au and 1.5 eV for Al employing electron spectroscopy with synchrotron radiation. These results were obtained from the vacuum annealed (111) - (1x1) surface of B doped type IIB diamond. Himpsel et al. pointed out that their barrier heights were closer to a second threshold in the internal photoemission measurements of Mead and McGill (64) corresponding to barrier heights of 1.25 and 1.0 eV for Au and Al, respectively.

Hicks et al. (66) have measured barrier heights of 1.13 eV for both Al and Au contacts on polycrystalline diamond films, using internal photoemission spectroscopy. Beetz et al. (34) performed internal photoemission experiments on Au contacts on CVD diamond to obtain two photoresponse thresholds, one at ~2 eV and the other at ~1.2 - 1.3 eV. It should be noted that barrier heights are difficult to determine from I-V or C-V data on these diamond films since the contacts deviate widely from ideal behavior as a result of the nature of the surface (see Secs. 3.1 and 3.2), the presence of electrically active deep-states and high bulk resistivity. Therefore, internal photoemission and other non-electrical techniques should be preferred for the determination of barrier heights of metal contacts on diamond films. Any pretreatment that affects the nature of the surface, e.g., its deposition history, passivation and the presence of adsorbates, will have an effect on barrier heights, saturation currents, reverse leakage, and device performance. In order to develop a better understanding, surface analytical studies must be conducted in conjunction with terminal electrical characterization of potential device structures.

4.0 ETCHING OF DIAMOND

Successful dry etching of diamond has been obtained by Efremow et al. (67) using a 2 keV Xe^+ ion beam and a directed flux of NO_2 molecules. The investigators speculated that the energetic Xe^+ ions catalyzed a reaction between adsorbed NO_2 at the surface and the underlying diamond forming CO and CO_2. An etch rate of 2000 Å/min was obtained for a substrate temperature between 0°C and 20°C.

Kobashi et al. (68) reported an electron-beam assisted etching technique employing a self-focused, intense electron beam enclosed by a plasma. These researchers employed a mixture of H_2, O_2 and He as well as H_2 and O_2. Selective etching was obtained using a 0.4 µm film of gold as a mask. A substrate temperature of ~800°C was obtained as a result of plasma heating.

Recently, an electron cyclotron resonance (ECR) plasma technique has been employed by Beetz and Lincoln (69) for etching diamond. The substrates were immersed directly into the ECR plasma and the reactive gases such as O_2 and NO_2 were employed in these experiments. Line features as small as 0.3 µm were delineated using a thin film mask of deposited Au. An etch-rate of several hundreds of Å/min was achieved.

These achievements suggest the feasibility of fabricating submicron devices in diamond. It should be emphasized that in developing a diamond device technology that realizes the full potential of the material, close attention must be paid to current concepts involved in the design and processing of small geometry silicon or GaAs devices.

5.0 DIAMOND SiO_2 INTERFACE

The SiO_2-diamond interface has been studied in detail by Geis et al. (70). The properties of this interface strongly depend on the cleaning procedure, the SiO_2 deposition temperature and the orientation of the diamond surface. For the cleaning procedure, these authors employed a plasma ashing technique with a 2% mixture of O_2 in He followed by a treatment in a solution of $(NH_4)_2S_2O_8$ in H_2SO_4 at 200°C and separate rinses in water and acetone. The SiO_2 was deposited at a temperature of 450°C from a mixture of SiH_4, Ar, N_2 and O_2. The electrical properties of the interface were obtained from measured capacitance data obtained between the semiconducting diamond and the Al electrodes deposited on the SiO_2. The measured C-V plot and a lower estimate of the electronic interface states at the SiO_2 and (100) oriented diamond surface as a function of energy in the band gap are shown in Fig. 3. These plots indicate that a device quality interface was obtained and that surface passivation and MOS device operation may be obtained with deposited oxide films.

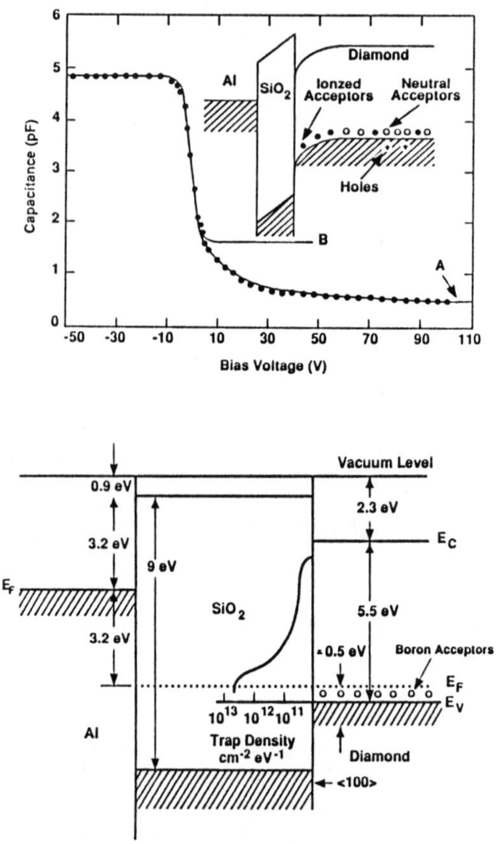

Figure 3. Metal/SiO_2/semiconducting diamond MOS system. (a) Capacitance-voltage characteristics (b) Distribution of traps in the band-gap. *Reprinted after Geis et al. (Ref. 70). By permission of IEEE (1991, IEEE).*

6.0 DEVICE STRUCTURES

The metal-semiconductor contact diodes reported in diamond probably cannot be considered 'devices' as such. These are essentially vehicles for assessment of the electrical quality of the material. However, the quality of these diodes in some cases is sufficiently high to warrant further development as high temperature rectifiers. Further development is also necessary if these structures are to be used as a gate element in Metal Semiconductor Field Effect Transistor (MESFET) structures. Due to surface related problems, all the FET devices reported to date (8)-(11), with the exception one (12), use an insulator or dielectric film between the metal gate and the channel region.

Probably the first reported three-terminal device was the bipolar transistor reported by Prins (44). This device was fabricated in a p-type semiconducting natural crystal substrate. The n-type areas of the structure were fabricated by using a radiation damage 'doping' effect obtained via carbon implantation. The device clearly demonstrated bipolar transistor action in diamond, although the current gain was very low. The schematic diagram of the structure and the electrical characteristics of the device are shown in Fig. 4.

The operation of a point contact transistor at 510°C was demonstrated by Geis et al. (45) in a synthetic B doped diamond crystal. The device used W probe point contacts as the emitter and the collector. The base contact was established by evaporating W on the side-faces of the crystal containing a high concentration of B. The schematic structure of the device and the reported characteristics are shown in Fig. 5. This device exhibited a power gain even at the highest operating temperature.

Geis et al. (71) also reported a vertical channel FET fabricated in a natural semiconducting diamond substrate. A schematic diagram of the device and its characteristics are shown in Fig. 6. This device had a transconductance of 30 µS/mm. Reduction in the high parasitic resistance associated with the substrate is expected to further improve the transconductance. Optimization of the design and the fabrication processes in diamond is expected to result in a device which can sustain a higher voltage and have a much higher cut-off frequency than in Si.

Gildenblat et al. (9) reported a FET device using a 1000 Å sputter-deposited film of SiO_2 as the gate dielectric. This device with a channel length and width of 60 µm, and 200 µm, respectively, was fabricated on a 1000 Å thick selected area B doped homoepitaxial diamond film grown by microwave plasma CVD on (001) insulating natural type Ia substrate. An Au/Ti metallization scheme was used for source and drain ohmic contacts with Au as the gate contact metal. The schematic structure of the device and the I-V characteristics measured at 300°C are shown in Fig. 7. A selected-area deposited Hall bridge pattern yielded a resistivity of 216 Ω cm, a hole concentration of 1 X 10^{14} cm^{-3}, and a mobility of 290 cm^2/Vs for representative films.

Shiomi et al. (8) recently reported the characteristics of a circular geometry MESFET (MISFET) device having a 20 µm gate-length. Schematics of the plan-view and the cross-section as well as the characteristics of this device are shown in Fig. 8. Synthetic type Ib substrates were used for the fabrication of these devices. Initially, a 0.90 µm diamond film with a B concentration of 1 x 10^{19} cm^{-3} was deposited via microwave plasma CVD. This was followed by the deposition of a 0.23 µm thick undoped diamond film. Titanium metal was deposited through a metal mask for the formation of ohmic contacts. The gate area was delineated photolithographically following the deposition of Al gate metal. The device structure was completed by etching Al to form circular gate areas. The drain current tended to saturate at a drain voltage of -50 V. The device had a transconductance of 2.0 µS/mm.

The fabrication and characteristics of a depletion mode diamond MESFET was reported by Tsai et al. (12). The device was fabricated on a type IIa insulating diamond crystal. The active channel region was formed by solid-state diffusion of B from a cubic BN source by rapid thermal processing (RTP) for 30 - 60 s at a temperature of 1400°C in an Ar ambient. The resulting B profile as obtained by SIMS is shown in Fig. 9. Annealed Au/Ti source/drain contacts at a separation of 700 µm and an as-deposited Ti/Au gate 150 µm in length and 1 mm in width were employed. A modulation of the drain current was obtained by applying a gate bias and pinch-off was observed at a high positive gate bias. The schematic and the characteristics of the device are shown in Fig. 10. An open channel current of 1.6 µA/mm at a drain bias of 10 V and a transconductance of 0.7 µS/mm were obtained.

An insulated gate field effect transistor was fabricated by Fountain et al. (10) using selectively grown B doped homoepitaxial films on type IA natural crystals. A RPECVD SiO_2 film was used as the gate dielectric. Titanium was used both for gate and source/drain metallization. As-deposited Ti provided ohmic contacts. The device had a gate length of 8 µm and width of 50 µm. The device could not be pinched-off and a transconductance of 38 μS/mm was obtained. Devices fabricated on the Li/B doped films showed better characteristics than those fabricated on B doped films (see Sec. 2), as shown in Fig. 11.

Zeisse et al. (11) reported an FET with an implantation doped active layer in a type IIa substrate and a deposited SiO_2 gate insulator. They employed a multiple B implant scheme (25 keV, 1.5×10^{14}; 50 keV, 2.1×10^{14}; and 100 keV, 3.0×10^{14}), that was conducted at liquid N_2 temperature, to obtain an approximately uniformly doped active layer of ~2100 Å in thickness. A subsequent anneal at 990°C contributed to room temperature carrier concentration of $\sim 5 \times 10^{15}$ cm^{-3}, with a mobility of ~30 cm^2/Vs. A bilayer Mo/Au metallization was used as source/drain contacts and Ti/Au as the gate contact. The device structure and the reported device characteristics obtained at room temperature are shown in Fig. 12. Current saturation and pinch-off was observed in these devices; a transconductance of 6.9 μS/mm was reported.

One of the primary reasons for the poor characteristics of the above mentioned devices, apart from the material quality of the currently grown films, is that the device design and process concepts used in their fabrication are yet to be optimized. Major improvements are expected with short gate lengths of the order of a micron and below. Reduction of source-channel and drain-channel spacings also will improve characteristics further by reducing series resistance of the device. Reduction of these resistances would lead to a better power handling capability and high frequency response.

High electric fields are required for avalanche multiplication in diamond, this attribute coupled with its high saturated carrier velocities, provide the potential for IMPATT operation. In conjunction with a high thermal conductivity and higher temperatures at which the junctions can be operated, diamond IMPATTS should be capable of handling much higher power at high frequencies. Simulation of this device has been reported by Trew et al. (7). The results predict a capability of producing 10 W of RF power with a 23% efficiency at 30 - 40 GHz. This represents a factor of five improvement in the power capability in comparison to IMPATT devices in Si and GaAs, although the conversion efficiencies are essentially similar. The RF performance is expected to degrade above 100 GHz and will be comparable to Si devices. However, IMPATT operation has not been demonstrated in diamond. Device demonstration would necessitate a better understanding of avalanche processes in diamond as well as the fabrication of optimized device structures.

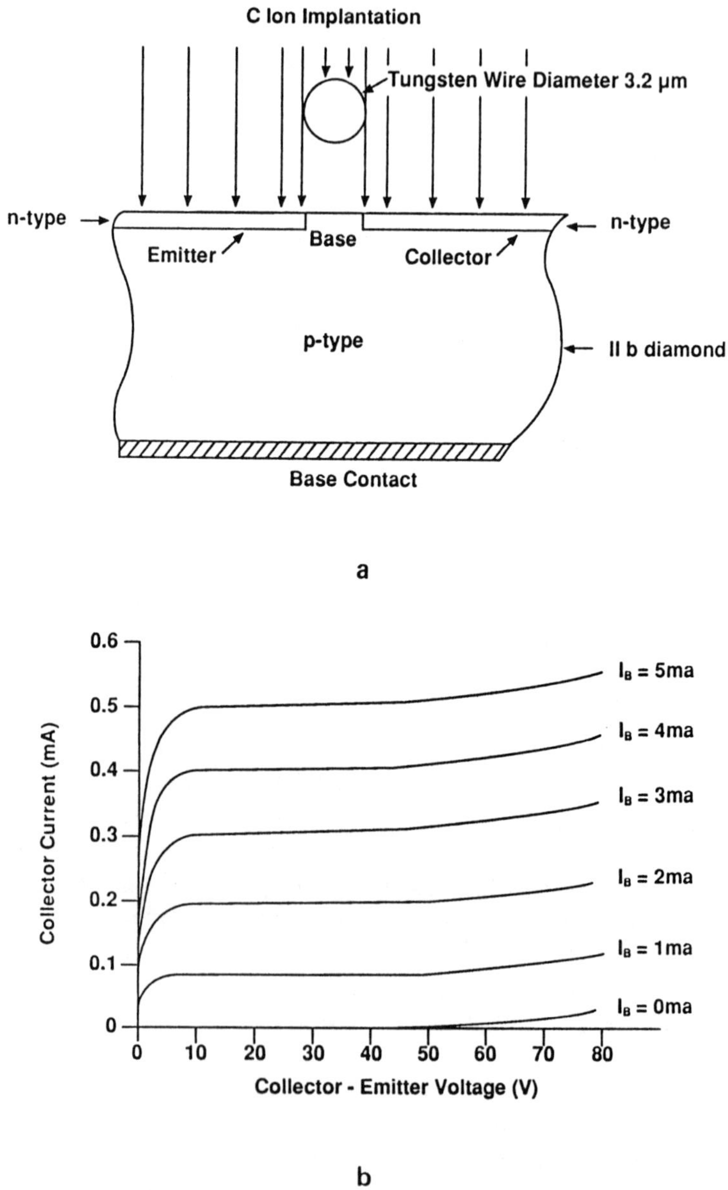

Figure 4. Implanted diamond bipolar transistor. Radiation damage effect of multiple energy C ion implantation was employed for n-type doping. The diamond substrate was held at 320°C during implantation. (a) Schematic of the fabrication process. (b) Current-voltage characteristics. *Reprinted after Prins (Ref. 44). By permission of AIP and the author (1982, IEEE).*

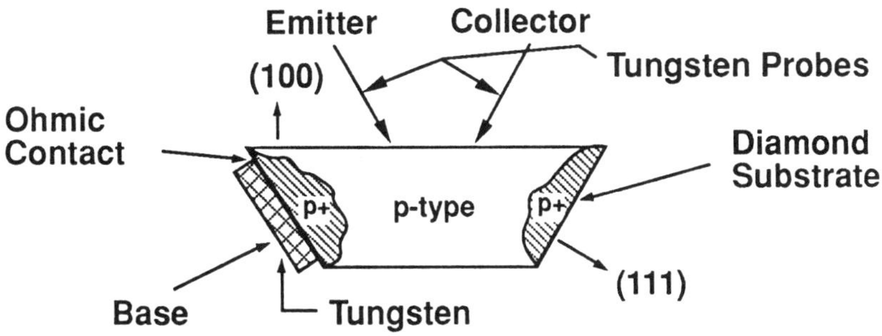

(a)

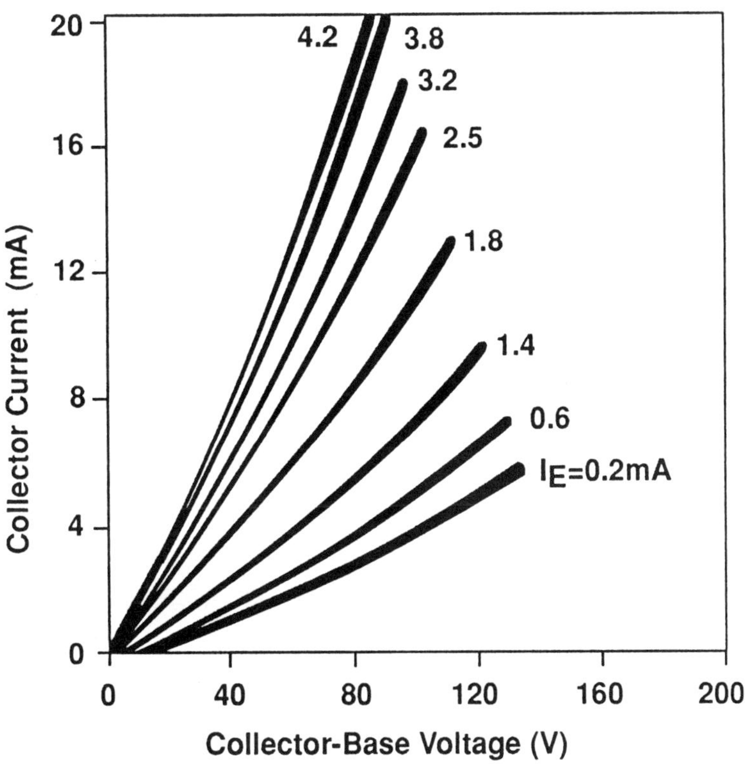

(b)

Figure 5. Diamond point-contact transistor. (a) Schematic device structure. (b) Current-voltage characteristics. *Reprinted after Geis et al. (Ref. 45). By permission of IEEE (1987, IEEE).*

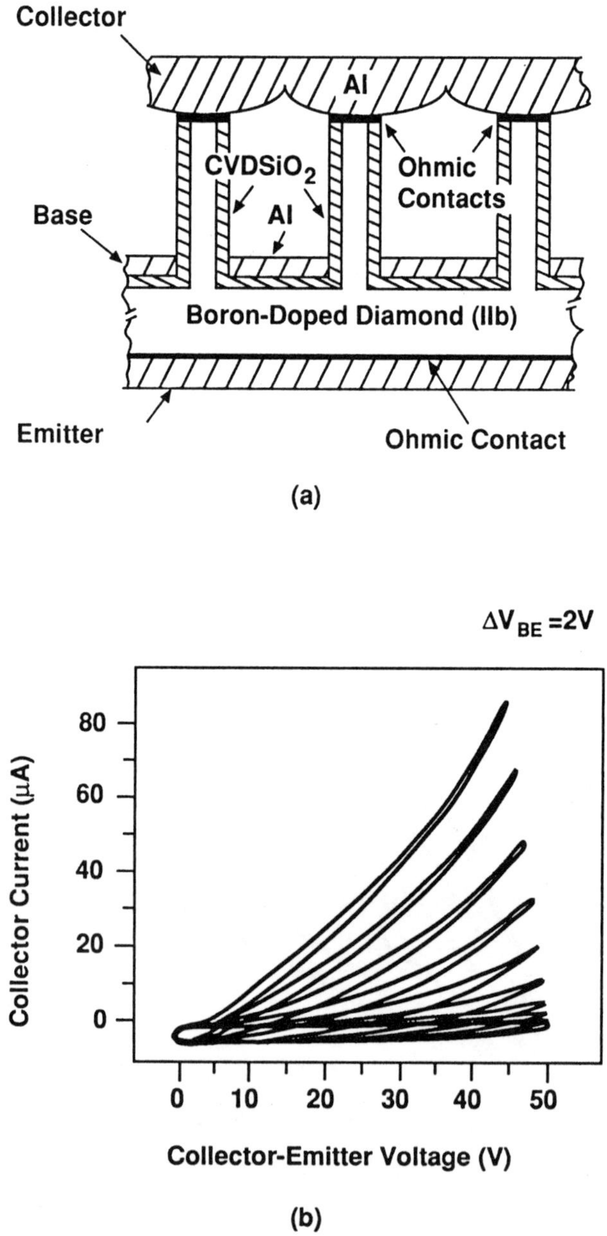

Figure 6. Diamond vertical FET. (a) Schematic device structure. (b) Current-voltage characteristics. *Reprinted after Geis et al. (Ref. 71). By permission of AIP and the authors (1988, AIP).*

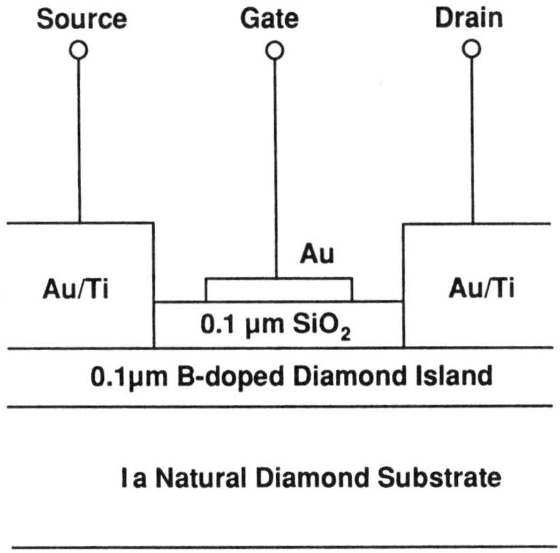

(a)

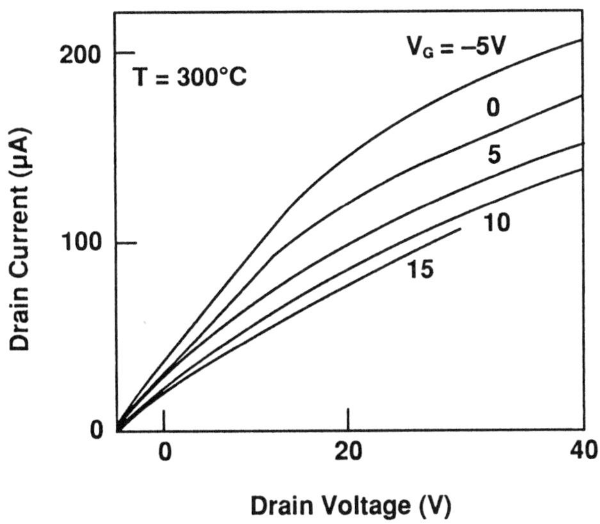

(b)

Figure 7. Insulated gate FET fabricated in selected area homoepitaxial film. (a) Schematic device structure. (b) Current-voltage characteristics obtained at 300°C. *Reprinted after Gildenblat et al. (Ref. 5). By permission of IEEE (1991, IEEE).*

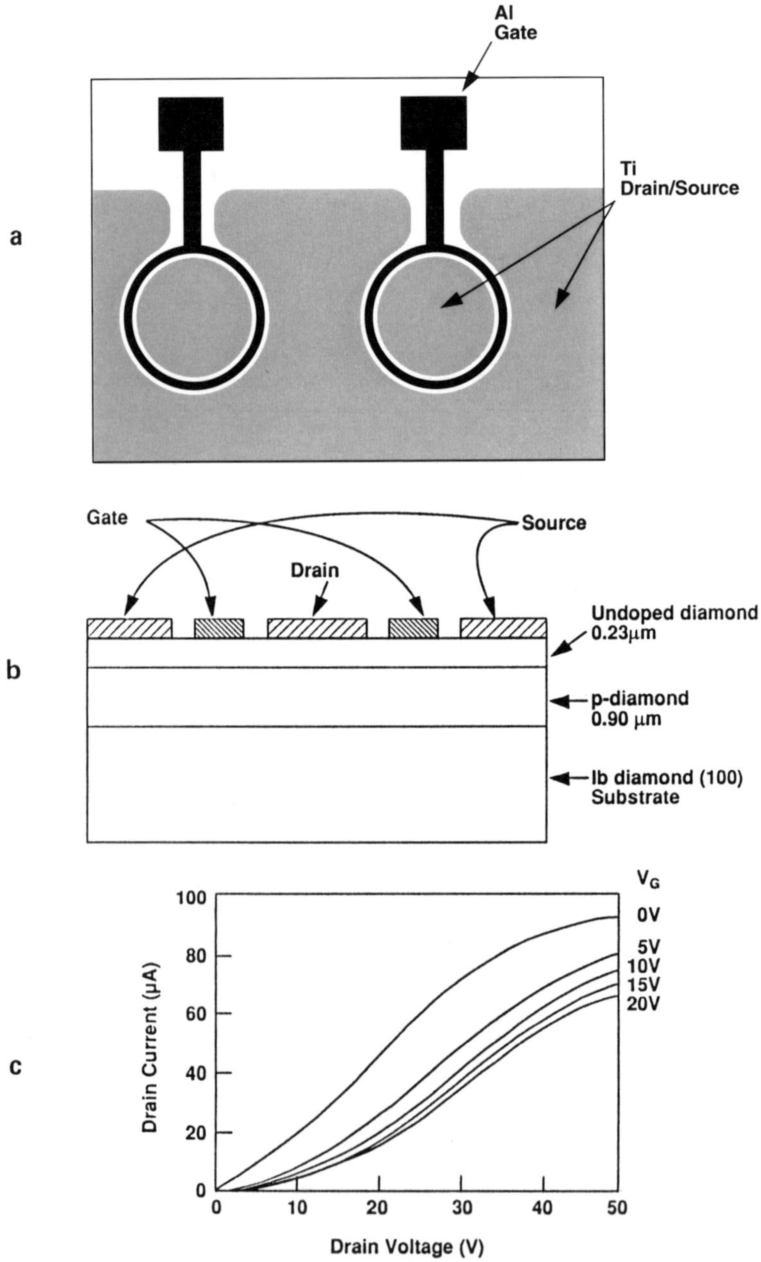

Figure 8. Metal semiconductor FET fabricated on undoped / B doped homoepitaxial diamond film. (a) Device plan-view. (b) Metal/undoped/B doped diamond homoepitaxial film device structure. (c) Current-voltage characteristics. *Reprinted after Shiomi (Ref. 8). By permission of the authors.*

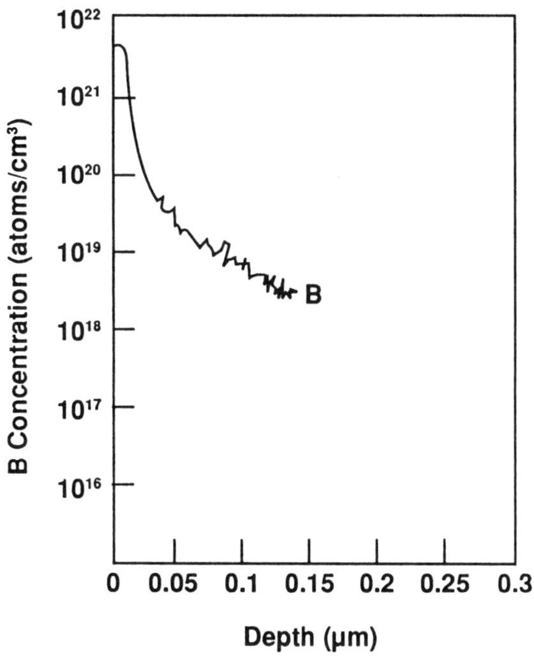

Figure 9. Thermally diffused B profile in type IIa diamond crystal as obtained by SIMS. *Reprinted after Tsai et al. (Ref. 12). By permission of IEEE (1991, IEEE).*

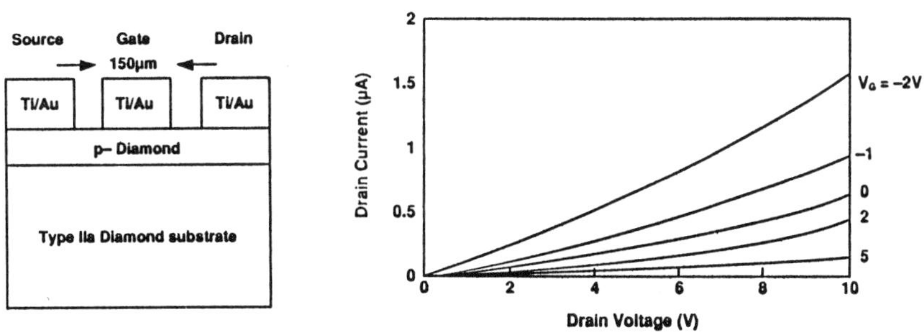

a b

Figure 10. Metal semiconductor FET fabricated in a thin B diffusion doped layer on a type IIa diamond substrate. (a) Device structure. (b) Current-voltage characteristics. *Reprinted after Tsai et al. (Ref. 12). By permission of IEEE (1991, IEEE).*

a

b

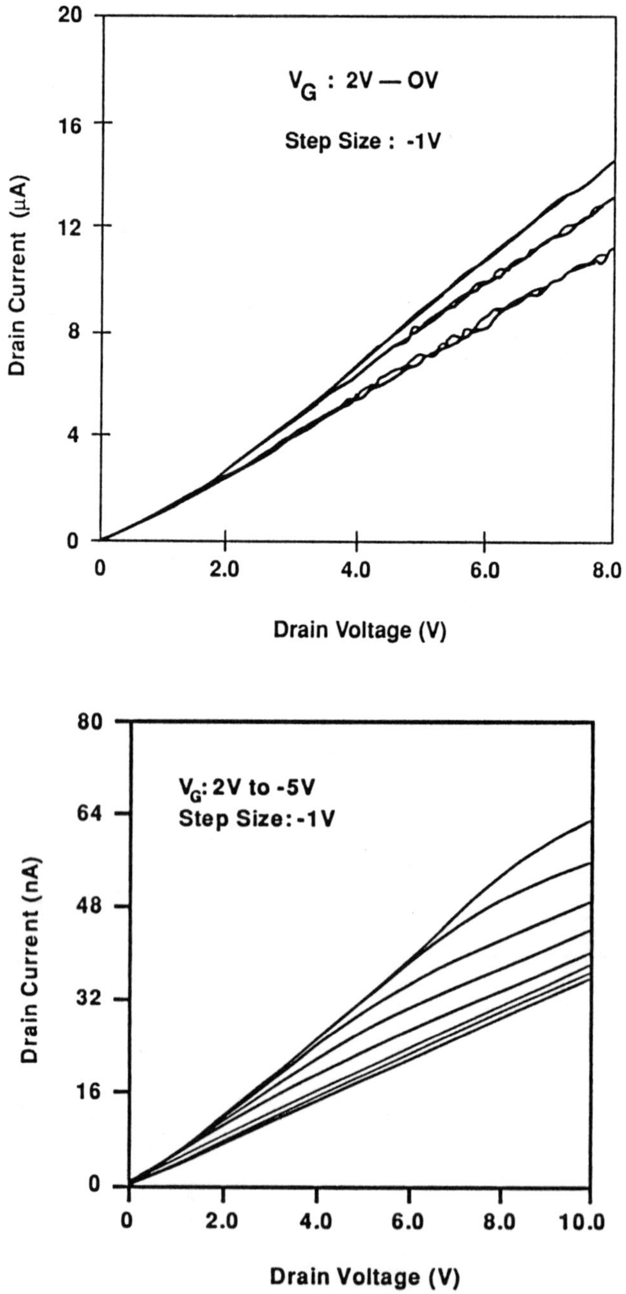

Figure 11. Current-voltage characteristics of insulated gate FETs fabricated on selectively grown homoepitaxial films (Ref. 10). (a) Boron doped film. (b) Lithium/B doped film.

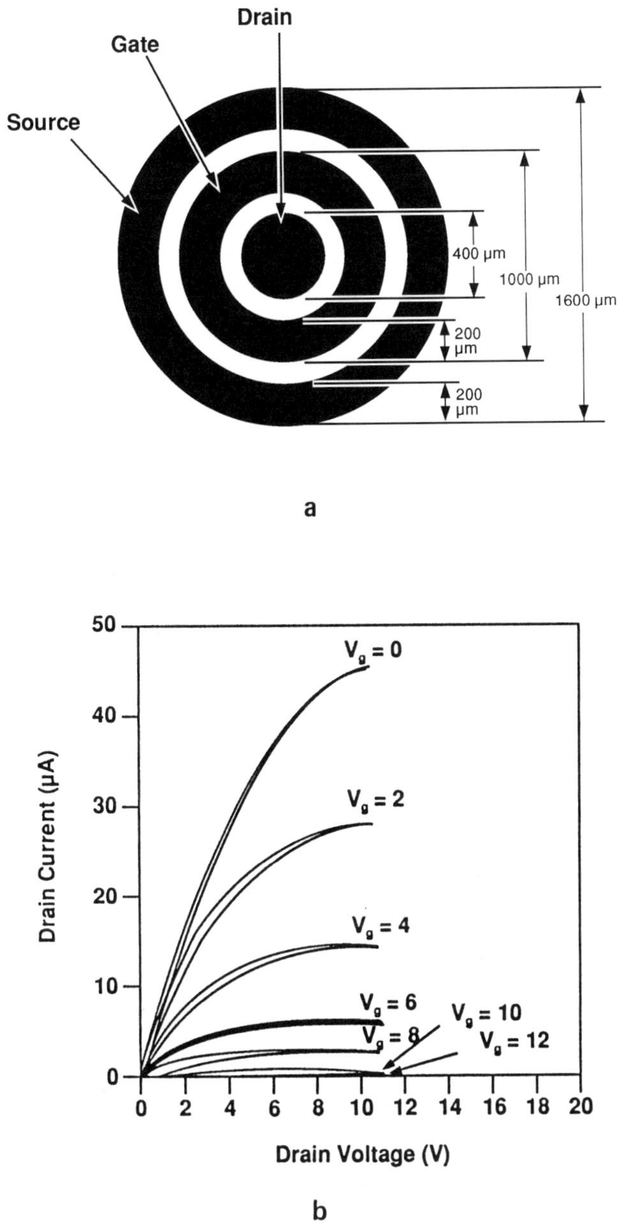

Figure 12. Insulated gate FET with the active layer formed by multiple B implantation. (a) Schematic device structure. (b) Current-voltage characteristics obtained at room temperature. *Reprinted after Zeisse et al. (Ref. 11). By permission of IEEE (1991, IEEE).*

7.0 CONCLUSIONS

In order to be able to take full advantage of the unique characteristics of diamond as a material for the fabrication of solid state devices, basic scientific understanding of the various experimental process techniques must be obtained. Areas of specific need include: introduction and activation of dopants, development of ohmic contacts with low contact resistivity, surface passivation, etching, and selected area processing. Attention also needs to be paid to proper design of devices incorporating novel features utilizing concepts and practices established in Si and GaAs device technology.

REFERENCES

1. Collins, A. T. and Lightowlers, E. C., in: *The Properties of Diamond,* (J. E. Field, ed.), Academic Press, San Diego (1979)
2. Bazhenov, V. K., Vikulin, I. M. and Gontar, A. G., *Sov. Phys. Semicond.,* 19:829 (1985)
3. Collins, A. T., *Semicond. Sci. Technol.,* 4:605 (1989)
4. Shenai, K., Scott, R. S. and Baliga, B. J., *IEEE Trans. Electron Dev.,* 36:1811 (1989)
5. Gildenblat, G. S., Grot, S. A. and Badzian, A., *Proc. IEEE.* 79:647 (1991)
6. Geis, M. W., *Proc. IEEE,* 79:669 (1991)
7. Trew, R. J., Yan, J. B. and Mock, P. M., *Proc. IEEE,* 79:598 (1991)
8. Shiomi, H., Nishibayashi, Y., and Fujimori, N., *Proc. 2nd Intl. Conf. New Diamond Sci. Tech.,* (R. Messier, J. T. Glass, J. E. Butler, and R. Roy, eds.), p. 975, Mat. Res. Soc., Pittsburgh (1991)
9. Gildenblat, G. S., Grot, S. A., Hatfield, C. W. and Badzian, A. R., *IEEE Electron Dev. Lett.,* 12:37 (1991)
10. Fountain, G. G., Rudder, R. A., Malta, D. P., Hattangady, S. V., Alley, R. G., Hudson, G. C., Posthill, J. B., Markunas, R. J., Humphreys, T. P., Nemanich, R. J., Venkatesan, V. and Das, K., presented at the Electrochem. Soc. Meeting, Washington, D. C., (May 6 - 10, 1991)

11. Zeisse, C. R., Hewett, C. A., Nguyen, R., Zeidler, J. R. and Wilson, R. G., *IEEE Electron Dev. Lett.,* 12:602 (1991)

12. Tsai, W., Delfino, M., Dodul, D., Riaziat, M., Chiang, Y., Reynolds, G. and Cooper, C. B. III, *IEEE Electron Dev. Lett.,* 12:157 (1991)

13. Nakahata, H., Imai, T., Shiomi, H., Nishibayashi, Y. and Fujimori, N., *Proc. 1st Intl. Conf. New Diamond Sci. Tech.,* (S. Saito, O. Fukunaga, and M. Yoshikawa, eds.), p. 285, KTK Scientific Publishers/Terra Scientific Publishing Company, Tokyo (1990)

14. Gildenblat, G. S., Grot, S. A., Wronski, C. R., Badzian, A. R., Badzian, T. and Messier, R., *Appl. Phys. Lett.* 53:586 (1988)

15. Mori, Y., Kawarada, H. and Hiraki, A., in *Diamond, Silicon Carbide and Related Wide Bandgap Semiconductors,* (J. T. Glass, R. Messier, and N. Fujimori, eds.), Mat. Res. Soc. Symp. Proc. 162:353 (1990)

16. Jeng, D. G. K., Tuan, H. S., Salat, R. F. and Fricano, G. J., *Proc. 2nd Intl. Conf. Electronic Mat.,* (R. P. H. Chang, M. Geis, B Meyerson, D. A. B. Miller, and R. Ramesh, eds.), p. 577, Mat. Res. Soc., Pittsburgh (1990)

17. Nishimura, K., Das, K. and Glass, J. T., *J. Appl. Phys.,* 69:3142 (1991)

18. Glover, G. H., *Solid St. Elect.,* 16:973 (1973)

19. Fujimori, N., Imai, T. and Doi, A., *Vacuum,* 36:99 (1986)

20. Geis, M. W., in: Diamond, *Silicon Carbide and Related Wide Bandgap Semiconductors,* (J. T. Glass, R. Messier, and N. Fujimori, eds.), 162:15, Mat. Res. Soc. Symp. Proc. (1990)

21. Vavilov, V. S., Guseva, M. I., Konorova, E. A. and Sergienko, V. F., *Sov. Phys. Semicond.,* 4:6 (1970)

22. Braunstein, G. and Kalish, R., *J. Appl. Phys.,* 54:2106 (1983)

23. Prins, J. F., *Phys. Rev. B,* 38:5576 (1988-I)

24. Sandhu, G. S., Swanson, M. L. and Chu, W. K., *Appl. Phys. Lett.* 55:1397 (1989)

25. de la Houssaye, P. R., Penchina, C. M., Hewett, C. A., Wilson, R. G. and Zeidler, J. R., *Proc. 2nd Intl. Conf. New Diamond Sci. Tech.,* (R. Messier, J. T. Glass, J. E. Butler, and R. Roy, eds.), p. 1113, Mat. Res. Soc., Pittsburgh (1991)

26. Vavilov, V. S., *Radiation Eff.,* 37:229 (1978)

27. Vavilov, V. S., Gukasyan, M. A., Konorova, E. A. and Milyutin, Y. V., *Sov. Phys. Semicond.,* 6:1998 (1973)

28. Vavilov, V. S., Gukasyan, M. A., Guseva, M. I., Karatygina, T. A. and Konorova, E. A., 8:471 (1974)

29. Kajihara, S. A., Antonelli, A. and Bernholc, J., in: Diamond, Silicon Carbide and Related Wide Bandgap Semiconductors, (J. T. Glass, R. Messier, and N. Fujimori, eds.), 162:315, Mat. Res. Soc. Symp. Proc. (1990)

30. Okano, K., Kiyota, H., Iwasaki, T., Nakamura, Y., Akiba, Y., Kurosu, T., Iida, M. and Kakamura, T., *Appl. Phys.* A 51:1731A (1990)

31. Okano, K., Kiyota, H., Iwasaki, T., Nakamura, Y., Akiba, Y., Kurosu, T., Iida, M. and Kakamura, T., to be published Solid St. Elect.

32. Vavilov, V. S., Gukasyan, M. A., Guseva, M. I. and Konorova, E. A., *Sov. Phys. Semicond.,* 6:741 (1972)

33. Okumura, K., Mort, J. and Machonkin, M., *Appl. Phys. Lett.,* 57:1907 (1990)

34. Beetz, C. P. Jr., Lincoln, B. A. and Winn, D. R., *Diamond Optics III,* (A. Feldman and S. Holly, ed.) 1325:241, SPIE (1990)

35. Landstrass, M. I. and Ravi, K. V., *Appl. Phys. Lett.* 55:975 (1989)

36. Landstrass, M. I. and Ravi, K. V., *Appl. Phys. Lett.* 55:1391 (1989)

37. Albin, S. and Watkins, L., *IEEE Electron Dev. Lett.* 11:159 (1990)

38. Parikh, N. R., Chu, W. K., Sandhu, G. S., Swanson, M. L., Childs, C., Mikrut, J. M. and McNeil, L. E., *Diamond Optics,* SPIE Proc., 969:95 (1988)

39. Harper, R. E., Johnston, C., Blamires, N. G., Chalker, P. R. and Buckley-Golder, I. M., presented at the 1st European Conf. on Diamond and Diamond-like Carbon Coatings, Crans-Montana, Switzerland (September 17-19, 1990)

40. Muto, Y., Sugino, T., Shirafuji, J. and Kobashi, K., *Appl. Phys. Lett.,* 59:843 (1991)

41. Miyata, K., Dreifus, D. L., Das, K., Glass, J. T. and Kobashi, K., presented at the Electrochem. Soc. Meeting, Washington, D. C. (May 6 - 10, 1991)

42. Venkatesan, V., Das, K., Dreifus, D. L., Glass, J. T., Fountain, G. G., Rudder, R. A., Posthill, J. B. and Markunas, R. J., presented at the Electrochem. Soc. Meeting, Washington, D. C. (May 6 - 10, 1991)

43. Grot, S. A., Gildenblat, G. S., Hatfield, C. W., Wronski, C. R., Badzian, A. R., Badzian, T. and Messier, R., *IEEE Electron Dev. Lett.,* 11:100 (1990)

44. Prins, J. F., *Appl. Phys. Lett.,* 41:950 (1982)

45. Geis, M. W., Rathman, D. D., Ehrlich, D. J., Murphy, R. A. and Lindley, W. T., *IEEE Electron Dev. Lett.,* 8:341 (1987)

46. Moazed, K. L., Nguyen, R. and Zeidler, J. R., *IEEE Electron Dev. Lett.,* 9:350 (1988)

47. Moazed, K. L., Zeidler, J. R. and Taylor, M. J., *J. Appl. Phys.,* 68:2246 (1990)

48. Fang, F., Hewett, C. A., Fernandes, M. G. and Lau, S. S., *IEEE Trans. Electron Dev.* 36:1783 (1989)

49. Geis, M. W., Rothschild, M., Kunz, R. R., Aggarwall, R. L., Wall, K. F., Parker, C. D., McIntosh, K. A., Efremow, N. N., Zayhowski, J. J., Ehrlich, D. J. and Butler, J. E., *Appl. Phys. Lett.,* 55:2295 (1989)

50. Nishimura, K., Das, K., Iwase, M., Glass, J. T. and Kobashi, K., in *Diamond, Silicon Carbide and Related Wide Bandgap Semiconductors,* (J. T. Glass, R. Messier, and N. Fujimori, eds.), Mat. Res. Soc. Symp. Proc., 162:347, Mat. Res. Soc., Pittsburgh (1990)

51. Childs, C. B. and Chu, W. K., US patent application (1989)

52. Childs, C. B., Sandhu, G. S. and Chu, W. K., presented at the 4th SDIO/IST-ONR Diamond Tech. Initiative Symp., Crystal City, Virginia, July (1989)

53. Sandhu, G. S., Ph.D. thesis, University of North Carolina at Chapel Hill (1989)

54. Prins, J. F., *J. Phys. D: Appl. Phys.* 22:1562 (1989)

55. Venkatesan, V. and Das, K., *IEEE Electron Dev. Lett.,* 13:126 (1991)

56. Kim, G. A., U. S. Patent number 4,643,161 (1987)

57 Lightowlers, E. C. and Collins, A. T., *J. Phys. D : Appl. Phys.,* 9:73 (1976)

58. Shiomi, H., Nakahata, H., Imai, T., Nishibayashi, Y. and Fujimori, N., *Jpn. J. Appl. Phys.,* 28:758 (1989)

59. Miyata, K., Matsui, Y., Kumagai, K., Miyauchi, S., Kobashi, K. and Nakaue, A., *Proc.2nd Intl. Conf. New Diamond Sci. Tech.,* (R. Messier, J. T. Glass, J. E. Butler, and R. Roy, eds.), p. 981, Mat. Res. Soc., Pittsburgh (1991)

60. Humphreys, T. P., LaBrasca, J. V., Nemanich, R. J., Das, K. and Posthill, J. B., *Jpn. J. Appl. Phys.,* 30:L1409 (1991)

61. Das, K., unpublished work (1990)

62. Humphreys, T. P., LaBrasca, J. V., Nemanich, R. J., Das, K. and Posthill, J. B., *Electron. Lett.,* 27:1515 (1991)

63. Sahaida, S. R. and Thompson, D. G., in Mater. Res. Soc. Symp. Proc., (T. D. Moustakos, J. I. Pankove and Y. Hamakawa, eds.) (1992)

64. Mead, C. A. and McGill, T. C., *Phys. Lett.* 58A:249 (1976)

65. Himpsel, F. J., Heimann, P. and Eastman, D. E., *Solid State Comm.* 36:631 (1980)

66. Hicks, M. C., Wronski, C. R., Grot, S. A. and Gildenblat, G. S., Badzian, A. R., Badzian, T. and Messier, R., *J. Appl. Phys.,* 65:2139 (1989)

67. Efremow, N. N., Geis, M. W., Flanders, D. C., Lincoln, A. and Economou, N. P., *J. Vac. Sci. Technol.* B3:416 (1985)

68. Kobashi, K., Nakaue, A., Glass, J. T. and Buckley-Golder, I. M., *Proc. 2nd Intl. Conf. New Diamond Sci. Tech.,* (R. Messier, J. T. Glass, J. E. Butler, and R. Roy,eds.), p. 975, Mat. Res. Soc., Pittsburgh (1991)

69. Beetz, C. P. Jr. and Lincoln, B. *Proc. 2nd Intl. Conf. New Diamond Sci. Tech.,* (R. Messier, J. T. Glass, J. E. Butler, and R. Roy, eds.), p. 975, Mat. Res. Soc., Pittsburgh (1991)

70. Geis, M. W., Gregory, J. A. and Pate, B. B., *IEEE Trans. Electron Div.,* 38:619 (1991)

71. Geis, M. W., Efremow, N. N. and Rathman, D. D., *J. Vac. Sci. Tech.,* A6:1853 (1988)

Index

Acceptor dopant 13, 56, 320, 385
Activation energy 13, 15, 382, 384, 385
Adhesion
  of in situ coatings 231
  of low pressure diamond 217
  of PCD layers 176
  of sintered diamond 18
  on solvent/catalyst substrates 150
  to substrates 344, 352
Adhesion energy 100
Amorphous carbon
  and benzene 132
  and nucleation 91, 98
  Auger peaks 265
  binding energy 259
  characterization by Raman
    spectroscopy 3, 251
  deposition of hydrogenated 324
  EELS spectrum 269
  evaporant source for PVD 168
  on polycrystalline Ni 103
  XPS-EELS spectra 271
Analysis techniques. *See: Chemical analysis, Fluorescence, Gas chromatography, Microscopy, Spectrometry, Spectroscopy*
  ex situ 111, 117
  in situ 111, 165, 324, 349
Applications
  abrasives 225, 343
  anvil 19
  automotive industry 225
  bearings 369
  biological 20, 27, 28, 369
  cutting
    tool 58, 187, 193, 198, 204, 366
  devices 381
  electro-optic 25, 281, 321
  electronic 15, 281, 318, 321
  electronic packaging 343, 372
  extrusion dies 365
  face-milling 209
  free-standing 231
  Fresnel lens 24
  guide surfaces 365
  heat sink 343
  heat transfer 346
  high
    temperature 14, 186, 372, 381, 395
  infrared 357, 375
  infrared coating 344
  jet engines 366
  long-term, mechanical 368
  long-term, optical 376
  long-term, passive 373
  machine tool coating 345
  mechanical 364
  medium-term, mechanical 366
  medium-term, optical 375
  medium-term, passive 372
  microdrills 198
  microwave 14, 15, 372
  millimeter wave 14, 15
  near-term 349, 372
  near-term, mechanical 364
  near-term, optical 374
  near-term, passive 371
  needle-like 19
  non-electronic 339
  optical 19, 71, 350, 354, 374
  oxidation resistant 357
  PCD tools 175, 189, 194, 197, 202, 209
  protective 9
  quarter-wave coating 22
  seals and bearings 366
  semiconductor devices 71, 174
  space borne 16

stripper-foil 21
surgical instruments 28
thermal 356, 357, 371
thin film 173, 183, 346
tool bits 71
tribological 14, 176
twist-drills 198
wear resistant 193
window 20, 342
x-ray 24
x-ray mask 71, 175, 374
Applications, Table of
CVD diamond 341
predictions 363
Atomic hydrogen
adsorption 313
by "electrical discharge" 158
detection by THG 116
effect on contaminants 81
in CVD 154
in DLC growth 86
in low pressure deposition 218
in OES 112
supersaturation 73, 89
Auger electron spectroscopy (AES). *See: Spectroscopy*

Ballas diamond 193
Bandgap 13, 26
"Boat" structure 3, 88, 89, 106
Boron
acceptor dopant 13, 56, 151, 173, 382
in type IIb diamond 11
profiles by SIMS 389
Boron nitride
superhard 229, 233
Boron-doped
Bragg
diffraction 303
team 1
Breakdown voltage 390
Bridgman
anvil 38, 40, 44
seal 38
Bridgman, P. W. 32, 33, 36, 38, 40
Bundy, F. P. 34, 36, 40, 43, 44, 64

Carbide
conductivity of 4
Knoop hardness of 8
sound velocity of 7
Carbon-12 118, 322
Carbon-13 118
Carbonado 36, 49, 58
high pressure 57
Carrier velocity 13
CARS. *See: Spectroscopy*
Cathodoluminescence (CL). *See: Microscopy*
Cemented carbides
as CVD substrates 184
as tool substrates 184
bonding to 176
brazing to 176
cost of tool inserts 210
endmills 205
microdrills 198
twist drills 198
"Chair" structure 2, 3, 88, 89, 106, 111
Chemical
characterization of composition 245
filament-assisted deposition 77
maps 317
mechanisms of growth 68, 108, 127, 271
polishing 256
potential 101
reactions, timescale of 68
transport reaction (CTR) 291
vapor deposition 68
Chemical analysis 258, 264
by AES 326
by EEELFS 326
by EXAFS 326
by SIMS 316
Chemical bonding 20
after argon sputter 265
characterization of 248, 258
detection of 253
of allotropes 244
of low pressure diamond 217
of PCD to cemented carbide 176
Chemical kinetics 169, 176, 177
mechanism 172
theory 148
Chemical properties 186, 229
Chemical vapor deposition 111, 119, 147, 150, 184
compared to PVC 187
plasma activated 218
plasma enhanced 158
thermally activated 218
with boron 382
Coating, in situ
batch 220
CVD 197
diamond sheets 222

impact on tooling 225
low pressure CVD 194
market value potential 224
multilayer 193
outlook 231
Coatings. *See: Films and coatings*
Coefficient
Friction, Table 10
Thermal Expansion, Table 6
Coefficient
expansion 341, 344, 369
friction 9, 341, 344, 348, 356
Hall 383
thermal 341
thermal
expansion 4, 6, 344, 372, 376, 377
Coherent anti-Stokes Raman scattering (CARS). *See: Spectroscopy*
Color
in synthetic 55
Composite powder 228, 232
Composition
analysis of 245
gas, for microwave 160
gas, for PVD and CVD 187
gas phase 119
of growth interface 110
Conductivity, electrical 21, 375, 382, 385
Conductivity, photo 384
Conductivity,
thermal 3, 5, 167, 343, 345, 346, 356
for optical switches 25
heat exchanger 15
in lasers 26
Table 4
windows 19
Corrosion 15
resistance 19, 369, 371
resistant 188, 193
Cross-sectional TEM. *See: Microscopy*
Crystal structure 1, 2, 230, 298, 324, 325 *See also: Lattice*
fivelings 53, 88, 293
MTP's 70, 275, 293
twins 53
twin planes 106
Cubic boron nitride
as nucleation enhancer 75
hardness 8
Cutting tools 175, 217, 364
by PVD 177
carbonado 58
economics of 174
endmills 205
future development 366
market value, annual 18
materials machined by 18
metal-cutting tool inserts 205
microdrills 198
multilayer coating of 193
not for ferrous material 15
superhard 230
TAB stamps 198
twist-drills 198
CVD coating
fine-grained 188. *See also: Films and coatings*
in situ 194, 197, 224
multilayer 193
on cemented carbide 184
on cutting tools 217
on hard metals 187

Dark field imaging
via TEM 273
Deep level transient spectroscopy (DLTS) *See: Spectroscopy*
Defects. *See: Crystal structure*
Density
column 113
current 91, 391
defect 71, 73, 75, 99, 275, 283
molar 3
nucleation 71, 104, 221, 256, 341, 342, 354, 364
of acceptor states 385
of states 12, 249, 265, 271, 272
particle 93
power 20, 169, 371
surface sites 109
Table 10, 376
Deposition. *See: Growth*
Derjaguin, B. V. (*alt. sp.: Deryagin*) 73, 150
Devices
bipolar transistor 395
field effect transistor (FET) 381, 395, 397
IMPATT 15, 397
insulated gate FET 396
MESFET 391, 395, 396
MISFET 396
point contact transistor 395
vertical channel FET 395
Diamond grit
as abrasive 343
cost of 61

equipment for production 63
for cutting, grinding, polishing 49
for nucleation 93, 104
friable 52
in PCD 176
size 51
used in pyrolysis 151
Diamonds
classification of 55
Dielectric
constant 14, 26, 341, 344, 371, 372
Dielectric loss tangent 14, 16
Diffraction
double 307
X-ray diffraction, Table 301
Diffraction, electron
LEED 259, 262, 298, 311
RHEED 298, 315
SAD 273, 303
XRD 298
Disks, magnetic 17
Donor dopants 13, 55, 320, 384, 385

EELS. *See: Spectroscopy*
Elastic properties 3
Elastic waves 7
Electrical properties 11
of diamond/SiO_2 interface 394
Table 13
Electrically active defects 13
Electrically insulating 15
Electron channeling microscopy. *See: Microscopy*
Electron diffraction pattern
via TEM 303
Electron energy loss spectroscopy (EELS). *See: Spectroscopy*
Electron microscopy. *See: Microscopy*
Electron mobility 11, 13
Electron velocity 14
Electronic properties 355
Ellipsometry
multispectral 349
spectroscopic 81, 267
spectroscopy 323
Environment
hot-filament 90
microwave 90
operating 9, 20, 344, 350, 352, 381
RF plasma 90
Equilibrium
gaseous 119
thermodynamic calculations 119, 125

Ex situ
analysis 117

Fedoseev, D. B. 73, 150
Films and coatings
adhesion 353
"backside" 293
by "alternating chemistry" 163
by ARE 168
by chemical transport reaction (CTR) 291
by combustion 171
by "continuous chemistry" 163
by DC plasma 165
by DC plasma jet 167
by ECR 162
by fluorocarbon pyrolysis 152
by hot filament 154
by hydrocarbon pyrolysis 154
by hydrogen pyrolysis 150
by low temperature deposition 355
by microwave plasma 86, 158, 292
by PVD 168
by RF plasma 165
by sputtering of graphite 168
characteristics 148
characterization of 244, 326, 345
composition of 317
conformal 18
contacts 386
continuous film nucleation 76
crystalline defects 281
CVD and PVD compared 168
deposition by CVD 147
deposition methods 147
diffraction intensities 299
electronic applications 381
functional characterization 350
future applications 364
growth conditions 92
growth phases on silicon 261
growth rates 362
growth texture 301
hot filament growth conditions 157
hydrogen concentration 321
industrial applications 339
IR 344
IR anti-reflection 22
large area deposition 359
luminescence of 318
market value, potential 224
mechanical properties. *See: Mechanical properties*

morphology 278, 291, 354
moth's eye 17, 22
multilayer 193
on IR windows 20
on PCD tools 197
on polycrystalline Pt 79
optical
21, 323, 324, 341, 344, 346, 353, 354
oxidation resistance. *See: Oxidation*
plasma enhanced 158
polycrystalline 4, 74
properties and applications, Table of 341
recommendations and predictions 176
reproducibility 358
semiconducting 151, 382
stripper foils 21
surface morphology 299
surface structure 315
textured, on non-diamond 103
thermal conductivity *See: Conductivity*
Fluorescence
photodissociation 117
ultraviolet 55
Fluorine 17, 154, 259
Fracture toughness 8, 9, 346
Frank-Van der Merwe. *See: Two-dimensional growth*
Free-standing structures
brazing to tools 176
impurities in 317
IR analysis of 254
on liquid metal substrates 157
Fresnel equations 323
Fresnel lens 23, 24
Fresnel reflections 376
Friction. *See: Tribological*
Future
of CVD diamond products 224
of low pressure diamond tools 227
of non electronic products 363

Gas chromatography 118
Gaseous kinetics 121
General Electric
"Compax" product 58
General Electric process
history 31, 32, 36, 40, 44, 46
present 61
Table 44
Getters 17
Graphite
Auger emission 264
C=C bonds 8
characteristics of 3
chemical bonds 70
crystallite orientation 281
detection by SAD 308
EELS peaks 269
epitaxial 103
form vs. diamond 34
friction coefficient 10
gas phase nucleation 91
in PVD 168
layer on Ni 314
molar heat capacity 6
nucleation 73
nucleation on (0001) 87
Raman peaks 249
role in diamond growth 149
thermal conductivity 4
thermal expansion coefficient 6
to diamond grit 51
XPS analysis 259
Graphitization 14
by high dose implant 389
damage 274
role in nucleation 80
threshold 341
Grinding 58, 209, 343
Grinding tools 18, 344
Growth
2-D 97, 100, 101
3-D 97, 100
dynamic balance 74, 93, 125
dynamic balance, model 132
vapor phase 148
Growth environment 70
characterization of 111
charged species in 109
flame 361
for CVD 68, 70, 73, 74
for FACVD 81
for homoepitaxial 108
gaseous 70
gaseous kinetics 121
hot-filament 92, 103
in situ analysis 111, 113
Li doping 385
microwave 92, 103
models 119, 125
plasma, plasma jet 91
rare gases in 113
thermodynamic equilibrium 119
torch 91
Habit
of diamond
crystals 93, 216, 278, 291, 325
rhombohedral 52

Hall
belt 40, 49
bridge pattern 396
coefficient 12, 383
effect 385, 386
mobility of holes 12
tetrahedral press 49
Hall, H.
T. 32, 40, 41, 43, 44, 46, 49, 57, 58
Hannay, J. B. 33
Hardness 7
and tribology 16
enhancement 356
Knoop, Table 8
of natural and synthetic 52
quantification of 347
superhard BN phases 229
Table 8
Vicker's 167
Heat exchange 15, 16
Heteroepitaxial
growth 102
single crystal diamond 71
substrates 103, 104, 263
Heteroepitaxy 100, 103
large-scale 96
on Si 103
High field electron velocity 14
High pressure techniques 32, 36
GE process 40
of carbonado synthesis 57
of diamond grit synthesis 49
of single crystal growth 53
of superhard PCD and PCBN 189
present status 61
High-resolution transmission electron (HRTEM). *See Microscopy*
Hole mobility 12, 14
Hole velocities 12
Homoepitaxial
Au contacts on 387
by RF induction PECVD 165
CVD 104
electrical resistivity 12
electronic devices 381
FET operation in 381, 396
Li doping of 385
metal contacts on 390
Homoepitaxy 100, 104
Hydrogen 105
abstraction 129
adsorption and desorption 313
and nucleation 73
barrier layers 369
chemisorbed 259
content in free-standing 255
detection by LIF 116
detection by OES 111
effect on nucleation 74
electrically active 255
etching 313
impurities 317
in CVD deposition 149
in CVD diamond 255
in CVD growth 108
in CVD growth environment 68
in diamond growth environments 118
in dynamic balance models 125
in kinetic models 132
in surface models 129
measurement of 321
plasma etching 354
tribology, contribution to 17
Hydrogen (solid)
density of 11
Hydrogen-terminated
growth surfaces 109
Hydrogen-termination 73

Impurities
absence of 55
characterization of 317
color 55, 56, 189, 326
detection by CL 320
detection by IR 253
electrically active 382
heavy mass 322
in carbonado 57
light mass 322
nitrogen 356
non-carbon 343
In situ
CARS 165
CVD coating 193, 197
diamond coatings 231
free-standing sheets 222
gas phase laser spectroscopy 165
monitor of deposition thickness 349
planarization techniques 354
Index of refraction
effective, IR 254, 257
for optical triggers 26
matching 21
moth's eye coating 22
of diamond 375
of diamond films 349
Table 11

Industrial applications. *See: Applications*
Interface
adhesion 217, 366, 368
anti-reflection coating 21
bonding 344, 364, 391
bonding to dummy 221
carbon etching 88
diamond/silicon 257, 277, 287
diamond/SiO_2 393
formation of W_2C and WC 81
gas/surface chemistry 108
interactions 185
mechanical stress 372
multiple twinning 90
nucleation sites 97
side of free-standing 293, 317
substrate/coating interactions 185
Interfacial evolution 76
Internal photoemission spectroscopy 392

Kinetics
chemical 148
gaseous 121, 125
gaseous reaction 119
modeling 124
reaction 156
surface 132

Laser 26
-induced fluorescence
(LIF) 116, 117, 156
damage 375
diamond 26
diode 113, 114, 124, 371
excitation 100
fluorescence 116
for hole cutting 18
host material 26, 376
mirror 20, 376
nucleation 84, 100
spectroscopy 165
surface modification by 388
window 19, 61
Laser Microprobe Mass Analysis
(LAMMA). *See Spectrometry*
Lattice. *See: "Chair" and "Boat" structure*
boron doping 382
cubic diamond 106
distortions 94
growth 71
heat transfer mechanism 5
hydrogen in 255
imaging 283
imperfections 273
match 83
nine-carbon model 129
softening by heat 70
structure 2. *See also: Crystal structure*
vibrations 248
Lattice
constant 3, 20, 102, 298, 301, 303
Lithium
donors 13
doping 384
Lonsdaleite 3
crystallized 311
lattice structure 106
vs. hydrogen partial pressure 127
Low energy electron diffraction (LEED). *See: Electron diffraction*
Low pressure techniques 132, 183
benefits of 227
for wear coatings 192, 194
of composite powder
production 228, 232
of cubic BN synthesis 229
problems 215
two concepts 222
Luminescence spectroscopy 318

Magnetic disks 17
Market potential
for diamond cutting tools 18
for diamond dies 18
of CVD diamond products 224
Mass spectrometry 117, 321
Mass spectroscopy 117
Mechanical properties 3, 347, 356
development issues 353
Membranes
free-standing on liquid metal 157
rim-supported 175
x-ray lithography masks 374
Methane
and crystal orientation 291
and defects 275
and graphitic inclusions 281
cauliflower morphology 278
equilibrium with acetylene 156
in CVD growth 68, 109
in hot filament growth 156
Microcrystalline carbon 3
Microscopy
ACFEM 326
electron 245. 272

electron channeling 326
HRTEM 283
HRTEM micrograph 93
HRTEM of fivefold 285
SEM 272, 287
STM 261, 326
TEM 268, 272
TEM of defect structures 275
TEM of multiply-twinned particles 275
TEM, plan-view 275
XTEM 285
Microwave 90. *See also: Growth environment*
Mobility 383, 384, 385, 397
adatom 292
carbon 83, 278, 291
electron 12, 13
Hall 12
hole 14
of carbon-containing species 74
of charge-carriers 26
of growth species 72
Model
(111) surface 127
band structure 265
chemical kinetic 169
computational 108
diamond growth 125
dynamic balance 125
equilibrium 127
"evolutionary selection" 302
for diamond growth 108
gaseous growth 119
multiply-twinned nucleation 90
non-equilibrium thermodynamics 127
quasi-equilibrium thermodynamics 127
sp^2 bonding 252
Modulus
bulk 3
elastic 184
Young's 8, 9, 26, 193, 376
Table 9
Moissan, H. 33
Molar density 3
Molar Heat
Table 6
Molar heat 5, 6
Molecular beam mass spectrometry 117
Molecular hydrogen
detected by REMPI 115
effect on nucleation 73
Moth's eye coating 23
Multilayer sheets 223
MTPs. *See:Crystal structure*

Needle-like 19
Nitrogen
carbon bond 8
donors 13
impurities enhance hardness 356
impurity, energy level of 384
in diamond types 11, 55, 253
reduces diamond conductivity 4
Nuclear magnetic resonance (NMR) 326
Nucleation 71, 83
analysis by AES 267
analysis by IR 257
analysis by SE 324
analysis by spectroscopy 258
analysis by XPS 260
analysis by XRD 302
by EACVD 83
by scratching 75, 216, 358
by scratching and seeding 92
conditions for 101
density 74, 75
enhancement 216
gas phase 90, 149
graphite 73
influence of interlayers 79
influence of substrate evolution 79, 216
kernels 105
kinetic aspects of 71, 122
mechanics of 73
mechanisms 70, 105
multiply-twinned 90, 106
on BN 104
on metals 104
on non-carbide 75
on non-diamond 70, 154
on Si substrates 83
on ß-SiC 76
on unscratched substrates 76, 78, 84
on WC substrates 83
process 74, 83, 215
related to crystal size 75
self-propagating 75
sites 76
spontaneous 215
thermodynamics of 71, 74
two-step process 76, 107

OES. *See: Spectroscopy*
Optical
absorption 11, 326
absorption centers 12, 26
coating. *See: Films and coatings*
emission 112

matching 21
mirror 20, 376
scatter 254, 256, 349, 364
switch 26
technique 386
transmission 374
transparency 162, 349, 357, 363
window 20
Optical emission spectroscopy (OES). *See: Spectroscopy*
Optical properties 11
infrared 253
Optically active centers 11
Optically triggered switches 26
Oxidation
for sample preparation 273
goals of research 341
of diamond 15
on {111} and {100} surfaces 280
resistance 357
resistant 353, 357
Oxidation barrier coating 374

Parsons, C. 33
Phase diagram 59, 127
Phosphorous 13, 312
Physical vapor deposition 147, 167, 177, 184
comparison with CVD 187
Soviet literature on 170
Planarization
in situ 354
Plasma emission actinometry 113
Polycrystalline diamond (PCD) 197
benefits of 227
Powder
for diamond grit 52
Powder, diamond
for use in carbonado 58
Powder, high purity carbon
in ARE 169
Powder, sintered diamond 18
Powder, superhard BN
in PCD 189
Powder, superhard composite 228
Properties. *See: Elastic, Electrical, Chemical, Mechanical, Optical*

Raman
application areas 346
CARS 165
correlated with IR spectroscopy 255
diamond "perfection" 354
films with no scattering 163
low temperature deposition spectrum 355
scattering (SERS) 82
spectroscopy 248
study vibrations 245
thermal conductivity character 357
with SEM 289
Raman signature
of amorphous carbon 3
of cubic diamond (carbon-12, -13) 2
of graphite 3
of Lonsdaleite 3
of microcrystalline carbon 3
Refractive index. *See: Index of refraction*
Resistivity
bulk 392
modification of 386
of DIA-JET diamond 167
of Hall bridge pattern 396
of heat sinks 343
of natural diamond 12
ohmic contacts 387
P-doped 384
wear 198
Resonance enhanced multi-photon ionization spectroscopy (REMPI). *See: Spectroscopy*
Ring
carbon bond 2

Saturated carrier velocity 13
Scanning electron microscopy (SEM). *See: Microscopy*
Scanning tunneling microscopy (STM). *See: Microscopy*
Secondary ion mass spectrometry (SIMS) *See: Spectrometry*
SEM. *See: Microscopy*
Semiconducting diamond
as x-ray target 25
boron-doped 151, 382
FET 395
ohmic contacts on 388
rectifying contacts on 389
type IIb 55
Semiconductor 11
bipolar transistor 395
devices 174, 363
See also: Devices
diodes 395
properties, Table 13

Semiconductor device
  requirements 71
Sheets, free-standing 193, 205
  benefits, drawbacks 222
  bonding of 194
Shock wave process 58
Single crystal
  dies 18
  film 289
  for semiconductor 71
  growth 49, 53, 55
  morphology 293
  substrate 75, 324, 381
Single crystal Cu 103
Single crystal diamond 34, 88
  adamantane leading to 61
  cost, tool life 204
  semiconducting 391, 392
Sodium 13, 15, 16
Solvent 36, 44, 49, 150
  carbon 313
  metal 36, 51, 53, 58
  metallic 36
Solvent/catalyst 150
Sound velocity 3
Spectrometry
  energy dispersive (EDS) 289
  LAMMA 318
  mass 117, 316
  Rutherford backscattering (RBS) 321
  SIMS 316
Spectroscopic ellipsometry. *See Ellipsometry*
Spectroscopy 156, 258, 349
  absorption 55
  AES 258, 263, 271
  AES, not for hydrogen detection 321
  CL 318, 320
  DLTS 386
  EEELFS 326
  EELS 249, 263, 267
  ESCA 258
  EXAFS 326
  FTIR 113
  in situ 165
  IR 245, 252, 321
  laser absorption 113
  OES 111
  photoemission 392
  Raman 245, 248, 346
  REMPI 114
  transmission infrared 81
  XAES 262
  XANES 87
  XPS 258, 271
  XPS-EELS 271
Spitsyn, B. V. 150, 168
Stranski-Krastonov 100
Structure. *See: Crystal structure*
Substrate adhesion 352
Surface
  analysis methods 258, 271
  bonding 267
  characteristics 193
    for CVD 149, 187, 188
  characterization 260, 263, 264, 298
  chemistry analysis 271
  dummy 221
  energy 176
  flatness 202
  graphitization 258
  large area deposition 232
  morphology 256, 292, 297, 299
  morphology and defects 278, 283
  mounting 198
  nucleation 215
  polishing 256
  RMS smoothness 175
  roughness 202, 210, 216, 254
  roughness-induced scatter 255
  states 259, 269
  structure 312
  superhard 221, 225
  topology analysis 289
Surface crystal
  vibrations 245
Surface damage
  by AES 264
"Surface layer grades" 232
Surface morphology analysis
  via SEM 289
Synthesis
  apparatus 171
  by combustion 172
  by hot filament 156
  by plasma reactor 165
  by shock wave 58
  history of 31
  low pressure 183, 193, 218, 230
  low-pressure 229
  of diamond film 147, 150, 168, 291
  of diamond grit 49
  of films 165
  of free-standing diamond 221
  of nanocrystalline diamond 216
  superhard 233
  Table of methods 219
  via halogenides 220
  via pyrolysis 151

Synthetic
  carbonado 57

Tennant, S. 1, 33
Tetrahedral press 48, 49, 63
Thermal conductivity. *See: Conductivity*
Thermal expansion 4, 6, 26
Thermodynamics
  equilibrium 127
  non-equilibrium 127
Third harmonic generation (THG) 116
Three-dimensional
  growth 101, 287
Three-dimensional shapes 191
  coating of 194, 222, 361
  fabrication of 192
Transistor
  GaAs power 16
Transmission electron microscopy
  (TEM). *See: Microscopy*
Transmission infrared spectroscopy 81
Tribological. *See also: Coefficient, friction*
  and hardness 16
  applications 14
  marine applications 20
  testing 348
Twins. *See: Crystal structure*
Two-dimensional growth 100, 101, 287
Two-dimensional structures 312
  biperiodical 310
Types of diamond 11

Vapor phase growth 148
Volmer-Weber 100
  See also: Three-dimensional growth
Water jet nozzles 17
Wear properties
  of coatings 183
  of cutting tools 205
  of low pressure diamond 193
  of natural diamond 17
  of sintered diamond powder 18
  of superhard BN 229
  of synthetic carbonado 58
  of tools 197
  testing of 202, 348
Wear resistance
  of TAB tools 202
Wear resistant
  applications 189, 192
  products, market value of 225
Wear resistant coatings
  by PVD 188
  deposition of 184
Wentorf, R.
  H. 40, 43, 44, 49, 53, 57, 64, 150
Windows 19
  Be 342
  high power microwave 372
  IR 9
  pressure leaks 350
  rim-supported 175
  rupture failures 350
  x-ray 175
  x-ray detector 341, 342
Wire dies 17
Work function 14, 94, 95

X-ray absorption near edge spectroscopy
  (XANES). *See: Spectroscopy*
X-ray photo-electron spectroscopy
  (XPS). *See: Microscopy*
X-ray topography 302
XPS. *See: Spectroscopy*
XTEM. *See: Microscopy*

Young's modulus. *See: Modulus*